PIG PRODUCTION:
BIOLOGICAL PRINCIPLES AND APPLICATIONS

Join us on the web at
Agriscience.delmar.com

PIG PRODUCTION: BIOLOGICAL PRINCIPLES AND APPLICATIONS

JOHN MCGLONE, PH. D.
Professor, Texas Tech University
Lubbock, Texas

WILSON POND, PH. D.
Visiting Professor, Cornell University
Ithaca, New York

THOMSON
DELMAR LEARNING

Australia   Canada   Mexico   Singapore   Spain   United Kingdom   United States

Pig Production: Biological Principles and Applications
by John McGlone, Ph.D., and Wilson Pond, Ph.D.

Business Unit Executive Director:
Susan L. Simpfenderfer

Executive Production Manager:
Wendy A. Troeger

Executive Marketing Manager:
Donna J. Lewis

Acquisitions Editor:
Zina M. Lawrence

Production Manager:
Carolyn Miller

Channel Manager:
Nigar Hale

Developmental Editor:
Andrea Edwards

Production Editor:
Matthew J. Williams

Cover Image:
Corbis

Cover Design:
Dutton & Sherman Design

COPYRIGHT © 2003 by Delmar Learning, a division of Thomson Learning, Inc. Thomson Learning™ is a trademark used herein under license.

Printed in the United States
1 2 3 4 5 XXX 06 05 04 03 02

For more information, contact Delmar Learning,
5 Maxwell Drive, Clifton Park, NY 12065-2919.

Or find us in the World Wide Web at
www.DelmarLearning.com or
www.Agriscience.Delmar.com

ALL RIGHTS RESERVED. No part of this work covered by the copyright hereon may be reproduced or used in any form or by any means—graphic, electronic, or mechanical, including photocopying, recording, taping, Web distribution or information storage and retrieval systems—without written permission of the publisher.

For permission to use material from this text or product, contact us by
Tel   (800) 730-2214
Fax   (800) 730-2215
www.thomsonrights.com

Library of Congress Cataloging-in-Publication Data
McGlone, John, 1955–
   Pig production: biological principles and applications / John McGlone, Wilson Pond.
      p. cm.
   ISBN 0-8273-8484-X
   1. Swine. I. Pond, Wilson G., 1930– II. Title.
   SF395 .M37 2002
   636.4—dc21   2002025742

NOTICE TO THE READER

Publisher does not warrant or guarantee any of the products described herein or perform any independent analysis in connection with any of the product information contained herein. Publisher does not assume, and expressly disclaims, any obligation to obtain and include information other than that provided to it by the manufacturer.

The reader is notified that this text is an educational tool, not a practice book. Since the law is in constant change, no rule or statement of law in this book should be relied upon for any service to any client. The reader should always refer to standard legal sources for the current rule or law. If legal advise or other expert assistance is required, the services of the appropriate professional should be sought.

The Publisher makes no representation or warranties of any kind, including but not limited to, the warranties of fitness for particular purpose or merchantability, nor are any such representations implied with respect to the material set forth herein, and the publisher takes no responsibility with respect to such material. The publisher shall not be liable for any special, consequential, or exemplary damages resulting, in whole or part, from the readers' use of, or reliance upon, this material.

Brief Contents

SECTION I	**ORIGINS OF THE MODERN PIG AND EVOLUTION OF MODERN PRODUCTION SYSTEMS**	**1**
CHAPTER 1	**Domestication and Early History**	1
CHAPTER 2	**Profile, Structure, and Complexity of Modern Pig Production and Marketing in the United States**	13
CHAPTER 3	**Societal Issues for Pork Production**	25
SECTION II	**BIOLOGY OF THE PIG**	**34**
CHAPTER 4	**Applied Anatomy and Physiology Related to Blood Sampling, Hematology, and Immunology**	34
CHAPTER 5	**Reproductive Biology**	51
CHAPTER 6	**Pig Genetics**	66
CHAPTER 7	**Modern Breeds of Pigs**	81
CHAPTER 8	**Growth, Development, and Survival**	88
CHAPTER 9	**Pork Composition and Quality**	110

SECTION III	**NUTRITION AND FEEDING**	**121**
CHAPTER 10	Nutrients, Nutrition, and Life-Cycle Feeding	121
CHAPTER 11	Feed Resources	148
CHAPTER 12	Formulating Diets for Pigs	175
SECTION IV	**HOUSING, ENVIRONMENT, AND NUTRIENT MANAGEMENT**	**194**
CHAPTER 13	Creating a Comfortable Microenvironment for Pigs	194
CHAPTER 14	Production Systems for Adult Pigs	207
CHAPTER 15	Production Systems for Growing Pigs	225
CHAPTER 16	Waste and Nutrient Management	259
SECTION V	**PIG PRODUCTION APPLICATIONS THAT MAKE BUSINESS SENSE**	**267**
CHAPTER 17	Management of the Breeding Herd	267
CHAPTER 18	Management of Sows and Piglets: Before, During, and After Farrowing	275
CHAPTER 19	Management of Growing Pigs	291
CHAPTER 20	Management of Pig Health	312
SECTION VI	**APPENDICES**	**346**
APPENDIX A	Human Resources and Pork Industry Directory	346
APPENDIX B	Glossary/Pig Dictionary	362
INDEX		383

Contents

Preface   xv

SECTION I   Origins of the Modern Pig and Evolution of Production Systems   1

Chapter 1   Domestication and Early History   1
Introduction   1
Phylogeny   2
Early Pig Production in Europe and Asia   4
History of the Pig in North America   5
Current Pig Numbers and Distribution Around the World   7
Disease Threats to Pork Exportation   9
Summary   10
Questions and Activities   11
Literature Cited   11
Internet Resources   11

Chapter 2   Profile, Structure, and Complexity of Modern Pig Production and Marketing in the United States   13
Introduction   13
National and International Competition   13
Modern Pork Production Farms: Farm Styles, Site Layouts, and Production Schedules   17
Summary   23
Questions and Activities   23

Literature Cited 24
Internet Resources 24

CHAPTER 3 **Societal Issues for Pork Production**   **25**
Introduction 25
Challenges and Issues 25
Summary 32
Questions and Activities 33
Internet Resources 33

SECTION II   BIOLOGY OF THE PIG   34

CHAPTER 4 **Applied Anatomy and Physiology Related to Blood Sampling, Hematology, and Immunology**   **34**
Introduction 34
The Pig in Biomedical Research 35
General Anatomy 35
Anatomical References 36
Anatomy and Sampling of the Pig Vascular System 40
The Pig Heart and Blood Supply 45
Immunology 48
Summary 49
Questions and Activities 50
Literature Cited 50
Internet Resources 50

CHAPTER 5 **Reproductive Biology**   **51**
Introduction 51
Anatomy and Physiology of the Pig Reproductive System 51
Behavioral Biology 61
Summary 64
Questions and Activities 64
Literature Cited 64
Additional Readings 65
Internet Resources 65

CHAPTER 6 **Pig Genetics**   **66**
Introduction 66
The Changing Pig 66

Additive versus Nonadditive Gene Action   67
Heterosis   69
Heritability Estimates   71
Genetic Improvement   72
Crossbreeding Systems   72
Molecular Genetics   76
Summary   78
Questions and Activities   79
Literature Cited   79
Internet Resources   80

Chapter 7 Modern Breeds of Pigs   81
Introduction   81
Major U.S. Breeds   83
Less Common U.S. Breeds   84
Uncommon or Specialized U.S. Breeds   85
Commercial Lines   85
Summary   86
Questions and Activities   86
Literature Cited   86
Internet Resources   87

Chapter 8 Growth, Development, and Survival   88
Introduction   88
The Nature of Growth   88
Prenatal Development   92
Postnatal Development   99
Summary   107
Questions and Activities   107
Literature Cited   108
Internet Resources   109

Chapter 9 Pork Composition and Quality   110
Introduction   110
Nutrient Composition of Pork   110
Carcass Measurements and Composition   114
Pork Quality   116
Pork Safety   118
Summary   119

Questions and Activities   119
Literature Cited   120
Internet Resources   120

SECTION III — NUTRITION AND FEEDING — 121

CHAPTER 10 — Nutrients, Nutrition, and Life-Cycle Feeding — 121

Introduction   121
Nutrients   121
Nutrition   136
Life-Cycle Feeding   140
Summary   145
Questions and Activities   145
Literature Cited   146
Internet Resources   147

CHAPTER 11 — Feed Resources — 148

Introduction   148
Energy Sources   148
Protein Sources   163
Summary   173
Questions and Activities   173
Literature Cited   174
Internet Resources   174

CHAPTER 12 — Formulating Diets for Pigs — 175

Introduction   175
Important Considerations in Diet Formulation   176
Mechanics of Diet Formulation   178
Effects of Feed Processing on Nutritive Value, Palatability, and Feed Consumption   180
Practical Swine Diets   181
Summary   191
Questions and Activities   191
Literature Cited   192
Internet Resources   192

SECTION IV — HOUSING, ENVIRONMENT, AND NUTRIENT MANAGEMENT — 194

CHAPTER 13 — Creating a Comfortable Microenvironment for Pigs — 194
Introduction   194
Components of the Microenvironment   195
Summary   205
Questions and Activities   205
Literature Cited   206
Internet Resources   206

CHAPTER 14 — Production Systems for Adult Pigs — 207
Introduction   207
Production Systems: Reasonable Choices   208
The Breeding Area   210
Field Data on Gestation and Farrowing Systems   211
Controlled Studies on Gestation and Farrowing Systems   214
Building Designs   219
Summary   219
Questions and Activities   222
Literature Cited   223
Internet Resources   223

CHAPTER 15 — Production Systems for Growing Pigs — 225
Introduction   225
Modern-day Changes in Growing Pig Standards   226
Building Styles   229
Alternative Finishing Systems   231
Building and Lot Layouts   235
Standard Pen Layouts   239
Loading Chutes on the Farm   254
Summary   256
Questions and Activities   256
Literature Cited   257
Internet Resources   258

CHAPTER 16 — Waste and Nutrient Management — 259
Introduction   259
Manure Management   259

xii   Contents

           Regulations and Potential for Pollution   263
           Manure Nutrient Collection Options   264
           Summary   265
           Questions and Activities   266
           Literature Cited   266
           Internet Resources   266

SECTION V   PIG PRODUCTION APPLICATIONS THAT MAKE BUSINESS SENSE   267

CHAPTER 17   Management of the Breeding Herd   267
           Introduction   267
           Areas of Opportunity   267
           Setting Production Targets—22 PPSY Today, 30 PPSY Tomorrow   271
           Setting Breeding Targets   271
           Seasonal Effects   272
           Summary   273
           Questions and Activities   274
           Literature Cited   274
           Internet Resources   274

CHAPTER 18   Management of Sows and Piglets: Before, During, and After Farrowing   275
           Introduction   275
           The Interdependence of the Breeding Staff and the Farrowing System   275
           Critical Production Variables   276
           Management of Physical Assets   278
           Management of Three Biological Assets: Sows, Piglets, and People   283
           A New Breed of Dedicated Stockperson   288
           Summary   288
           Questions and Activities   289
           Literature Cited   289
           Internet Resources   290

CHAPTER 19   Management of Growing Pigs   291
           Introduction   291
           Management of the Modern Grower/Finisher Unit   291
           Normal Pig Behavior   292
           Daily Pig Management   297

　　　　　Behavioral Problems   301
　　　　　Handling Pigs   306
　　　　　Transporting Pigs   308
　　　　　Summary   309
　　　　　Questions and Activities   309
　　　　　Literature Cited   310
　　　　　Internet Resources   311

CHAPTER 20   **Management of Pig Health**   312
　　　　　Introduction   312
　　　　　Approaches to Ensuring Pig Health   313
　　　　　Herd Health Program   316
　　　　　Infectious and Noninfectious Diseases   320
　　　　　Summary   342
　　　　　Questions and Activities   343
　　　　　Literature Cited   343
　　　　　Internet Resources   345

SECTION VI   APPENDICES   346

APPENDIX A   **Human Resources and Pork Industry Directory**   346

APPENDIX B   **Glossary/Pig Dictionary**   362

INDEX   383

Preface

This book has been a labor of more than a decade, during which the U.S.A. and world swine industries have undergone monumental changes. What once was a large farm is now redefined; what once was a pig is now redefined; and what once was a pork producer is now redefined, especially in the United States.

WHY THE BOOK WAS WRITTEN

Modern pork production has developed and changed dramatically in the past few years. This book was written to include modern concepts in commercial pork production for the benefit of students and pork producers.

The U.S. pork industry has undergone an unprecedented, major restructuring. In 1950, a large farm had 300 sows. In 1975, a large farm had 1,000 sows. In 2003, large farms are found in many locations and have over 100,000 sows. At one location, the largest farms now have 25,000 sows and the growing pigs are at a distant site (not on the same property).

When the industry went through its most significant restructuring over the last decade, the educational needs for students changed as well. Excellent 2-year programs and the traditional 4-year Bachelor of Science programs are available in the United States, and 3- and 5-year programs are offered in Europe. At the same time, the educational needs of college students in the science, business, and management of pigs have changed. Advanced students and on-farm supervisors need skills that were not required in the past. Pork production is a business that needs people with logical, business sense.

Worker training and worker turnover is a problem that was less obvious 20 years ago. Today, human resource issues are reported by many farmers to be their number one concern and a significant cause of the variation in biological output of their farms. Although management of human resources is a recognized problem on commercial farms and although the solutions seem obvious, the situation seems to be only worsening. Five years ago, a survey of local, large-scale pork production units found a 55% turnover rate among pig farm workers. Today the number is estimated to be over 80%. The solution, it seems, is not so obvious after all and we must recognize, discuss, and attempt to resolve this real-world problem.

Animal Science students have changed as well during the past decades. Students are less often from a farm background and even less so from a pork production family. Graduates from an urban background who go to work at commercial pork production units can be successful. Indeed, workers from all walks of life find themselves in the pork industry. The scenario of the family pork producer sending his son or daughter to the university to learn about the family business is still found among students, albeit much more rarely than in the past.

Animal Science students who find themselves on a commercial pork production unit are often very well-trained in pig biology and management compared with their co-workers. One recent graduate found himself working with a collection of co-workers, including a farm manager who had worked in a coal mine, a worker who did not speak English, an overall site manager who came from the construction side of the business, and a company president who was recruited from a Fortune 500 company that was not an agricultural enterprise. It is the Animal Science graduate who can lead and train the other workers. It is the Animal Science graduate who holds the special knowledge and training that can help such a pig production enterprise succeed.

TARGET AUDIENCE

This book was written for a wide audience that includes primarily undergraduate students with some previous background in the Animal Sciences. Secondly, this text was written for workers on commercial farms who might have some on-site training, but who lack an understanding of pig biology. Most especially, this text is written for the advanced swine student at the college, university, or company, who wants to learn and who may one day train others on a commercial farm.

BOOK SECTIONS

The book is divided into two main sections. The first section is directed toward pig biology. While techniques (i.e., artificial insemination or necropsy) can be taught to students without a basic understanding of pig biology, those who have such a background will improve the outcome of the technique.

The second major section is dedicated to applications on the farm. Applications are, of course, site specific and must be modified to each site. An attempt was made to include more explanations of why pig biology drives us to use each application. The underpinnings of why practices are performed is important for professional development among pig workers.

TERMINOLOGY

Terminology is always a challenge in this industry, starting with what to call those who work in this industry. In the midwestern United States, they are called hog farmers. Among scientists, they are swine producers. Among national leaders, they are pork producers, and among the international community, they are pig farmers.

"Hog" is a term that is problematic because in the United Kingdom, a hog is a castrated male. "Swine" is an unflattering term in parts of Europe. The ultimate product is pork, so many people lean toward "pork production" as the most descriptive term for this industry. However, pork is not managed on the farm—the pigs are managed. To refer only to pork production reduces the importance attached with management of a biological animal—the pig. The term "pig" is the most widely accepted word for this very important agricultural and biomedical species and, thus, the most common term used in this text is "pig" rather than "swine" or "hog". Interestingly, the same problem exists in Latin America where there are several terms for the pig, including merannos, cerdo, and puercos—each having a unique use in different Spanish-speaking countries. An industry-validated glossary of terms is included at the end of the book to help us speak the same language.

USEFUL WEB SITES

Students and members of the pork production community will find the Internet full of useful information about pigs and pork production. Some useful links that students should check regularly include the following:

Scientific and nonprofit organizations with information on pigs:
http://www.asas.org
http://www.fass.org
http://www.animalagriculture.org/home.asp
http://www.aif.org

Selected university sites with pig information:
http://www.pii.ttu.edu (our Texas Tech University pig site)
http://www.oznet.ksu.edu/dp_ansi/swine/swine.htm (Kansas State University)
http://www.extension.iastate.edu/ipic/ (Iowa State University's Pork Center)
http://www.ansi.okstate.edu/ (Oklahoma State University)
http://www.ansci.uiuc.edu/ (University of Illinois)
http://www.cals.ncsu.edu/an_sci/home/home.html (North Carolina State University)

The National Pork Producers Council and the Pork Board:
http://www.porkboard.com
http://www.nppc.org/indexchoose.html

List of links to market prices:
http://www.nppc.org/PROD/marketinglinks.html

News about pigs and pork:
http://www.agriclick.com (links to National Hog Farmer magazine)
http://www.porkmag.com (Pork magazine)
http://www.porknet.com (pig and pork news)
http://www.meatingplace.com (meat industry news)
http://www.thepigsite.com (news with a focus on health)

http://www.pighealth.com (pig health information)
http://www.agriculture.com (Successful Farming magazine)

Pigs are the most important agricultural animal on earth. Humans eat more pork than any other meat. The pig and pork production industries around the world are experiencing the same sort of restructuring and continued growth that is happening in the United States. The world needs more people trained in pig biology with an understanding of how to use their biological knowledge to make wise business decisions. The goal is to manage pigs with an eye toward efficient pork production. Practitioners of biologically based pig management will create a profitable and personally rewarding production environment for all who are fortunate enough to work with this special animal. The authors of this book are committed to those people who dedicate their lives to working with pigs. It is important that the needs of the pigs are met in a cost-effective manner because, after all, it is the pigs that in the end give the ultimate sacrifice in service to humans.

INTERNET DISCLAIMER

The authors and Delmar Learning affirm that the Web site URLs referenced herein were accurate at the time of printing. However, due to the fluid nature of the Internet, we cannot guarantee their accuracy for the life of the edition.

INSTRUCTOR'S MANUAL

There is an Instructor's Manual to accompany this text available online at *www.agriscience.delmar.com* under "Resources." Written by the authors of this text, the Instructor's Manual contains chapter summaries, discussion topics, and answers to the end-of-chapter Questions and Activities.

ACKNOWLEDGMENTS

The authors and Delmar Learning wish to thank the following reviewers for their time and content expertise:

Kevin J. Rozeboom
Minitube of America
Verona, WI

Gary L. Cromwell
University of Kentucky
Lexington, KY

Bob Goodband
Kansas State University
Manhatten, KS

In addition, the authors thank the many students and pork producers who have experienced commercial pork production and who have helped shape our desires to teach pig production.

About the Authors

John J. McGlone was born in New Rochelle, New York from the son of Irish immigrants. He graduated from Holy Family High School in Huntington, NY and traveled to Washington State University where he obtained his B.S. in 1977. While conducting undergraduate research he won the undergraduate research award in 1977. Inspired by Dr. Keith W. Kelley, he studied behavior, immunology and physiology and obtained his M.S. in Animal Science in 1979. From 1979–1981 he studied under the leader in pig environmental management—Dr. Stanley E. Curtis at the University of Illinois. His PhD program included co-advisors Dr. Curtis and Dr. Ed Banks (deceased) from the Department of Ecology, Ethology and Evolution and work on pig behavior and neuroscience. Following graduation in 1981 from the University of Illinois, he spent 3 years on faculty at the University of Wyoming. His responsibilities included teaching, research and extension in swine. During that time, he received an award from the H. F. Guggenheim foundation for studies of pig aggressive behavior. In 1984 he began a faculty position at Texas Tech University where he has taught animal physiology, swine production and contemporary issues. His research relates to pig stress and includes a multi-disciplinary approach. His research, education and economic development projects have been funded by a variety of sources. He has spoken on the subject of pig management, behavior, welfare and on the future of the pig industry in the USA, Mexico, South America, Europe, Asia and Australia. He is a member of the American Society of Animal Science, the International Society for Applied Ethology, Animal Behavior Society, Society for Neuroscience, New York Academy of Sciences, Sigma Xi and Gamma Sigma Delta. He is married to Barbara who teaches elementary school reading and has two beautiful daughters in college studying music.

Wilson G. Pond is a native of Minnesota. He graduated in 1948 from Bloomington High School and received the B.S. degree from the University of Minnesota in 1952 with a major in Animal Husbandry and a minor in Agricultural Economics. His graduate training was at Oklahoma State University where he received the M.S. degree in 1954 and the PhD in 1957 in Animal Nutrition and Physiology. He served in the U.S. army in 1954–1956. From 1957–1978 he served on the faculty of the Department of Animal Science at Cornell University where he taught and did research in swine nutrition. In 1978 he was appointed Research Leader of the Nutrition Unit at the R.L. Hruska U.S. Meat Animal Research Center, Clay Center, NE. He served there until 1990 when he transferred to the USDA Children's Nutrition Research Center (CNRC), Baylor College of Medicine, Houston, TX, where he used the pig as an animal model for research in human infant nutrition. Since 1997, when he retired from CNRC, he has been a Visiting Professor, Department of Animal Science, Cornell University. He is author or co-author of more than 300 research papers on animal nutrition published in refereed journals, and author, co-author or editor of numerous book chapters and 13 books related to agriculture, has served as Editor-in-Chief, Journal of Animal Science (1975–1978), President of the American Society of Animal Science (1981), and is a member of American Society of Nutritional Sciences, American Society of Animal Science (Fellow), American Association for the Advancement of Science (Fellow), and other scientific societies related to animal agriculture.

SECTION I

ORIGINS OF THE MODERN PIG AND EVOLUTION OF PRODUCTION SYSTEMS

1

DOMESTICATION AND EARLY HISTORY

INTRODUCTION

Pigs have been a part of most human cultures since even before they were first domesticated. Archeological records indicate that their earliest domestication was underway in southwest Asia 9,000 years ago. By 6,000 to 7,000 years ago, pig domestication had spread into ancient Syria, Sudan, and Egypt, and then westward into Greece and southeast Europe. It later spread to western Europe (Porter, 1993). In the meantime, domestication was underway in eastern Asia. Archeological evidence indicates that pigs have been domesticated in China for 7,000 years (Porter, 1993). Most early civilizations have included pork in their diets since the domestication of the pig. Exceptions include Jews and Muslims, whose religious beliefs prohibit pork in the diet.

European explorers carried pigs with them when they settled in the New World. These pigs would seed the New World for generations and influence the importance of pork in the diets of North Americans.

Chapter 1 describes the origin of the domestic pig; its phylogenetic classification; the early development of pig production; and the history, current status, and future of pig production in North America and throughout the world.

PHYLOGENY

Pigs have a well-defined role in the animal kingdom. Paleontology evidence indicates that pigs originated about 40 million yr ago and were present in Europe, Asia, and Africa. Their wide geographic distribution at these early stages of their evolution was associated with wide variation in their food supply, the climatic environment, and their general characteristics. Presently existing wild pigs of Africa (bushpig, giant forest pig, warthog), Asia (babirusa, bearded pig, Javanese warty pig, Sulawesi warty pig), and Europe (Eurasian wild boar) probably resemble their early ancestors of a millenium or longer ago (Towne and Wentworth, 1950; Mellen, 1952; Porter, 1993) (See Figures 1–1, 1–2, and 1–3). Wild pigs currently roaming many areas of the United States were likely introduced by early explorers and settlers. It appears unlikely that North America was the site of origin of any of the contemporary wild pigs found in the United States. The peccary—not a true pig—is native to the Americas. The peccary is not in the same family, genus, or species as the domestic pig.

FIGURE 1–1
Warthog from Africa.

FIGURE 1–2
Babirousa Found in SE Asia.

FIGURE 1–3

European Wild Boar, the Ancestor of the Domestic Pig.

Pigs (wild and domestic) are even-toed ungulates, a characteristic shared with ruminant animals (e.g., cattle, deer, goats, sheep). Even-toed ungulates belong to the order Artiodactyla. Pigs are nonruminant Artiodactyla. A separate suborder, Suina, includes hippopotamuses, peccaries, and pigs. All true pigs are members of the Suidae family. Domestic pigs are descendants of the European wild pig, *Sus scrofa*, and the eastern Asiatic banded pig, *Sus vittatus*, a descendant of *Sus crystatus* (see Table 1–1).

TABLE 1–1

Phylogenetic Classification of the Pig.

KINGDOM	ANIMALIA
Phylum	Chordata
Class	Mammalia
Order	Artiodactyla (even-toed ungulates)
Suborder	Suina (hippopotamuses, peccaries, pigs)
Family	Suidae
Genus[a]	*Sus*
Species	*barbatus* (bearded pig of Malaya, Sumatra, Borneo)
	celebensis (warty pig of Sulawesi)
	crystatus (Indian crested pig or Asiatic wild pig of India)
	salvanius (pygmy pig of Southeast Nepal, Assam)
	scrofa (wild boar of Europe, Asia); domestic pig
	verrucosus (Javanese warty hog of Java, Sulawesi, Philippines)
	vittatus (Banded pig of Malay Archipelago)

[a]A total of five genera exist that can be called pigs. *Sus* contains seven species, including *scrofa*, the domestic pig (also sometimes incorrectly called *Sus domesticus*). The other four genera are *Babirusa*, *Hylochloerus*, *Phacochoerus*, and *Potomochoerus*. These four genera are not discussed here.

Source: Adapted from Towne and Wentworth (1950); Mellen (1952); Porter (1993); Pond and Mersmann (2001).

The Suidae includes animals from Africa and Asia that are pig-like, such as the warthog of Africa and the Babirusa of Indonesia, and many others. Members of the Suidae have sparse hair coats and show behavioral thermoregulation (swimming or rolling in mud and wallows). The *Sus* genus includes seven distinct species, including the domestic pig, *S. scrofa* (see Box 1–1). Domestic pigs vary greatly in size, color, body shape, ear carriage, behavior, prolificacy, and other traits. Some species of the genus *Sus* interbreed, so they may not be technically distinct species. Chinese and European pigs were intermixed several centuries ago, indicating that contemporary domestic pigs share many common genes. The domestic pig has 38 chromosomes (Gimenez-Martin et al., 1962) whereas the European wild pig has only 36; the hybrid pig has 37. McFee and Banner (1969) reported no evidence of reduced fertility associated with these different chromosome numbers.

\tBox 1–1 \t Domestic Pigs Are Known by Their Full Scientific Name as:					
Phylum	Order	Suborder	Family	Genus	Species
Mammalia	Artiodactyla	Suina	Suidae	*Sus*	*Scrofa*

EARLY PIG PRODUCTION IN EUROPE AND ASIA

European and Asian communities bred pigs that were unique to specific regions. During the height of the Persian, Egyptian, and Chinese civilizations, around 3000 to 4000 B.C., pigs figured prominently in the culture and diets of the time. Pig production in western Asia and the Mediterranean then spread north and west through Europe. Pig production in China and other parts of eastern Asia continued to expand.

The pig was an important part of the Roman Empire. Romans refined techniques for breeding and growing pigs. After the fall of the Roman Empire, breeding techniques and selection programs were lost, but the pig herds had already moved to the British Isles. The Celtic people also bred large numbers of pigs. Pigs became fully incorporated in the northern European cultures as well. Some cultures, including Jewish, Islamic, and Hindu allowed pork consumption in early times, but later forbade pork consumption. These cultures considered the meat unclean because: (1) many ancient cultures were nomadic and pigs did not herd easily; (2) pigs were scavengers that ate unhealthy carrion and waste products; and (3) the meat could contain parasites or bacteria that could cause illness.

By the time explorers set sail for the New World, pig breeding was well established in Europe. Game preserves held wild pigs for hunting and farmyards held organized pens of pigs. Most pigs were of one of two types developed by the Romans (large body size used for lard, drooping ears or small body size with less body fat, erect ears). Pigs of both types were propagated and improved in local areas.

Around the year 1500 A.D., most domestic pigs were dark colored, like the European wild boar. The Celtic type of pig represented pigs of northern Europe, whereas the Iberian pig represented pigs of the Mediterranean region. Celtic pigs were larger and generally leaner and more heavily muscled than Iberian pigs. Early explorers from the Mediterranean countries brought back Asian pigs to increase prolificacy. Celtic pig producers later added genes from Chinese pigs to their herds.

Pigs accompanied seafaring explorers on their voyages for three purposes: (1) for food; (2) to supply seed stock for future herd development; and (3) to be hunted on later visits. Pigs brought for food or herd development were the domestic pig types of the day, whereas pigs brought to hunt were more like the European wild boars prevalent on the continent.

HISTORY OF THE PIG IN NORTH AMERICA

PHASE I. THE EARLY EXPLORERS AND SETTLERS

Salt pork was a staple in the diet of seafaring explorers of the New World; fresh pork was a luxury. The tradition among explorers who intended to settle in the New World was to bring some pigs with them to develop an agricultural base. Christopher Columbus carried a famous group of eight pigs on his second voyage. These eight pigs, when left in Cuba, soon became feral (took on the behavior of wild animals, though domesticated). These pigs and those that followed on later explorations by de Soto, Ponce de Leon, and Cortez were probably Iberian-type pigs. Pigs arrived in California in the 1500s and many became feral.

John Cabot, the discoverer of New England, carried pigs, probably of the Celtic type, to that area of the United States. In the 1600s, during the establishment of the colonies along the East Coast of the United States, a complement of three sows (and apparently at least one boar) was brought to Jamestown, Virginia. Within 18 mo, the herd had reached more than 60 pigs and 2 yr later, it had swelled to 500 to 600 pigs.

The pilgrims who landed at Plymouth Rock received livestock brought by Edward Winslow. In the tradition of kings and landlords of cultures around the world, the animals were propagated for 4 yr and were then divided equally among the colonists. More pigs were added over time and many became feral, although their survival was impaired by the cold northern climate. As pig numbers increased, their noses were ringed to prevent rooting and destroying crops. Farms gradually extended westward to open more cropland for cultivation and livestock feeding.

In the early to mid 1600s, colonists could buy pigs from either the Virginia settlement or from the Caribbean. By the late 1600s, German, Dutch, English, and Swedish settlements in Pennsylvania were well established. The Pennsylvania practice of fattening pigs on corn spread north and south along the Atlantic seaboard.

As the population of pigs grew with their spread along the Atlantic seaboard, excess pigs were often released into the woods to fend for themselves. This practice resulted in a sizable feral population that remains a problem today. To the early settlers, however, pigs were wonderful animals because they seemed to thrive even with limited feed resources. East Coast pig farmers developed a booming export business through the 1700s. Massachusetts specialized in exporting barrels of cured pork and Virginia exported bacon. By the end of the American Revolution, pigs were the most numerous livestock species in North America.

PHASE II. POST-REVOLUTIONARY TIMES THROUGH THE CIVIL WAR

After the Revolutionary War from the late 1700s through the early 1800s, many farmers moved west over the Appalachian Mountains to establish new farmlands. The core of East Coast pig production remained from Virginia through Pennsylvania, New Jersey, and Massachusetts. Concentrated areas of pork production remain today in Pennsylvania and Virginia.

The westward expansion of agriculture led to the settlement of Ohio. A series of breeding programs by farmers in Ohio, as well as in Pennsylvania and Massachusetts, resulted in the formation of several breeds in Ohio, namely the Chester White, Poland China, and the Spotted Poland China (now named the Spot). Cincinnati, Ohio became the center of pork processing, where local pigs were slaughtered and shipped over the Appalachian Mountains to the large population centers along the East Coast. Salted, cured, and pickled pork allowed for transport over long distances to the east, making pigs the preferred livestock for long-distance marketing.

PHASE III. LATE 1800s TO LATE 1900s

The corn belt grew to the west through Indiana, Iowa, Missouri, southern Minnesota, southern Michigan, southern Wisconsin, and into eastern Kansas and eastern Nebraska. Rich, fertile soil, an abundance of rainfall, and more open plains resulted in a massive development of farmland. The production of corn and cereal grains—the crops of choice—soon far exceeded the needs of human consumption, and the pig proved to be an efficient converter of these feed resources to meat for human consumption. The pig soon became known as the "mortgage lifter" in Midwest agriculture and the region was soon referred to as the "corn belt," where grain production is greater than that of any other region of the United States.

Along with pork production, pig slaughter and processing were developed in Kansas City, St. Louis, and other Midwest cities, including Chicago, the stockyard giant for many years. The logic behind the location of processing facilities near grain and pig production was that pigs should be raised and slaughtered where the grain is grown rather than shipping grain to population centers of the East. It is more expensive to ship 3 or more lb of grain than to ship 1 lb of live pig to the markets, so the argument went. However, recent shifts in the cattle and poultry industries, and now in the swine industry, have cast doubt on the reliability and economics of this concept.

Farmers have traditionally sold a seasonal crop with major marketing in the spring and fall. This seasonal variation in the pork supply has moderated in recent years with the trend toward larger and fewer swine farms. In 1940, Iowa produced the most grain of any state. The grain was fed to cattle, pigs, and poultry. Indeed, Iowa was the number 1 producer of grain-fed beef, chicken, and pork. From 1955 to 1990, the beef cattle feedlots shifted to the feeding region extending from west Texas through Nebraska. At the same time, poultry production shifted from the Midwest to the Southeast. Now pig production, though still concentrated in the Midwest, has expanded into the Southeast, principally North Carolina, and into the arid high plains of the Southwest and beyond. Accompanying these shifts in pig distribution, pork slaughter and processing plants have appeared in these expanded areas of pork production. Large grain shipments now enter these newer areas of pork production, representing a shift from the earlier concept of concentrating pork production in grain production areas.

PHASE IV. LATE 1900s AND BEYOND

In 1990, Iowa ranked first in pig numbers, Illinois was second, and North Carolina was seventh. By 1995, North Carolina had passed Illinois and is now second in number of pigs produced. The trend continues, with more and more pigs produced out of the core of the midwestern United States and into the rural areas near eastern population centers, the arid western plains, and even into the deserts of Utah and Arizona. Reasons for the redistribution of the pig population include environmental concerns, government regulations, and a desire to produce pigs in isolated areas to control pathogens and avoid encroachment by urban sprawl.

CURRENT PIG NUMBERS AND DISTRIBUTION AROUND THE WORLD

Pigs are found throughout the world, but their distribution is not uniform. Pigs are found in great numbers in most of Asia, Europe, and North and South America. However, cultural and religious differences among regions of the world are partially responsible for differences in distribution patterns. Table 1–2 shows the amount of pig meat produced around the world in 2000, along with the human population and the amount of pork available per capita per yr in each country or region, based on population in 1999 (FAOSTAT, 2001).

TABLE 1–2
Human Population and Pig Meat Production Around the World, 2000.

COUNTRY OR REGION	HUMAN POPULATION (BILLIONS)[a]	PIG MEAT PRODUCTION (T)	PIG MEAT/ CAPITA/YR (LB)
Asia (incl. China and Japan)	3.634	50,371,208	28.2
China	1.274	43,053,600	74.4
Africa	0.767	565,036	1.6
European Union	0.375	17,565,100	103.0
Eastern Europe	0.121	4,899,593	88.9
South America	0.341	2,962,215	19.1
Canada	0.031	1,675,000	119.1
Japan	0.127	1,270,000	22.1
Mexico	0.097	1,034,906	23.3
Republic of Korea	0.047	940,836	44.4
Russian Federation	0.147	1,250,000	18.7
United States	0.276	8,532,000	68.0
World Total	5.978	91,030,043	33.4

[a]1999 FAOSTAT (2001) figures for human population.
Source: FAOSTAT (2001) http://apps.fao.org/

TABLE 1-3
World Meat and Milk Production in 1970 and 2000.

SOURCE OF MEAT OR MILK	AMOUNT PRODUCED MT/YR (MILLIONS)		PERCENT INCREASE, 1970 TO 2000
	1970	2000	
Pig	35.8	91.0	254
Beef and veal	38.4	57.1	149
Chicken	13.1	56.9	434
Turkey	1.2	4.8	400
Sheep	5.5	7.6	138
Goat	1.3	3.7	285
	–	–	–
Total Meat	100.4	233.2	233
Cow milk	359.3	484.7	135

Source: FAOSTAT (2001). http://apps.fao.org/

On a worldwide basis, more pork is produced and eaten than any other meat (Table 1–3). In 2000, the world consumed 59% more pork than beef, its closest competitor. However, chicken consumption and production are rising rapidly in the United States and around the world. If the current trend continues, pork and chicken consumption will be equal by about the year 2020. However, food consumption patterns vary over time for reasons that include changes in economic status, political unrest, and many others.

PIG NUMBERS IN EACH COUNTRY

In a given region or country, the amount of pork served in restaurants and homes may vary from none to an abundance, for ethnic or religious reasons. Muslim, Hindu, Jewish, and several other religions forbid pork consumption. Many ethnic groups in India and parts of South Asia do not eat pork. However, peoples of central and northern parts of Asia produce more pork and eat more in total (not per capita) than peoples of any other region of the world. Latin Americans eat more pork than any other meat overall, but peoples of some countries (e.g., Argentina) eat beef in the largest amounts.

Europe has an interesting distribution of meat consumption. Overall, Europeans eat more pork than any other meat, but in some countries, beef, lamb, or poultry lead the way. People in Mediterranean countries, except for Spain and Portugal, tend to eat less pork, whereas northern Europeans tend to eat more pork than other meats.

Nearly all pork produced in most countries is consumed locally. Notable exceptions are Taiwan and Denmark. On a per capita and historic basis, Taiwan produces more pork than any other major Asian country. Taiwan has more than 22 million people and produces about 128 lb of pork per person. A great volume of pork has been targeted for export, especially to Japan. Taiwan, whose limited land area precludes feed production to satisfy a large pig industry, imports grain and soybeans for feed formulation to support its pork export business. The high pig density in Taiwan creates a

TABLE 1–4
Pig Meat Production and Consumption for the 15 Member Countries of the European Union and for the United States, 1999.

COUNTRY	POPULATION (MILLIONS)	PIG MEAT PRODUCED T	PER CAPITA (LB)	PIG MEAT CONSUMED T	PER CAPITA (LB)
Austria	8,177	499,000	134	547,600	147
Belgium/ Luxembourg	10,579	1,004,700	209	350,382	73
Denmark	5,282	1,641,800	684	336,416	140
Finland	5,165	181,860	77	176,427	75
France	58,886	2,377,000	89	2,220,788	83
Germany	82,178	3,979,800	107	4,576,973	123
Greece	10,626	138,300	29	295,597	61
Ireland	3,705	250,700	149	139,324	83
Italy	57,343	1,471,702	56	2,052,024	79
Netherlands	15,735	1,711,000	239	960,632	134
Portugal	9,873	349,601	78	403,773	90
Spain	39,634	2,893,000	160	2,371,381	131
Sweden	8,892	325,400	81	2,347,492	86
United Kingdom	58,974	1,047,000	39	1,526,129	57
TOTAL	375,039	17,860,863		16,204,899	
United States	276,218	8,532,000	68	8,475,192	67

challenge, as in other small countries such as Denmark and The Netherlands, in dealing with the high volume of manure.

In Denmark, as in Taiwan, a high proportion of the pork produced is exported, although, as in Taiwan, Denmark's per capita pork consumption is high. Denmark takes pride in exporting high-quality pork products. It has major markets in the United Kingdom (a net pork importer) and Asia, particularly Japan. About 80% of Danish pork is exported. Denmark controls about 20% of the world trade in pork. Pork production in the 15 member countries of the European Union (EU) is shown in Table 1–4. Denmark, Belgium/Luxembourg, The Netherlands, Spain, and Austria have the highest per capita production of pork in the EU.

Prior to 1996, the United States was a net pork importer. Pork exports from the United States in 1997 were valued at about $1 billion. Pork exports from the United States are expected to increase in the future as a means of adding value to domestic feedstuffs—an alternative to exporting them directly.

DISEASE THREATS TO PORK EXPORTATION

The threat of spreading contagious diseases from one country to another is a major factor in determining export patterns of pork and other animal products. A recent example of such an occurrence is the 1997 crisis in Taiwan when the foot-and-mouth disease virus was confirmed. In 1996, the Taiwan pig industry was valued at more than

$3 billion, with about one-half of the production exported. The disease outbreak caused a voluntary halt in pork exports because other countries feared infection from this serious and highly contagious virus. Mainland China and other Asian countries are known to have foot-and-mouth disease in their pig populations. In 1997, The Netherlands had an outbreak of hog cholera that stopped its pork exportation. In early 2001, foot-and-mouth disease was diagnosed in cattle in the United Kingdom and in continental Europe, causing deep concern among other countries that the disease would spread. Newly infected countries placed severe restrictions on all exports of animals and animal products and the United States and other countries imposed severe quarantine and inspection measures to minimize the further spread of foot-and-mouth disease. Not only are cattle, sheep, goats, and swine susceptible to foot-and-mouth disease, but several wild animal species (e.g., deer) are also vulnerable. The introduction of foot-and-mouth disease, hog cholera, African swine fever, and numerous other contagious diseases into the United States would deal a devastating blow to the livestock industry and to the entire economy. These and other diseases are discussed further in Chapter 20.

THE FUTURE OF PORK PRODUCTION

Who will produce pork for the growing world demand in the future? No one can predict accurately, but there may be an indication about the potential for future production by looking at costs of production around the world. The cost of pork production in some less-developed countries is very low. For example, in China and Brazil, feed resources are generally available, but the infrastructure is inadequate at the present time to allow a massive export effort. However, China, Brazil, and other countries, particularly those in Asia and South America, should be watched as they continue to develop pork production and marketing systems.

Canada and the United States have the lowest cost of production among the pork export countries of the Western Hemisphere. As the future unfolds, countries with low production costs will have an advantage in the export market, but the ultimate determinant of total world pork production will be consumer acceptance of pork. Consumer acceptance is based on such factors as palatability, price, and perceptions of safety and environmental friendliness of the pork production and processing systems.

SUMMARY

Chapter 1 sets the stage for understanding modern pork production and how pork production contributes to contemporary society. It begins with a brief history of domestication of the pig and its place and phylogenetic classification in the animal kingdom. The general origins of domestication and the early development of pig production in Europe and Asia are introduced, followed by a brief history of pig production in North America, including the activities of the early explorers and settlers through the 1700s and extending through the 1800s and 1900s into the 21st century. The chapter ends with a snapshot of current pig numbers and distribution in the world and the emerging globalization of pig production through export marketing of pigs, pork, and pork products.

QUESTIONS AND ACTIVITIES

1. From an evolutionary perspective, the Artiodactyla originated as a large-bodied (cow size) nonruminant that looked like a giant pig. Specialization developed over time—some animals developed a rumen while others did not. What are the major advantages of having a rumen for plains and forest-dwelling animals? What are some advantages a smaller-bodied animal like a pig might have for not developing a rumen?
2. Why do you think early explorers carried domestic pigs on their ships (say in the 1500s) rather than a similarly sized ruminant (like a sheep or goat)? Why not cattle?
3. Why do some cultures and religious sects forbid pork consumption?
4. Examine the pig numbers in selected countries like The Netherlands, Canada, Denmark, Japan, Taiwan, and the United States over the past decades. Why are pig numbers changing in these countries? What do the data suggest about how pig numbers may change in the future? The FAO and USDA Web sites will be of help.

LITERATURE CITED

FAOSTAT. 2001. Agriculture Data, United Nations, Rome, Italy. Available at: *http://www.fao.org*.

Gimenez-Martin, G., J. F. Lopez-Saez, and E. G. Monge. 1962. Somatic chromosomes of the pig. J. Hered. 29:281.

McFee, A. F. and B. W. Banner. 1969. Inheritance of chromosome number in pigs. J. Reprod. Fertil. 11:161–163.

Mellen, I. M. 1952. The Natural History of the Pig. Exposition Press, New York, NY.

Pond, W. G. and H. J. Mersmann. 2001. Biology of the Domestic Pig. Cornell University Press, Ithaca, NY.

Porter, V. 1993. Pigs: A Handbook to the Breeds of the World. Cornell University Press, Ithaca, NY.

Towne, C. W. and E. N. Wentworth. 1950. Pigs from Cave to Cornbelt. University of Oklahoma Press, Norman, OK.

INTERNET RESOURCES

General pig information:
http://www.porkboard.org/Home/default.asp
http://www.nppc.org/

http://www.porkscience.org/default.asp

Pig trivia and history may be found at:
http://www.mnpork.com/education/swine.php3
http://www.geocities.com/TheTropics/Shores/7484/index.html
http://www.northcanton.sparcc.org/~orchard/pigs/pigs.htm
http://members.ozemail.com.au/~rroutley/

2

PROFILE, STRUCTURE, AND COMPLEXITY OF MODERN PIG PRODUCTION AND MARKETING IN THE UNITED STATES

INTRODUCTION

Pork production was a side-line business on the majority of U.S. family farms 100 years ago. Today, the majority of pork in the United States comes from just a few farms. Pig farmers today must compete in the low-cost and fluid commodity markets or they must develop niche markets based on some special element of their production method. While the majority of U.S. pigs come from highly uniform production methods, producers can choose from a wide diversity of production methods and production unit sizes. Chapter 2 describes a snap-shot in time of modern, U.S. pork production.

NATIONAL AND INTERNATIONAL COMPETITION

Who will produce pork for the growing world demand in the future? No one can predict with accuracy; however, an indication about the potential for the future can be seen by looking at cost of production around the world. The cost of production in some less-developed countries is very low; for example, China and Brazil have some available feedstuffs (especially Brazil), but the infrastructure is inadequate at this time to allow for a massive export effort. These countries and others in Asia and South America are expected to develop pork production and export capabilities.

The costs of production in Europe, Canada, and the United States (representing the major pork exporting countries in the world) vary considerably. As shown in Table 2–1, Canada and the United States have the lowest cost of production among the pork exporting countries. With the United States being a major world

TABLE 2-1
Cost of Production in Selected Countries and Regions in North America and Two Exporting Countries of Europe.

COUNTRY	COST OF PRODUCTION, $/MARKET HOG[a]
United States—Western corn belt	78
United States—Eastern corn belt	80
Canada—Western prairies	79
Canada—Eastern prairies	69
The Netherlands	150
Denmark	145

[a]Market weights are heavier in North America than in Europe.
Source: Data were collected by Martin and Kruja (1999).

FIGURE 2-1
Map of U.S. Grain-producing Areas.

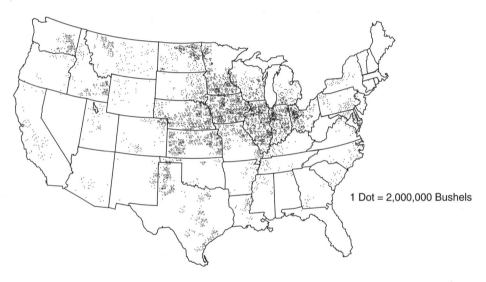

1 Dot = 2,000,000 Bushels

competitor in potential pork production, where will the industry be located and how will it be structured?

Pig distribution in the United States has traditionally been centered around the grain-producing areas in the Midwest, the region that produces the highest amount of grain and pork (see Figure 2–1). A recent shift in growth of the U.S. pig industry has seen gains in pig numbers in the Southeast, especially in North Carolina. Recent growth regions include Oklahoma, Colorado, Texas, and Arizona.

The top 17 states in pig inventory are listed in Table 2–2. Iowa has the greatest number of sows and market animals; however, in 1996 the numbers of piglets farrowed was greater in North Carolina than in Iowa. This reflects a higher overall efficiency of North Carolina sow herds than of those in Iowa.

TABLE 2–2
Numbers (thousands) of Breeding Pigs and Growing (market) Pigs and Total Numbers of Pigs in Inventory in 1997 in the Leading U.S. States. Table Values Are in Thousands of Pigs.

STATE	BREEDING	MARKET	MARKET: BREEDING ANIMAL RATIO	TOTAL
Arkansas	110	750	6.8	860
Georgia	105	695	6.6	800
Illinois	550	3,850	7.0	4,400
Indiana	450	3,150	7.0	3,600
Iowa	1,300	10,800	8.3	12,100
Kansas	180	1,180	6.6	1,360
Kentucky	80	520	6.5	600
Michigan	130	820	6.3	950
Minnesota	600	4,400	7.3	5,000
Missouri	460	2,840	6.2	3,300
Nebraska	450	3,100	6.9	3,550
North Carolina	1,050	8,450	8.0	9,500
Ohio	200	1,250	6.2	1,450
Oklahoma	180	1,190	6.6	1,370
Pennsylvania	120	850	7.1	970
South Dakota	160	970	6.1	1,130
Wisconsin	115	675	5.9	790
Other states	560	3,610	6.5	4,170
United States Total	6,800	49,100	7.2	55,900

The United States has about 7.2 times as many market animals as breeding animals (Table 2–2). A low ratio of market-to-breeding animals reflects a low breeding herd productivity or a large number of feeder pigs shipped to other states, or both.

Midwestern states account for 11 of the top 17 states in pig inventory. Only Arkansas, Georgia, North Carolina, Kentucky, Pennsylvania, and Oklahoma are out of the core Midwestern states, and only Pennsylvania and Oklahoma are out of the Midwest and Southeast centers of pork production. The growth in pig inventory in Oklahoma, Texas, and Colorado suggests that these states will be significant centers of production in the future due to the region's dry climate and rural setting. Pennsylvania's proximity to heavily populated areas, rivers, and waterways makes the future of the pig industry in this state questionable. The industry has the same problem in the southeastern United States and, to a large extent, the Midwest. This leaves major U.S. expansion to the Southwest and West.

HOW THE STRUCTURE OF THE INDUSTRY IS CHANGING

The United States Department of Agriculture (USDA) tracks pork producers using the criterion that a pig producer is anyone with a pig on a farm at any point in a given year (See Figure 2–2). Using this criterion, 61% of U.S. farms had only 3% of the pigs in the United States in 1996 (roughly 1.5 million head). At the other end

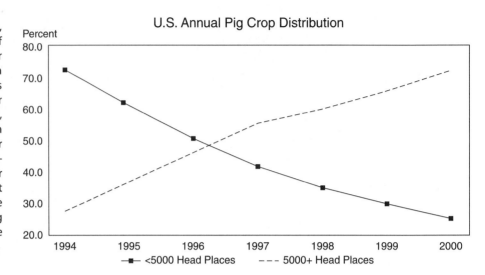

FIGURE 2–2
Data from USDA (USDA, NASS, 2001) on Percentage of Pig Farms with Less than or More than 5,000 Pigs/Farm. In 1994, Less than 25% of all Pigs Were Raised on Farms of Over 5,000 Pigs; However, by 2000, 73% of all Pigs Were Raised on Such Large Farms. Well Over Two-Thirds of all Pigs Are Produced on Farms Having Over 5,000 Pigs. These Recent USDA Figures Point to the Rapid Industry Restructuring That Took Place through the 1990s in the United States.

of the spectrum, larger farms with over 1,000 head of pigs represented only 7.5% of the farms, but had over two-thirds of the pigs.

In 1980, the United States had just over 600,000 pig operations (each containing one pig or more). In 1990, the number of hog operations dropped to less than 300,000. The number of pig operations was less than 200,000 in 1996. In 1999, there were 98,460 U.S. pig operations; of this number, 7,125 operations (7.2% of all operations) produced 68.5% of all U.S. pigs. The number of operations in several size categories (less than 100 head through greater than 5,000 head) and the change from 1994 are shown in Figure 2–2. The same trend occurs in Europe, Asia, and Australia. The pig industry is consolidating with fewer, but larger, farms.

In the 1990s, the U.S. pig industry experienced a trend toward growth of so-called "megafarms"—farms with over 10,000 sows and over 200,000 pigs marketed per year (Table 2–3). The number of megafarms is growing rapidly; there were 43 pig units with over 10,000 sows in 1996, but there were 53 such megafarms identified in 1997. These 53 farms had a single-year increase in sow numbers of 18% and they now represent just over 30% of the U.S. sow herd. In 2001, the top 10 farms had 2.5 million sows on 25 farms with an average sow herd size of about 99,000 sows and an annual average pig output of over 1,800,000 pigs/farm. While megafarms represent about 30% of the U.S. sow herd, they produce over 40% of the finished pigs because they are more efficient, *on average,* than smaller farms. Yet, megafarms are not as efficient as the very best of the smaller farms. Average U.S. hog farms produce about 17 pigs per sow per year. While megafarms produce from 18 to 22 pigs per sow per year, some smaller units (and the occasional large unit) can produce 24 to 25 pigs per sow per year. Sow herd productivity is not the only measure of success. If the cost of production is lower, the profit potential is greater. Large farms have low costs due to economy of scale (i.e., buying in large quantities at a lower price). Small farms lower their costs by supplying their own grain or having lower labor costs.

TABLE 2–3
Top 10 Pig Farms in the United States in 2001.

Rank	Farm	Location(s)	# Sows
1	Smithfield Foods[a]	NC, MO, OK, IL, UT, VA	710,000
2	ContiGroup Companies[b]	MO, NC, TX	201,000
3	Seaboard Corporation	KS, CO, OK, TX	185,000
4	Triumph Pork Group	OK, NC, CO, IA, WI, IL, MN	140,000
5	Prestage Farms	NC, MS	122,000
6	SMS of Pipestone	MN, SD, IA, NE, OK, IN, OH	120,000
7	Cargill	NC, AR, OK	109,000
8	Tyson Foods	NC, MO, AR, OK	107,000
9	Iowa Select Farms	IA	100,000
10	Christensen Farms	MN, NE	80,500
		Total	1,874,500[c]

[a]Smithfield now includes Carroll Foods and Murphy Family Farms.
[b]Includes Premium Standard Farms and Continental Grain.
[c]An increase in 15% over the previous year. This represents over 35 million market hogs produced/year, or about one-third of U.S. pig numbers. The top 25 farms in 2001 had 2.5 million sows and produced about 50% of U.S. market hogs.
Source: Successful Farming (October, 2001).

MODERN PORK PRODUCTION FARMS: FARM STYLES, SITE LAYOUTS, AND PRODUCTION SCHEDULES

Modern pork production units are located in regions and locales that are very different than those of past years. Historically, pig farms were located in regions or on farms that produced a large amount of grain. The pigs were used as a way of "walking the grain off the farm," or adding value to the grain. Rather than selling $1.00's worth of grain, the farmer could generate a gross market value of $2.00's worth of live pig.

At the same time that farmers were adding value to their grain, pig farms were getting larger and fewer in number. As farms grew in size, new challenges arose, the most significant in the areas of pig health and waste/nutrient management. Producers who wished to protect their pigs from neighbor pigs and wildlife pathogens moved their farms to remote locations in Arizona, Utah, and the southern high plains. Larger farms had greater waste management issues and, thus, needed a more remote location.

LEGAL REQUIREMENTS

Modern pig farms, which, by definition, are larger than farms of the past, must meet certain legal requirements. Many of these requirements were not on the books even 10 years ago. Legal requirements can be found in areas of:

- Set-back distances from other farms or houses
- Water or well requirements
- Waste treatment rules and regulations

- Groundwater protection
- Air quality protection
- Fire, flood, and disaster considerations

Set-back requirements are the distance a farm must be from other houses or structures. State regulations vary from no set-back requirement to complex requirements. In the state of Illinois, farms with less than 50 animal units (125 pigs over 50 lb) have no set-back requirements. Farms with over 7,000 pigs over 55 lb must be 1/2 mile from nonfarm residences and 1 mile from populated areas. The more pigs, the greater the set-back distance required. In some states, farms must be a regulated distance from people and must meet requirements to protect groundwater.

Before site selection can be finalized, the pork producer needs to know certain other features of the farm and the location. Among the considerations are the:

- Available capital
- Through-put of pigs required
- Environmental situation (laws, regulations, farm situation)
- Nearest neighbors and the need to reduce offensive odors
- Need to capture manure nutrient for crops
- Skills of the stock people
- Availability of bedding
- Availability and cost of power

Clearly, a different site layout is required if the farm is indoors versus outdoors. Getting roads and power to the site is a major expense for farms in remote locations. With a drive toward more sustainable production systems, integration of the pig farm with a cropping system is desirable. The skills of the available workforce may determine which type of production system is built (indoor versus outdoor). The availability of bedding will impact the choices for production system as well. Some locations do not have a large supply of bedding (e.g., The Netherlands) and because bedding is a low-value, bulky substance, transportation costs to move it great distances favor its use locally. When wheat straw, for example, is transported hundreds of miles, the transportation costs can exceed the cost of the material.

SITE LAYOUTS

Farms must accommodate seven major production phases:

- Breeding
- Gestation
- Farrowing
- Nursery
- Growing
- Finishing
- Isolation of replacement breeding stock

The farm site may have from one to seven "sites"—a collection of barns or rooms that are separated by some distance.

In recent decades, pig farm layouts have undergone major changes. In the 1970s, production units used one-site production, in which the isolation facility was a wing off to the side and all of the other buildings were under one roof or were a collection of outdoor pens connected by short walkways.

With one-site production, workers have an easy time moving from one room to another. The main disadvantage of one-site production is that disease organisms flow from pigs of one age group to the other groups freely. Sows infect the piglets and then the growing pigs can re-infect the sows. Once a pathogenic organism enters the barns, it is nearly impossible to get it out, short of depopulating the farm. Disease organisms are common on commercial farms, so the one-site production model is quickly becoming a thing of the past.

At the very least, the sow herd should be located in a separate site from the growing pig herd. Breeding and gestation can be located in separate buildings, but they can also be combined under one roof (see Figure 2–3). The main difference between modern farms and those of the past is the number of sites the growing (nursery, growing, and finishing) pigs occupy. Three models are available on pig farms today, but the current move is toward the wean-to-finish facility (pigs in a single building or site from weaning until market). If the farm has a breeding/gestation/farrowing complex as one site, the growing pigs are either in one or two sites. Thus, the farm would be said to be a two-site (breeding/gestation/farrowing + nursery/growing/finishing) or three-site (breeding/gestation/farrowing + nursery + growing/finishing) facility.

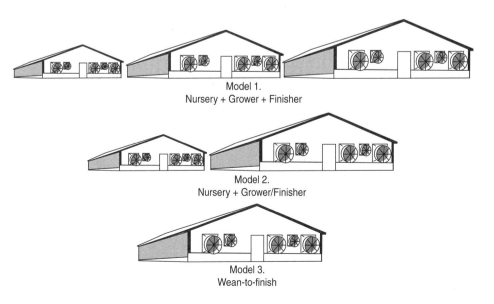

FIGURE 2–3
Three Models of Facilities for Growing Pigs. Model 1 (three buildings) Was Popular in the Early 1980s. Model 2 Was Popular in the Early 1990s. In the Late 1990s, the Third Model Became a Popular Scheme for Taking Pigs, in a Single Building, from Weaning (generally at less than 21 days) to Finishing (slaughter weight in the United States of over 260 lb).

PRODUCTION SCHEDULES

Modern pig farms have a different look and organization than old-style, small-scale farms. Even modern farms built on a small-scale should incorporate the improved features of modern farms, the production schedules and site layouts.

Hardy (1998) presented two diagrams that illustrate the modern-day pork production unit. Figure 2–4 shows the life cycle of the pork production unit. Conception to processing requires 305 d, but phases within the production cycle require different facilities and management.

Figure 2–5 shows Hardy's (1998) organization of a modern, 80,000-sow production unit (producing 1.6 million market hogs/yr). The unit is self-contained, with genetic multiplication, feed manufacturing, and pork processing all in the same company. This production size (80,000 sows) is less efficient compared with production units that process 4 million pigs/processing plant (the offspring from 200,000 sows).

Like the site layouts, the production schedule for modern farms is very different than it was a few decades ago. The primary reason why production schedules changed is that farms became much larger. When a pork producer had 100 sows, it was more efficient to group breedings, which results in batch farrowing and batch marketing. A 200-sow farm that had four breeding groups could farrow 50 litters in a batch and thus market 400 to 500 pigs at a time. Semi-trucks in the United States hold about 200 pigs, so a single breeding/farrowing batch would work well with two truckloads going to market. Older sources such as the Pork Industry Handbook (Purdue University) gave examples of 1, 2, 5, 7, and 10 batch farrowing production schedules.

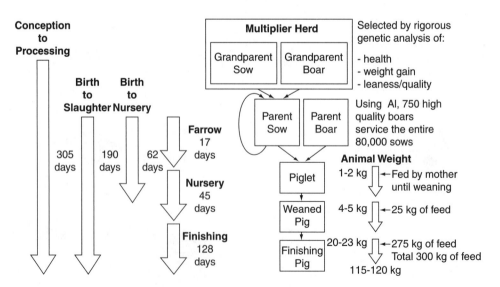

FIGURE 2–4
Life Cycle of a Pork Production Unit.

Source: Adapted from Hardy (1998).

Regardless of the production schedule, if farms are producing a high number of pigs weaned per sow per yr, it is the weekly output of pigs/wk that is important. A number of factors can determine the numbers of pigs produced per wk, including:

- The size of trucks taking pigs to market—if a truck holds 200 pigs, then marketing 200 pigs/wk is an efficient use of transportation. Thus, 500 sows are needed to produce one truckload/week.
- If the finishing barns hold 1,000 pigs of each gender, then the sow units need to produce 2,000 pigs/wk—this level of production requires 5,000 sows.
- If the packing plant wants to process 16,000 pigs/d, then the plant needs 4 million pigs/yr, or the output of 200,000 sows.

Table 2–4 shows the numbers of sows that represent different numbers of pigs marketed per year and per week.

FIGURE 2–5
Organization of a Modern, 80,000-Sow Production Unit.

Source: Adapted from Hardy (1998).

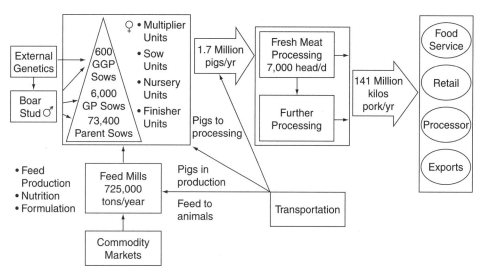

TABLE 2–4
Approximate Numbers of Pigs Marketed per Year and per Week when the Farm Markets 20 Pigs per Sow per Year.

NUMBER OF SOWS	APPROXIMATE NUMBERS OF PIGS MARKETED	
	PER YR	PER WK
500	10,000	200[a]
1,250	25,000	500
2,500	50,000	1,000
5,000	100,000	2,000[b]
10,000	200,000	4,000

[a]Represents one truckload/wk.
[b]Represents 1,000 pigs of each sex each week.

Modern farms are often on a weekly production schedule. With the weekly production schedule, the farm production cycle is divided into 20 groups or weeks. The farm might breed sows each week and farrow each week. The following is an example snap-shot of the weekly system:

Week	Status of Group
1	Breeding
2	Early pregnancy
3	Early pregnancy
4	Pregnancy check by boar and ultrasound
5	Gestation
6	Gestation
7	Gestation
8	Gestation
9	Gestation
10	Gestation
11	Gestation
12	Gestation
13	Gestation
14	Gestation
15	Gestation
16	Gestation
17	Farrowing, week 1
18	Farrowing, week 2
19	Farrowing, week 3
20	Weaning, return to estrus

On the weaned pig side of the farm, the production system would require about 20 to 24 wk to reach a market weight of about 114 kg (250 lb) starting at a weaning weight of 5 to 7 kg (11 to 15 lb) at 21 days of age.

PIG FLOW

With 20 breeding groups of sows, the farm would breed, farrow, wean, and market a batch of pigs each week. The flow of pigs from the unit is determined by the productivity in such measures as numbers bred, conception/farrowing rate, pigs born alive, and mortality. Based on a modest productivity of 20 pigs marketed per sow per yr, a farm of 1,250 sows would market 25,000 pigs/yr. A farm of 2,500 sows would market 50,000 pigs or about 1,000 pigs/wk—500 males and 500 females. In farms where producers finish pigs in single-sex barns with 1,000 pigs each, the best pig flow is obtained with 5,000 sows marketing about 2,000 pigs each wk (1,000 of each sex).

Table 2–4 lists the weekly numbers of pigs from farms of various sizes. One can see that the 1970's estimate of the proper size farm (500 sows) is based on marketing one truckload of pigs per week. Current pig flow requires at least 5,000 sows marketing 100,000 pigs/yr.

Another estimate of a good-sized farm is to market 8,000 pigs/wk, so as to provide one shift's kill at a packing plant. This is attractive to both the producer and the processor (around 20,000 sows producing 400,000 pigs/yr). To fill all five days/wk at a pork processing plant requires about 100,000 sows (2 million pigs/yr). The new economy of scale has 200,000 sows (or a bit more) to produce 16,000 pigs/d, processing five days per week using two (8h) shifts/d (the third shift is for clean-up and equipment maintenance).

With the drive toward larger-sized farms, consolidation of the larger farms to capture greater amounts of the processing capacity of large-scale plants is expected. If the processing plants grow in size, the production units may do the same—probably in multiples of 5,000 sows. Thus, over the last 25 years, the size of dedicated, single-pork production units has effectively increased from 500 to 5,000 sows. The next jump may be to much larger farms that capitalize on economies in pork processing.

SUMMARY

The U.S. swine industry is rapidly changing in structure and complexity. Chapter 2 describes these changes and characterizes the trends toward larger, integrated systems of production. Site layouts, production schedules, legal requirements, and pig flow through these facilities are addressed. Consolidation of large farms to capture a greater portion of the processing capacity of large-scale slaughter and processing facilities may be expected as the industry moves toward greater integration of feed processing, pig production, and meat processing.

QUESTIONS AND ACTIVITIES

1. The USDA publishes a quarterly report called the Hog and Pig Report. The December issue reports data on pig numbers in each state. Certain states have changed dramatically over the past 20 yrs in pig numbers. On a graph, plot the pig numbers each 5 yr for the past 20 yr for Iowa, North Carolina, and Oklahoma. What are the trends?

2. Kansas passed an anti-corporate farming law in the early 1980s. The law was partially repealed in the mid 1990s to allow county-by-county approval or disapproval of pork production units. Track the sow herd numbers in Kansas from 1980 through the present year.

3. Modern-day American pork processing plants process about 16,000 pigs/d and they work for 250 d/yr. Assuming that 20 pigs are marketed per sow per year, how many sows are needed to process the following numbers of pigs/d:
   a. 16,000 pigs/d
   b. 8,000 pigs/d
   c. 4,000 pigs/d
   d. 500 pigs/d

4. Assume a farm has 10,000 sows and produces 200,000 pigs/yr. How many pigs might be produced/wk? If pre-weaning mortality suddenly increases from 15% to 25% due to a disease outbreak and this continues for 4 wk, how many fewer pigs will be marketed when those pigs are finished? At a $100 value per market pig, how much will this disease cost in terms of acute loss of marketed pigs?

INTERNET RESOURCES

These pages give updated statistics about the pig industry:
http://www.nppc.org/PorkFacts/pfindex.html
http://www.usda.gov
http://www.agriculture.com/sfonline/sf/2001/october/0111pork powerhouses.html

Breeds of pigs:
http://www.ansi.okstate.edu/breeds/swine/

Example regulations:
http://www.epa.gov/
http://www.ipcb.state.il.us/statutes/compiled statutes.htm

LITERATURE CITED

Hardy, B. 1998. Management of large units. pp 561–581. In: Progress in Pig Science, edited by J. Wiseman, M. A. Varley and J. P. Chadwick. Nottingham University Press.

Illinois Department of Agriculture. Available at: *http://www.agr.state.il.us/IMAGES/Aglogo.gif.*

Successful Farming. Available at: *http://agriculture.com/sfonline/.*

USDA. 2001, June. U.S. Hog Breeding Herd Structure. Agricultural Statistics Board, NASS.

3
SOCIETAL ISSUES FOR PORK PRODUCTION

INTRODUCTION

A century ago, many people were familiar with agriculture production practices and most had hands-on experiences producing food and fiber. Today, fewer than 2% of the U.S. population is directly involved in agriculture production. Not only are people unfamiliar with agricultural production practices, they often do not like some production practices when they become aware of production techniques. The U.S. population has a heightened concern for pollution, food safety, and care of pigs on commercial farms. This chapter addresses these concerns and issues. In dealing with societal issues, unlike traditional hard-science disciplines (e.g., nutrition, genetics, etc.), the pig industry must address issues of ethical values and consumer perception. Although animal scientists work in the world of science, they cannot ignore consumer thinking and behavior.

CHALLENGES AND ISSUES

Modern-day pork production faces a major challenge in dealing with animal activists. Most activists who oppose pig production practices have good intentions, but others exaggerate the issue to cause change. Propaganda is produced on several important issues—some truthful and some not. Distinguishing myth from fact is the first step toward developing a logical position on any issue. How many myths can be listed for domestic pigs? A few that come to mind are "facts" like:

- Sweating like a pig [pigs do not sweat].
- Your room is like a pig sty [pigs are, in fact, clean animals when given a chance].
- Pigs stink [when cared for properly, pigs do not smell any worse than other animals].

Commercial pork producers find it very difficult to deal with societal concerns when the public holds so many myths. Informed decision making begins by understanding the facts.

Pigs and pig production units are subjects of societal concern. Some of the concerns are warranted and some are clearly not. Unfortunately, in today's society, if someone or some group has a concern about a particular farm, the farm has a problem with which it must deal. Entire industries have been wrecked from scares that lack validity—a prime example is the Alar scare that unjustly caused a brief decimation of the apple industry. Apples were said to be "contaminated" with Alar, which, in fact, was a harmless chemical.

> If the public even perceives a problem with the industry, then the industry has a problem with which it must deal.

If the concern is real and based on facts and reason, the pig industry must act to improve the situation. If the concern is only perceived by society, the pig industry must still act to educate the public about the situation.

This chapter raises a number of issues in case studies designed to spark discussion. Each issue is first presented from the point of view of concerned citizens. Then, facts and descriptions are given to support the pig industry. Each issue is presented as a polarized, societal concern. More than two views may be held on each issue.

Readers should be able to relate to each side of each issue (pro and con) because sooner or later, most people are likely to confront individuals or groups who hold these views. The advantage held by students of pig biology—and other informed citizens—is that they can gather and interpret pig-specific information from a variety of diverse sources and synthesize the information into a rational, informed argument that supports their point of view.

Many of these issues cause heated debate in some circles. Discussion and debate are healthy ways of learning and moving opinion in a logical direction. For the most part, people who hold extreme views are not "bad people." Those who hold opposing views should find ways to agree to disagree at the end of a heated debate and should also try to look at the issue from the other person's point of view.

In addressing societal issues, individuals or groups should adopt the following sequence of learning:

- Read about the issue from at least two sides.
- Listen carefully to the opposing view.
- Study the opposing side's arguments.

- Organize the arguments on paper.
- Speak on the topic from a given position.

The sequence that leads to action saves speaking on each issue for the last step. When individuals do finally speak, they should remember the following words, paraphrased from President Abraham Lincoln:

It is better to remain silent and be thought a fool, than to open your mouth and remove all doubt.

Case Study #1—Pig farms smell!

In one Nebraska county, National Farms[a] had a cluster of farms that had about 17,000 sows. Neighbors claimed that on a normal day, buildings emitted an offensive odor. When the lagoons were emptied and the old manure was stirred, the farms emitted a highly offensive odor. Neighbors sued National Farms because they claimed their quality of life was reduced by being exposed to the offensive odor of the pig farm.

The community benefited greatly from many high-paying jobs and from community service by their corporate neighbor. The farms were in place for more than 10 yr by the time the neighbors complained. Still, certain neighbors brought a lawsuit against National Farms because of the offensive odor and the court ruled against National Farms. National Farms has since sold that farm, in no small part due to neighbors complaints.

- What would you have done as a member of the jury in the case?
- What should the community do, if anything?
- What should National Farms do?

In the minds of some homeowners, nothing lowers their property values like having a pig farm built next door! Most people would not like to live next to a hog farm because everyone knows pig farms smell bad. Try answering these questions:

- Would you like to have your home next to a pig farm?
- Why are pig farms not inside the city limits of most towns?
- How is it that some pig farms, for example those in the Orient, are actually inside the city limits?
- If a pig farm did not smell, would you then like to live near one?

Older agriculture instructors (and even some younger ones) may say to their students, when the smell of the pigs reaches their noses on the first day of swine class at the farm, "Take a deep breath everyone; that is the smell of money!" The saying is still true, but rather than manure odor being the sign of a profitable agribusiness, it is more often associated in the popular press with lawsuits and money paid to neighbors.

---

[a] National Farms exited the pig industry in the late 1990s.

Most pig farms do emit an odor that many people find offensive. Pork producers do not mind the odor too much or they would be in another business. Farms do not have to emit offensive odors. Today's modern technologies provide for several solutions to the problem of off odors. However, most solutions to air pollution problems do have an added cost.

Case Study #2—Pig farms were here first!

In some parts of the United States, farms were built 1 or 2 miles from town. As the town grew, the population moved closer and closer to the farms. The urban sprawl eventually reached the pig farms. What should happen? Should the people not move closer to the farms? Or should the pig farms move or close down?

Two conflicting principles apply here:

- The pig farms were there first.
- The pig farms allow odor pollution to leave their borders and pollute the air of their neighbors.

The community is most often the winner in this situation. Being first at a location does not give the original farms the right to pollute the air or water.

Case Study #3—Pigs suffer on farms!

Several news magazines ran full-page ads depicting the apparent suffering of sows on commercial farms. They are housed alone, pumped full of drugs, and lead a life that is so stressful it drives them mad. They show neurotic behaviors such as biting the bars. They are trapped.

Many sows, in fact, are housed alone. Female pigs are social animals, but as a consequence of being a social animal, they have strong dominant-submissive relationships. In such relationships, the submissive animals may suffer health and reproductive problems. The submissive sow is probably better off alone, and the dominant sow does as well alone as she does in a social group. Boars are usually solitary animals as adults and are often individually penned on well-managed commercial farms.

- How much space do you think sows need? Enough space to turn around? Enough space for social interactions? Enough space to stand up and lie down comfortably without touching the sides of the pen?
- Should sows be given the freedom to interact socially?
- Should sows be prevented from full social interaction to protect the submissive sows?

Case Study #4—You can get worms from eating pork!

A man was at the barber one day, shortly after he moved to town. The woman cutting his hair asked him what he did. He told her that he was a pork producer. The woman then said—quite firmly—that she did not eat pork because it has those little worms and she doesn't like to eat worms. She was talking about the *Trichinae* parasite that causes the disease trichinosis in people.

The incidence of *Trichinae* infection of pork is very low in the United States. Recent estimates of the incidence of infected carcasses are less than 0.6%.

If the incidence is so low, why doesn't the pork industry try to eradicate the parasite? Attempts have been undertaken, but eradication is incomplete. Why is pork not tested and labeled as "*Trichinae* free"? The main reason is an age-old marketing problem. To be labeled "free" of something, each carcass must be tested—this adds considerable expense, although at least one pork producer certifies its product as "*Trichinae* free." If two packages of pork are in the retail case—one labeled "*Trichinae* free" and the other bearing no label—what conclusion will consumers draw? They conclude the package without the label has the *Trichinae*, even though its odds of being infected are very low.

- Should pork be required to be labeled as "free" or "not free" from *Trichinae* infestation?
- Should the industry eradicate trichinosis from the U.S. pig herd?

Case Study #5—Pork is full of cancer-causing chemicals!

We not only place carcinogens in our processed pork products, we create new ones when we cook pork. Most food products contain additives, and anything added to meat must have disadvantages. Food additives are either clearly unnatural or unhealthy, or both.

Preservatives are one class of food additive. Adding a preservative reduces the chance of microbial growth and potential food-borne illness. The effects of acute food poisoning can be anything from uncomfortable to deadly.

A balance must be struck, on the side of reason and public safety, to prevent food-borne acute illness while not adding chemicals that increase the risk of long-term carcinogenicity.

- Should additives be used in meat products?
- Would reducing acute deaths from food-borne bacteria by 90% be worth adding 10% to the long-term rate of cancer during a 20-yr exposure? What percentages would be acceptable?
- What advanced technologies could be used or developed to solve the double-sided problem of acute microbial and chronic carcinogenicity?

Case Study #6—Pig farms pollute the water!

Modern-day pig farms are fewer in number but much larger in animal inventory. With thousands of pigs on a single site, the potential for pollution is enormous. If 100 pigs are on 100 farms (total of 10,000-pig inventory), a spill of waste on any one farm with 100 pigs would not have very serious consequences. But if a single farm has 10,000 pigs and this farm releases waste, the consequences could be very serious.

Pig farms are considered nonpoint sources of pollution by the U.S. Environmental Protection Agency (EPA). This designation means that pig farms do not have permission to discharge waste into waterways (rivers, streams, lakes, or groundwater sources). The only reason a pig farm may discharge into a waterway is by accident, not by design.

A large farm, like the 10,000 pig farm example, would pollute in a significant manner as a result of a natural disaster, such as very heavy rainfall leading to flooding. A larger farm would be able to afford greater and safer containment facilities.

- Should farms of all sizes be required to spend whatever is necessary to prevent water pollution?
- Should farms be located in geographic regions that are far enough away from waterways so as to minimize natural disasters from causing pig farm water pollution?
- Should pig farms be located adjacent to large rivers (like the Mississippi) that sometimes overflow their banks?
- In what regions in the United States could large, environmentally friendly pig industries be located? What added costs would be associated with these regions?

Case Study #7—Pigs eat grain that people could eat!

Both breeding animals and growing pigs eat and metabolize 3 to 4 lb of grain-protein mixtures to put on 1 lb of live weight gain. The actual edible meat might be only 0.5 lb of meat; thus on the order of 8 lb of feed are required to produce 1 lb of edible pork. Apart from some fish and poultry, the pig is the most efficient of nonaquatic meat animals.

The diets fed to pigs are usually nutritionally complete and could just as well sustain normal growth and development of humans. So why not feed these diets to hungry people? In theory, the feed used for 100 million pigs could feed over 50 million people. The logic goes like this:

1. Nearly 100 million pigs/yr are marketed in the United States alone.
2. Pigs require about 6 mo to reach 250 lb.
3. During the growth period, the average pig might eat 3 lb of complete feed/lb of live weight gain.
4. Each year, 100 million pigs eat 1,000 lb of feed each (100 billion lb of feed/yr).
5. If an average person eats 5 lb of compete pig feed/d, or 1,825 lb/yr (365 × 5), the 100 billion lb of pig feed would feed just over 54 million people (100 billion lb/1,825 lb per person).

If people have a choice, would they eat only corn-soybean diets formulated with vitamins and minerals? This question was once posed to a group of philosophy students. They said they would not like to eat such a bland diet themselves, but perhaps starving people would like such food. In fact, vegetarians who eat a diverse vegetable diet do not choose their diet based on consumption of the most efficient plants. They eat lettuce, tomatoes, pinto beans, wheat, oats, etc., but not the bland diets fed to farm animals—grain plus protein sources and micronutrient supplements.

People in developing countries that gain affluence show a common change in behavior. One of the first uses of added family income is to buy animal products to provide a higher plane of nutrition for the family. Meat consumption goes up in proportion to income in developing countries.

Meat consumption does not follow the ideas of activists in developed countries. Rather, meat consumption follows the agricultural resources of the country and cultural preferences of the people. Pigs are a significant part of the culture of the people of the Orient, and even when feed resources are at a premium, pork consumption continues. Why does pork consumption remain high in the Orient?

1. Forages are at a premium.
2. If grains (such as rice products) are to be fed, they would be less efficiently utilized by ruminants; grains are directed toward pigs and poultry.
3. Cultural developments favor pork in the diet.

What are other reasons why pork might be eaten, even in poor countries?

Case Study #8—Large corporations are taking over the industry and killing the family farm!

Farms have increased steadily in size in recent years. The average pig inventory/farm has increased as the number of farms has declined. Smaller, family-style farms are being lost in rural America at an alarming rate. These farms are being replaced by large corporations that have different values than those on the family farm. The corporate view is directed toward profits and not toward individual workers. Corporations channel resources toward areas that enhance profits, with little concern for people. As corporate America takes over the agricultural community, a part of America is lost.

Other industries grow in size and limit competition as the industries mature. The number of companies that make microcomputers has dropped over 80% in the last 10 yr. The microcomputer industry is consolidating, and as it does, there are fewer, but larger manufacturers.

Corporations, in fact, must be concerned about their workers. If they are to succeed, agricultural corporations must be concerned about their workers and the communities in which they reside. Megafarm corporations, by virtue of their large size, can actually afford to apply a small amount of money/pig to protect the environment. The old-style family farm could dump manure in the river without regulatory consequence, but modern-day corporations are held accountable to environmental standards. As the industry becomes more vertically integrated, it must be more concerned about its public image. A scandal is much more financially damaging for a vertically integrated company than for a small family farm.

Efficiencies have increased as corporations expanded their role in pork production. This economy of scale and heightened environmental concern means more people can be fed in a less expensive manner while protecting the environment.

Some midwestern states have forbidden or discouraged corporate ownership of farrow-to-finish production or have forbidden vertical integration of livestock farms and meat processing plants. It seems as though the more restrictive the state legislation, the greater the harm to the state's agriculture.

One of the best examples of anti-corporate farming laws that hurt the pig business is in the state of Kansas. Kansas outlawed corporation ownership of more than 20 acres of land for farrow-to-finish pig units in the early 1980s. The result was a steady decline in both pig numbers and pork processing in Kansas. By 1990, pig inventory was down over 40%, compared with 10 yr earlier, and all pork processing had left the state. The lesson is clear: Restrictive legislation hurts family farms by ripping apart the industry infrastructure as the corporate units grow in other regions, taking the allied industries with them.

The same consolidation that has taken place in live pig production has taken place in pork processing, feed manufacturing, equipment manufacturing, and all other allied industries. As the pig industry evolves, further consolidation in the primary industry and allied industries is likely.

SUMMARY

This chapter discussed societal issues—issues that society considers important, but might not have direct and immediate economic impact (such as changing growth rates or reproductive rates). Pork producers historically have not focused their attention on societal issues, but today's society demands that pork producers focus attention on and solve these issues. Important societal issues were discussed, including animal welfare, environmental issues, food safety, corporate versus family farming, feeding grain to pigs rather than directly to humans, and several variations on these issues. Pig farmers and students should be fully aware of societal concerns and be able to discuss them in a logical and rational manner to work toward solutions that are compatible to the producer and the consumer.

QUESTIONS AND ACTIVITIES

1. Look on the Internet for information on the pig Industry. Find the home page of the National Pork Producers Council. Look for information on societal issues being addressed by companys and organizations.
2. Pick an issue and try to argue from the point of view that is most opposite to your personal view.
3. Write a position statement for a controversial issue to be used by:
    a. a family pig farm
    b. a corporate pig farm
    c. a community
    d. a government

INTERNET RESOURCES

Pro-Commercial Pig Production:
http://www.nppc.org/
http://www.porkboard.org/Home/default.asp
http://www.soundagscience.org/

Anti-Commercial Pig Production:
http://www.keeper.org/hogfight/
http://www.hsus.org/ace/352
http://www.hfa.org/index_fla.html
http://www.awionline.org/
http://www.hogwatch.org/

SECTION II
BIOLOGY OF THE PIG

4

APPLIED ANATOMY AND PHYSIOLOGY RELATED TO BLOOD SAMPLING, HEMATOLOGY, AND IMMUNOLOGY

INTRODUCTION

The domestic pig resembles humans in anatomy and physiology. Some pigs have organ weights and sizes similar to those of any size human. A 60-kg (132-lb) human and a 132-lb pig have similarly sized hearts, lungs, livers, kidneys, and spleens. In addition, the functioning of the pig's heart is nearly similar to that of the human heart. The digestive system, kidney, liver, pancreas, skin, and other organs also function similarly in pigs and humans. The structures and functions of some of these organs and tissues are described in greater detail in Chapters 5, 6, 8, 9, and 10. This chapter describes some of the salient features of pig anatomy and physiology related to procedures and information-gathering activities that might arise during commercial pig production. Uses of the pig in biomedical research are also described. Some commercial pig farms sell domestic or specially bred miniature pigs for biomedical research. These pig producers have a different market than the pork market, but their production is likewise important.

This chapter serves as a reference chapter for students, owners, and workers in commercial pig production. The two primary needs for anatomical information involve blood collection and necropsy. Blood is obtained from pigs as a part of routine herd health surveillance. Necropsies are performed to assess potential causes of death or illness in animals. Pigs that die on commercial farms without a known cause should be necropsied; information in this chapter will help the pig biologist prepare for a necropsy. Also, pig health is managed through the use of a sanitation program, vaccinations, and biologically active drugs such as antibiotics. While the applied aspects of pig health are discussed in Chapter 20, this chapter provides some background information on the pig immune system.

THE PIG IN BIOMEDICAL RESEARCH

The pig is used as a model system in modern biomedical research directed toward benefiting human health. Pigs are utilized as a model species in cardiovascular, skeletal, urinary, nervous, and digestive system function and in dermatology, alcoholism, teratology, ethology, reproductive biology, immunology, anesthesiology, surgical studies, and developmental biology.

The similarities between human and pig anatomy and physiology are so striking that work is underway to develop the technology to transplant pig organs into humans. This process—called xenotransplantation—usually fails because the immune system of the recipient species attacks and kills the organ from the donor species. Two major efforts are underway to improve the success rate of xenotransplantation. The first involves suppressing or changing the human recipient's immune system. The second, more favorable, method is to change the donor organ of the pig, through genetic manipulation, to be immunologically compatible with the recipient human. Human use of pig hearts, dopamine-producing brain cells, livers, kidneys, and the pancreas may become practical in a few years because human organs for transplantation are in short supply. The use of pig organs for xenotransplantation would mark a major breakthrough in human medicine.

GENERAL ANATOMY

The pig (like other animals) has three planes associated with three dimensions. Anatomical points that are located more toward the head are identified as anterior, superior, or rostral. Those located more toward the tail are posterior, inferior, or caudal.

Body regions, especially parts of the limbs, are said to be proximal if they are closer to the body core or distal if they are farther away from the core. Any given point on the pig's body has an infinite number of other points that are proximal to (close) or distal from (far) that point. If a point is away from the center of the body, it is said to be lateral. If it is nearer the center, the point is said to be medial. If a point on the body and an organ are on the same side of the animal, it is said to be ipsilateral. If the organ or tissue of interest is on the other side of the body, it is said to be contralateral. Thus, thinking of the kidneys; each kidney has in close proximity an adrenal gland. The left kidney has an ipsilateral adrenal gland, but the right adrenal gland is contralateral to the left kidney.

The pig's body can be divided into three planes: medial, transverse, and frontal (see Figure 4–1). The medial plane divides the pig from head to toe, down the center. The exact medial plane is also called the midsaggital section. An infinite number of saggital sections (cuts) may be made lateral to the medial plane.

Transverse planes are at a right angle to the medial plane. The transverse, also called cross section or coronal, plane is equivalent to cutting or viewing the pig in sections from left to right anywhere from head to tail.

The frontal plane, or the z-axis plane, is the least common plane used in anatomy. The frontal plane is at a right angle to both the medial and transverse planes.

Other anatomical terms are frequently encountered. If a site is dorsal to another site, it is closer to the spinal cord in contrast to a more ventral site (closer to the navel). A midventral laparotomy cuts open the belly of the pig at its center line.

FIGURE 4–1
Anatomical Planes Common in Pig Anatomy. The Three Planes, Representing an X, Y, and Z Plane, Are Median, Transverse (or cross section), and Frontal.

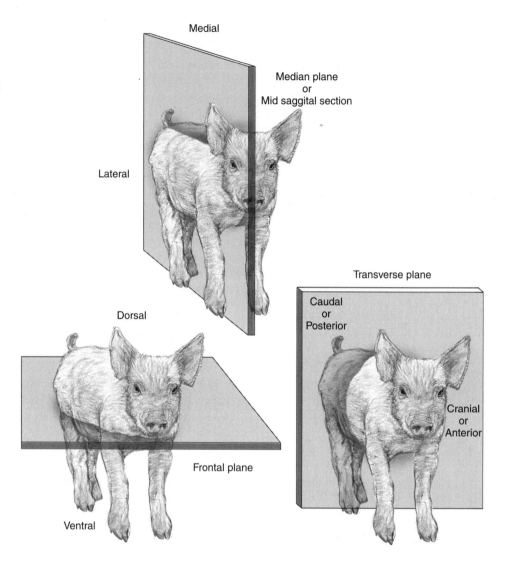

"Central" has two meanings in anatomy—it can refer to nervous tissues encased in bone (brain and spinal cord), in contrast to peripheral tissues. The central nervous system is encased in bone. The peripheral nervous system travels throughout the body and is not encased in bone. In some cases, "central" refers to inside a vascular (blood) vessel. Thus, a pig can be said to have a central line, which means it has a tube inside a major blood vessel.

ANATOMICAL REFERENCES

This book includes some anatomical materials that pig producers can use as a reference when they attempt to solve problems that arise on the farm. They must understand anatomical sites and terms to deal with pig injury and disease. Bones have features like a head or a shaft, but for the purposes of this chapter these will not be described. As an example, if a pig were to injure its tibia, one can see that this bone is on the pig's back leg.

Figure 4–2 shows the 32 major bones of the pig. Injuries to the limbs are more common than are injuries to the body core. Injuries to the joints between bones are also common.

Figures 4–3, 4–4, and 4–5 show the internal organs of the pig as they appear in dissection upon necropsy. One must understand normal pig anatomy to be able to recognize an abnormal state. Those who work with pigs should compare anatomical drawings with a healthy pig that is euthanized humanely. This educational experience can be performed on an injured pig whose suffering may be relieved by euthanasia.

Figure 4–3 provides a view of the organs when the ventral aspect of the peritoneum (cavity containing intestines and other organs) is opened. The primary view is of the intestines, which indicate if the pig had been eating and if the intestinal contents are healthy in appearance. Intestinal contents that appear swollen, contain liquid, or are bloody indicate an enteric problem. The stomach should be removed and its contents examined to see if the pig had been eating and if the stomach might have ulcerations of its lining. The liver should be examined for signs of milk spots or white areas that indicate the presence of parasites such as *Ascarid*.

Figure 4–4 shows a lateral view of the body, including the lungs and intestines. The lungs and heart should be removed and examined. The lungs should be a bright pink color and have a spongy texture. They should float in water. If the tips of the lungs appear dark red, the pig may have had areas of lung tissue that were not exchanging oxygen. These can be cut out to see if they float. Healthy, air-filled lung tissue will float, while previously or actively infected lung tissue may sink in water. The heart should also be examined at this time.

FIGURE 4–2
Skeleton of the Pig.

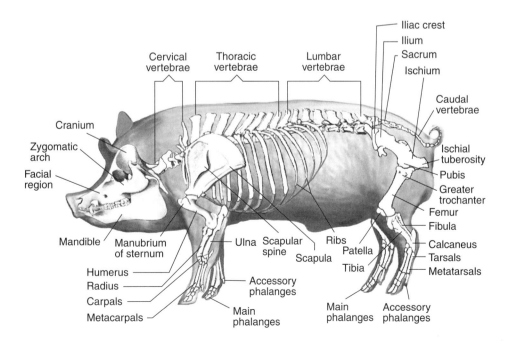

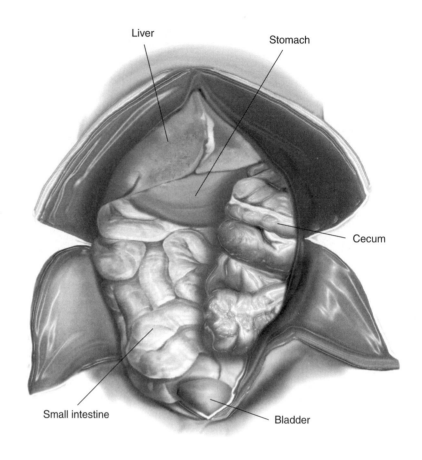

FIGURE 4–3
Abdominal Organs from the Ventral View.

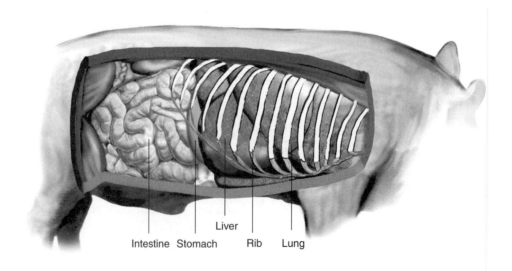

FIGURE 4–4
Lateral View of the Body Including the Lungs and Intestines.

Figure 4–5 shows the deep organs of the peritoneal cavity when the ventral organs are removed. Note especially the rich blood supply along the dorsal wall of the peritoneum. Note also the kidneys that lie against the dorsal body cavity. Removal of the kidney with its associated kidney capsule will remove the adrenal gland on the anterior aspect of the kidney. Note, upon dissection, that the yellow-colored adrenal gland is triangular in shape.

FIGURE 4–5
Deep Organs of the Peritoneal Cavity When the Ventral Organs Are Removed.

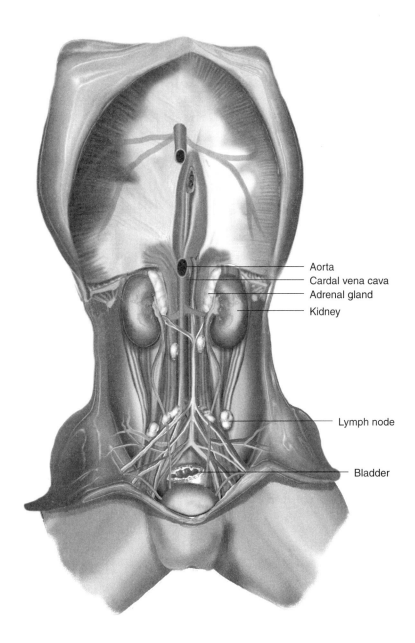

ANATOMY AND SAMPLING OF THE PIG VASCULAR SYSTEM

Blood is collected from pigs to assess individual pig health or to estimate the health of the herd. A blood sample from a single pig can indicate its nutrient level or the presence of an antibody to a specific pathogen, but a single sample provides information only about that one animal. To understand the herd's health status, blood samples should be obtained from a number of pigs. A representative sample of pigs will give a better idea about herd health than will a single sample.

Blood is a convenient tissue to sample in the evaluation of the health of the pig herd. Those who sample blood must know what sort of blood is needed (plasma, serum, or whole blood), how much is needed, whether any blood preparations are required (some assays require certain anticoagulants, for example), if the blood must be kept cold or at room temperature, how the sample is to get to the laboratory, acceptable transport time, and how to obtain the blood sample.

BLOOD COLLECTION

Pig arteries and veins of the pig are much like those of other mammals. Placement of major vessels varies among species. It is a challenge to find convenient places to sample blood in the pig. For example, the jugular vein is close to the surface and easy to access in horses, cattle, and sheep, but in the pig, the jugular vein lies deep in the tissues of the neck, essentially making jugular venipuncture a "blind stick" in the pig (see Figure 4–6).

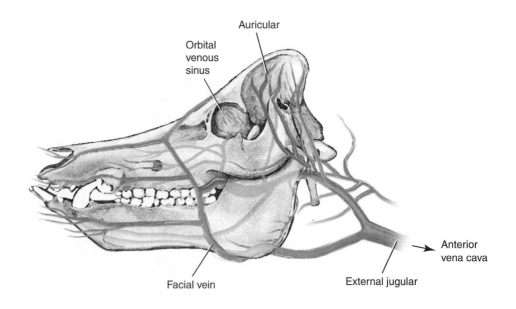

FIGURE 4–6
Anatomy of Veins of the Head and Neck of the Pig. These Veins Are Useful Sites for Blood Collection.

Chapter 4 Applied Anatomy and Physiology

TABLE 4–1
Common Blood Collection Tubes (tubes with * are commonly used).

Color of Tube Top	Fluid Type	Anticoagulant	Example Uses
Red*	Serum	None	Antibodies, minerals, other proteins
Blue	Either	Na heparin or none	Special blood chemistries
Brown	Plasma	Na heparin	Lead determinations & others
Black/light blue	Plasma	Na citrate	Coagulation studies
Gray	Plasma	Glycolytic inhibitors	Glucose determinations
Green*	Plasma	Lithium heparin	Na-, Ca-sensitive assays
Yellow	Plasma	Sodium citrate	DNA extraction
Purple/Lavender*	Plasma	EDTA	Clotting factors

When blood is sampled for laboratory or diagnostic purposes, it is often drawn in a manner to sample one or more of its fractional components (see Table 4–1):

- Serum (fluid minus all cells and clotting factors)
- Plasma (fluid including soluble clotting factors)
- Red blood cells (obtained only from plasma)
- White blood cells (obtained only from plasma)

The blood is sampled for some specific purpose, such as determining the presence of fluid or a particular cell type. Once the type of sample is determined, the correct sampling method must be employed. The two most common methods of obtaining blood are (1) in a syringe of sizes ranging from 1 mL to 60 mL or (2) in a vacutainer (a vacuum tube with a double-sided needle inserted).

RESTRAINT OF PIGS FOR BLOOD SAMPLING AND OTHER PROCEDURES

Pigs may require temporary restraint for a number of procedures, but, most commonly, pigs are restrained to sample blood. Blood is typically sampled from the venous system because the veins have lower blood pressure and thinner walls. Pigs will heal quickly from a venipuncture. Arteries have a much higher blood pressure and thicker walls. Sampling from arteries can result in excessive, unintentional blood loss. Vein sampling sometimes results in a puncture to a nearby artery; this is usually not a serious problem for the pig.

Blood sampling (see Table 4–2) can lead to formation of a hematoma, a blood clot associated with a damaged blood vessel (usually a vein). The hematoma resembles a bruise if it is near the surface. Formation of a hematoma is not life threatening, but it may reduce the value of a carcass if the site does not heal before slaughter. Venous blood flows toward the heart. Thus, if a hematoma is present in a given place, it may reduce blood flow from the hematoma back to the heart. Sampling blood from the same vein from the hematoma toward the heart has a low chance of success. The same vein can be sampled distal from the hematoma (away from the heart). If the hematoma prevents blood return from a major region of the body, other veins will have to compensate. During this process, tissue nutrient delivery may be inefficient.

TABLE 4–2
Points to Sample from or Inject into the Venous System of Pigs.

	PIGLETS	YOUNG PIGS	ADULTS
1. Ventral neck			
Vena cava	✓	✓	✓
External jugular	✓	✓	✓
Lingual-facial	Difficult	✓	✓
Facial	Difficult	✓	✓
2. Orbital sinus	✓	✓	NA
3. Ear vein	Difficult	Difficult	✓
4. Tail vein	Difficult	Difficult	✓
5. Heart (cardiac puncture)	NA[a]	NA[a]	NA[a]
Preferred method of restraint	On back	On back	Standing, snared

[a]Not advised except as a last resort or in the case of euthanasia.

Serum or Plasma?

When left in a tube or syringe, whole blood will clot in a matter of minutes. Thus, technicians must make an early decision to determine if the blood sample is better suited to become serum or plasma. Serum is the fluid left after whole blood has clotted. The clot contains red and white blood cells and several clotting proteins that are tied up in the clot (fibrin is a major protein that forms the clot). The serum does not contain any cells or any of the major clot-forming proteins.

Plasma is the fluid left after the red and white blood cells are removed. To get plasma, an anti-clotting factor must be used. If the objective is to collect red or white blood cells, the technician must collect plasma. The anticoagulant allows red and white blood cells to float down (sediment) toward the lower part of the tube (see Figure 4–7).

Common anti-clotting factors include heparin, sodium citrate, sodium flouride, and ethylenediaminetetraacetic acid (EDTA). Heparin, a natural protein found in the lungs (and other organs) of animals, prevents clotting in the lungs where red blood cells squeeze into small spaces. The trauma and agitation of blood moving through very small capillaries normally causes blood to clot. Local heparin prevents these clots. Heparin has a co-metal attached—either sodium or lithium. The availability of two forms of heparin is convenient because some assays to be performed on the blood sample are sensitive to either sodium or lithium in the blood collection tube.

Sometimes the anticoagulant interferes with the assay to be performed. Sodium citrate adds sodium and lowers the pH of the tube (citrate is citric acid). EDTA binds calcium in a strong bond that might interfere with calcium determinations or calcium-dependent assays. Sodium fluoride is the appropriate anticoagulant for use in measuring plasma glucose because fluoride prevents glycolysis (breakdown of glucose).

Syringe or Vacutainer?

Blood can be collected easily by syringe or vacutainer. Serum is collected by syringe and needle, by inserting the needle into the vein (or artery) and drawing on the syringe. The technician has control over the force of draw (negative pressure on the vein). Too much pressure causes the vein to collapse and no blood will enter the syringe. With too

FIGURE 4–7
Drawings of Whole Blood Collected into Glass Tubes as Either Serum (without anticoagulant) or as Plasma (with anticoagulant).

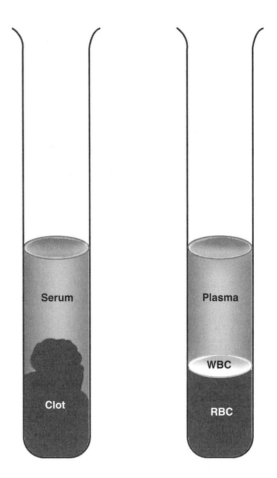

little pressure, blood will not flow. The syringe is the preferred method to learn blood collection because the technician can feel the vein pressures directly.

If plasma is required, any of the anticoagulants can be added to the syringe. A tube that contains anticoagulant should be available. The syringe can be emptied into this tube.

There are several steps in bleeding by syringe:

1. Prepare the syringe (possibly add anticoagulant).
2. Wipe the skin with alcohol or disinfectant that does not interfere with the intended assays.
3. Insert the needle while drawing back gently on the barrel of the syringe.
4. When a blood vessel is entered, stop moving the syringe and allow the syringe to fill by applying gentle pressure.
5. Gently add the contents of the syringe into an appropriate tube.
6. Apply gentle pressure on the puncture mark with gauze to help stop bleeding.

Collection into a vacutainer is easier than collection into a syringe because the glass tube already has the appropriate negative pressure needed to draw blood. Also, for plasma collection, the anticoagulant is already present in the tube. The vacutainer is more convenient than the syringe because bleeding can be performed easily with one hand. The needle is inserted toward the vein and the vacutainer holder and tube are pushed forward to cause a vacuum on the needle. The needle is thrust in, and when the vein is entered, the tube fills with blood.

The steps in bleeding by vacuum tube are:

1. Gather the appropriate vacuum tube, needle, and vacutainer holder (vacutainer with or without anticoagulant).
2. Wipe the skin with alcohol or disinfectant that does not interfere with the intended assays.
3. Insert the needle into the animal's skin, toward the vein.
4. Place the vacuum tube on the other end of the needle to allow negative pressure on the vein.
5. As the needle is pushed in and pulled out gently and the vein is entered, blood will fill the tube.
6. Hold the tube steady while the blood is entering. More vacuum tubes may be attached to draw greater volumes or aliquots of blood.
7. Withdraw the needle and the vacuum tube, allowing air to enter the vacuum tube (or not, depending on the objective).
8. Apply gentle pressure on the puncture mark with gauze to help stop bleeding.

Blood Sampling Young Pigs

The easiest method to sample blood from piglets and smaller pigs is to lay the pig on its back and insert the needle in the vena cava or other vessels of the neck. Pigs may be laid on their back on any flat surface, but construction of a V-shaped trough makes bleeding much easier.

The V-shaped trough should be constructed using a 90°-angle, turned on its side (see Figure 4–8).

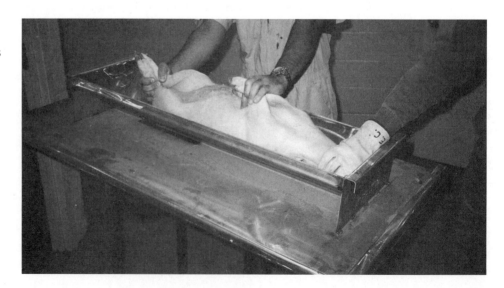

FIGURE 4–8
Diagram of a Bleeding Trough for Young Pigs. Pigs from Birth to Over 20 kg (40 lb) Can Be Bled in This Trough. The Trough Operates Well with Four Legs with Locking Wheels.

Young pigs provide access to the vena cava, external jugular, and linguo-facial veins in their neck. All the veins can be reached with a 1.5-in needle (see Figure 4–9).

Blood Sampling Older Pigs

The most common way to sample blood from an adult pig is to restrain the pig by use of a snare, a convenient tool that holds a sow or a boar. The snare is either a rope or a wire attached to a pole. Figure 4–10 shows a wire snare with an attached metal pole designed specifically for restraining adult pigs. The operator places the wire in the pig's mouth and closes the wire by pulling back and tightening the wire. The pig will show two behaviors: it will scream loudly (an alarm signal) and will pull back from the operator. The rare pig will lunge forward—the operator should be prepared for this possibility. A second person can obtain blood or otherwise handle the restrained pig.

While snared, the pig can have its vena cava, jugular, facial, ear, or tail vein sampled. The vena cava and external jugular veins require a 4-in needle to reach the vessel. Extremely large sows or boars may require a 6-in long needle. The facial vein of the neck may be reached with a 1.5-in long needle. This smaller vein requires more practice to sample consistently.

THE PIG HEART AND BLOOD SUPPLY

The pig heart is similar in size and function to the human heart. These anatomical similarities have led to the use of pig heart valves as replacements for diseased human heart valves. The valves, made of cartilaginous material, do not seem to be rejected after transplantation like other soft tissues.

Pigs develop heart diseases in manners similar to humans, making them a valuable model species for atherosclerosis. Pigs fed a high-fat diet will develop atherosclerotic plaques in blood vessels. Continued and developing occlusion of the coronary artery

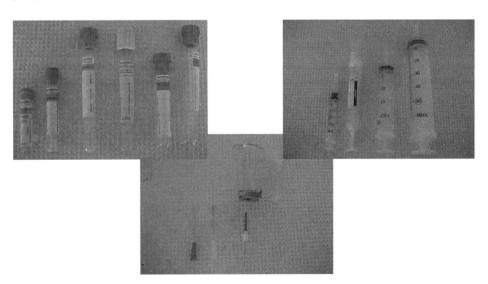

FIGURE 4–9
Blood Collection Supplies.

FIGURE 4–10
Wire Snare.

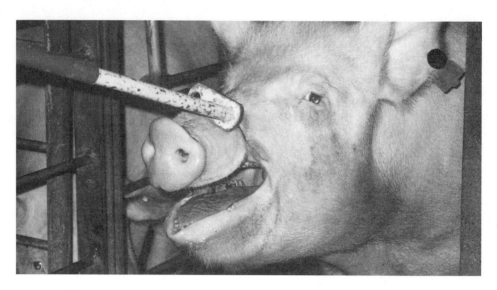

will lead to heart disease in older pigs. Heart failure is a significant cause of death in older pigs (as it is in humans).

Pigs with the genetic condition Porcine Stress Syndrome (PSS) have a mutation on a muscle protein, including on the heart muscle. Pigs may have one or two copies of the Halothane (Hal$^+$ or Hal^{++}) gene. The condition was named after the anesthetic Halothane, which induces the often-fatal condition in pigs and humans called malignant hyperthermia. Both Hal$^+$ and Hal^{++} pigs are more susceptible to acute illness and death when they are even mildly stressed. After induction of the condition by handling or stress, pigs will increase blood flow to their skin and their body temperature will rise very quickly. Cardiac arrest will follow in many cases.

Normal values for some blood measures are given in Table 4–3. A normal pig has about 40% of its blood volume in red and white blood cells. As pigs become dehydrated, the percentage of packed cell volume (PCV; also called *hematocrit*) goes up. A PCV or hematocrit value is the percentage of the whole blood that is composed of red blood cells. A high PCV indicates dehydration; a low PCV may be associated with a metabolic disease state such as iron-deficient anemia.

A minute or two after a pig takes a large drink of water, the PCV goes down (as water enters the blood). During states of anemia, the PCV will go down. Anemia can be caused by lack of ability to synthesize red blood cells (RBCs) or by a greater loss of RBCs.

Pigs have a plasma glucose concentration of about 100 g/dL. Plasma glucose rises after a meal in an acute manner. Blood glucose also can increase while pigs are stressed.

The normal blood supply of pigs is estimated to be about 8% of body weight. A complete bleeding of pigs at slaughter yields only 5% to 6% of body weight in blood. Some blood remains in the vessels as they collapse.

When technicians obtain a blood sample, they should not expect to remove enough blood to cause signs of anemia. As a general rule, technicians should take less than 10% of the blood volume for a given sample. Table 4–4 gives the recommended maximum blood volume to be drawn in a single blood collection. Most determinations require much less blood than the pig has available.

TABLE 4–3
Normal Blood Values for Pigs.

MEASURE	VALUE
Blood volume, % of body weight	8
Total RBC, $10^6/\mu L$	6–8
Diameter of RBC, μm	6
Diameter of WBC, μm	8
Packed cell volume, %	40
WBC, $10^3/\mu L$, birth to 2 weeks old	10–12
WBC, $10^3/\mu L$, older pigs	11–15
Neutrophils, adult, %	45
Lymphocytes, adult, %	50
Monocytes, adult, %	3
Eosinophils, adult, %	2
Basophils, adult, %	<1
Glucose, mg/dL	80–120
Cholesterol, mg/dL	60–200
Albumin, g/dL	3.2–4.0
Globulin, g/dL	3.4–4.0
Gamma globulin, g/dL	2.5–3.0
Fibrinogen, g/dL	0.2–0.4

Source: Adapted from Dukes' Physiology of Domestic Animals (1984) and research from our laboratory.

TABLE 4–4
Total Blood Volume and Safe Bleeding Volumes for Pigs of Various Sizes.

AGE & WEIGHT OF PIG	TOTAL BLOOD VOLUME, mL	MAXIMUM DRAW, mL[a]
Newborn, (3 lb) (1.4 kg)	110	10
3–4 wk (15 lb) (6.8 kg)	544	50
8–9 wk (35 lb) (16 kg)	1,280	120
24 wk (240 lb) (109 kg)	8,720	800
Adult (440 lb) (200 kg)	16,000	1,280

[a]One-time draw with the expectation the animal will live and grow normally. The estimate is rounded from about 10% of total blood volume. If repeated blood samples are needed, the volume of blood drawn should be reduced considerably.

White blood cells (WBCs) number 10,000 to 15,000/μL (10 to 15 million/mL). When pigs have an active infection, the number of WBCs increases dramatically, commonly two-fold. Stress also releases neutrophils from their attachment on the vascular epithelium by a process called demargination. During stress, the percentage of neutrophils increases to well above 50% and the ratio of neutrophils to lymphocytes reverses from being predominantly lymphocytes to being predominantly neutrophils. The neutrophil:lymphocyte (N:L) ratio is normally less than 1:1, with an average in the range of 0.4:1 through 0.7:1. If the N:L ratio is above 1.0:1, adult pigs might be experiencing stress. Young pigs, however, normally have a greater percentage of neutrophils than lymphocytes.

IMMUNOLOGY

The immune system is the pig's natural mechanism to defend itself against microbial, parasitic, and physical insults. The immune system is aided by the skin and mucous membranes, which provide physical and chemical barriers to entry of foreign material and organisms (for information on applied pig health, see Chapter 20). The immune system of the pig is divided into three main components: innate (or maternal), humoral (soluble in blood), and cellular.

Piglets are born with an immature immune system. The pig placenta does not allow significant amounts of antibodies to pass from the mother to the piglet. Also, in utero and in early life, the piglet cannot yet build antibodies in the face of antigenic challenge.

To help the piglet cope with environmental pathogens, it eats milk (colostrum) that is rich in all types of immunoglobulins (antibodies). The piglet's gut is "open" for the first 48 h of life and, thus, antibodies in the milk can be absorbed into the piglet's blood during these unique first hours. After the gut "closes," large proteins like immunoglobulins will not pass through the small intestine into the blood, but they can provide some local protection against gut pathogens.

Maternal antibodies in the colostrum are a significant source of plasma antibodies for the piglet. At birth, piglets have less than 1 mg/mL of immunoglobulin. After consumption of colostrum over the first h of life, piglet blood immunoglobulin rises to over 20 mg/mL. These maternal antibodies decline over a period of weeks. The piglet typically does not make significant antibodies of its own until after 3 wk of age, and even from 3 to 6 wk of age, the ability to synthesize immunoglobulin is still developing. Thus, from birth through about 6 wk of age, maternal antibodies absorbed by the piglet shortly after birth and coating the gastrointestinal tract while nursing are the primary sources of protective antibodies.

The ability to synthesize immunoglobulins in response to antigen exposure occurs in the humoral (blood) arm of the immune system. Antigens are foreign molecules, primarily proteins, against which the pig will build antibodies. The immunoglobulins (Ig), or gammaglobulins, are subdivided into the following classes:

- IgM: The first antibody formed when white blood cells are initially exposed to an antigen; IgM is a pentomer molecule with five linked antibodies. The affinity of IgM for the antigen is weaker than other classes.
- IgG: When exposed to an antigen a second time, the pig builds very high levels of antibodies, mostly in the class of IgG. White blood cells switch from synthesizing IgM to synthesizing IgG after continued antigen exposure.
- IgA: Also referred to as secretory antibody, IgA is found at higher concentrations in the fluids of the mucous membranes (respiratory, gastrointestinal, and reproductive tracts, and eyes). IgA is a dimer; and this structure resists breakdown by enzymes found in the mucous membranes.
- IgD and IgE: These immunoglobulins are less studied in pigs than in some other species. IgD and IgE function in allergic and anaphylactic responses.

Antibodies coat and, in some cases, disable the pathogen. More importantly, the presence of antibodies aids in the ability of cellular components of the immune system to kill pathogens.

The cellular arm of the immune system is very complex and interacts extensively with the humoral arm. Immune cells (WBCs) originate in the bone marrow. The cellular arm of the immune system includes the following cell types and their functions:

- Lymphocytes: These cells cytes are found in great numbers in the blood. They kill cells and synthesize antibodies. Lymphocytes look like a round, dense cell—they are primarily nucleus and have a few granules in their cytoplasm (with some exceptions). Some lymphocytes live for years—they are generally long-living.
- T-helper cells: These lymphocytes help B cells make antibodies; also called CD4 cells.
- T-cytotoxic suppressor cells: These cells have two functions: to kill cells coated with antibodies and to suppress antibody synthesis (in a negative feedback manner); also called CD8 cells.
- B cells: B lymphocytes synthesize and secrete antibodies.
- NK cells: These natural killer cells kill foreign cells, cancer cells, and viral-infected cells. Antibodies are not required for cell killing (although presence of an antibody heightens killing). NK cells are a first line of defense against viral-infected and malignant cells. One genetic line of mini pig develops skin tumors starting at birth. Until the NK cells develop (~4 to 8 wk of age) the tumors are present. As the NK cells come on line, the tumors are killed.
- Neutrophils: These cells are short-lived, and have a convoluted nucleus and granules in their cytoplasm. Neutrophils are phagocytic—they phagocytize (eat or engulf) bacteria and debris. The granules are used to enzymatically digest the engulfed cell or particle.
- Basophils: These granular leukocytes kill other cells. Basophils are involved in allergic reactions. Among all WBCs, pigs have less than 1% basophils.
- Eosinophils: These cells are involved in killing of parasites. Unless there is a parasitic infection, eosinophil numbers are usually very low.

SUMMARY

This chapter summarizes useful anatomy and physiology information for the domestic pig and provides reference anatomical terminology, landmarks, and anatomical sites to observe during necropsy. Access to blood vessels is discussed, along with methods of blood collection.

The similarities between human and pig anatomy and physiology are so striking that work is underway to develop the technology to transplant pig organs into humans—a process called xenotransplantation. This success of this ongoing research would mark a major breakthrough in human medicine.

The pig's immune system is divided into three main components: innate, humoral, and cellular. Piglets are born with an immature immune system that is fortified by the sow's colostrum, which is rich in all types of antibodies. The humoral component of the immune system synthesizes immunoglobulins in response to antigen exposure. The cellular arm of the immune system produces a variety of cells that attack foreign cells.

Pig producers must understand pig anatomy, physiology, and terminology to deal with injury and disease on the farm.

QUESTIONS AND ACTIVITIES

1. If a bullet passes anterior to the eye in a dorsal to lateral direction, 10 cm from the anterior tip of the cranium, where might it be traveling?
2. Carefully dissect a euthanized pig. Indicate where the following organs are located: heart, lungs, spleen, liver, stomach, small intestine, large intestine, cecum, kidney, adrenals, and pancreas.
3. Write up a standard operating procedure for the following, to be used on a commercial farm:
   - Blood collection from a 10-kg (22-lb) pig
   - Blood collection for a sow

LITERATURE CITED

Pond, W. G. and H. J. Mersmann. 2001. Biology of the domestic pig. Cornell University Press. Ithaca, NY.

Straw, B., J. A. Roth, and L. J. Saif. 2001. Basics of Immunology. PIH-122. Pork Industry Handbook. Purdue University Extension Service.

Swenson, M. J. 1984. Dukes' physiology of domestic animals. Cornell University Press. Ithaca, NY.

INTERNET RESOURCES

Fetal pig dissection:
http://mail.fkchs.sad27.k12.me.us/fkchs/vpig/

Anatomy of the heart:
http://www.tmc.edu/thi/anatomy.html

5
REPRODUCTIVE BIOLOGY

INTRODUCTION

Just as reproductive biology is fundamental to the survival of the species, the efficiency of the reproductive process is vital to the economic survival of pork producers. Those producers who understand and manage pig reproductive biology have the greatest chance of success at both causing pregnancy and generating the greatest economic benefit from the herd.

Breeding (also called a "service") requires sperm and egg. Today, in vitro fertilization can be used to create embryos in the culture dish. The levels of participation of the live boar and sow now vary from virtually no animal participation, to insemination, through completely natural service (see Table 5–1). With the development of cloning techniques, great excitement is expected in the field of farm animal reproductive biology in the next few decades.

Reproductive pig biology, for both the male and the female, consists of three levels: anatomy, physiology, and behavioral biology. This chapter explores each level.

ANATOMY AND PHYSIOLOGY OF THE PIG REPRODUCTIVE SYSTEM

THE BOAR

In the United States, the male pig is called the *boar* and the castrated male pig is called a *barrow*. In some parts of Europe, the male castrated pig is called a *hog* (in the United States, hog is a generic term).

TABLE 5–1
Levels of in Vitro and in Vivo Reproduction.

Event	In Vitro	In Vivo
Ovarian and follicular maturation	Partially successful	Yes
Ovulation	In the laboratory only	Yes
Sperm maturation	In the laboratory only	Yes
Fertilization	Yes	Yes
Implantation	Yes	Yes
Pregnancy	No	Yes

DEVELOPMENT OF THE BOAR REPRODUCTIVE ORGANS

The boar develops its sex organs starting about 20 d and ending about 90 d after fertilization (Figure 5–1). The early fetal period ends around d 85 to 90 of pregnancy when the testicles descend from the body cavity into the scrotum. Failure to descend results in either unilateral or bilateral cryptorchidism. This early period of fetal development may shape lifetime male reproduction in ways that are not fully understood.

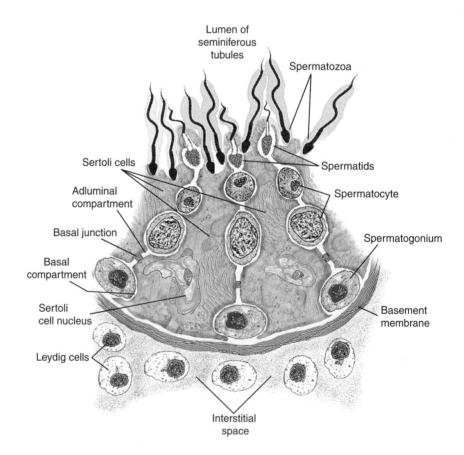

FIGURE 5–1
Drawing of a Histological Section of a Boar Testes.

FIGURE 5–2
Diagram of Boar's Reproductive Tract.

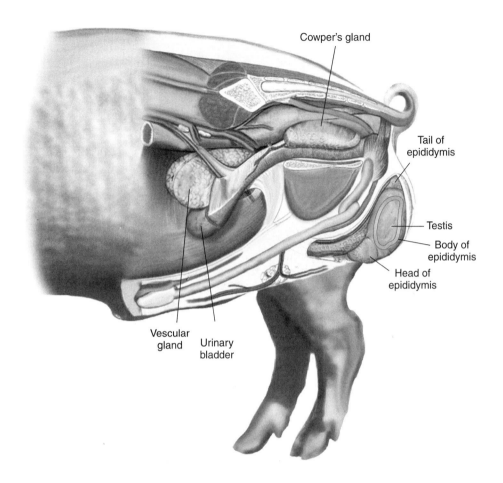

The two primary endocrine cells of the testes are the Leydig cells and the Sertoli cells (Figure 5–2). The perinatal period of male sexual development is from about d 90 of gestation to around d 21 of postnatal life when the germ cell numbers increase and the Leydig cells differentiate.

The pubertal period begins around d 30 to 70 of age when the Sertoli cells slow down proliferation (increase in cell number), cell junctions appear, germ cell differentiation begins, and spermatocytes and spermatids are observed. Leydig cells are developed by d 120, and puberty is said to start around d 160 (Table 5–2). However, behavioral expression of breeding takes place among males at a very early age, including before d 30 of age. Effective and fertile breedings would not be possible until much later.

The Leydig cells are located in the testicular interstitium. Leutinizing Hormone (LH) is produced and secreted by the pituitary and binds to specific receptors on the Leydig cells. Upon receptor binding by LH, the Leydig cells synthesize and secrete androgens, primarily testosterone. Testosterone and other androgens stimulate (1) male sexual behavior, (2) development of sperm, and (3) development of secondary sex glands.

Follicle-stimulating hormone (FSH) stimulates Sertoli cells to produce androgen-binding proteins, inhibin, and enzymes. Inhibin enters the blood system and inhibits FSH secretion from the pituitary.

The rete testis is a series of tubules that extends from the seminiferous tubules, which are lined by sertoli cells. Leydig cells are located in the interstitial spaces among the seminiferous tubules. The area is rich in blood supply, lymph, and nerves.

SPERMATOGENESIS

Spermatogenesis is the process of producing sperm. Spermatozoa travel from the seminiferous tubules to the rete testes to the epididymis. The stages of spermatogenesis are shown in Figure 5–3.

The epididymis, which stores sperm, is attached to the testis. Upon ejaculation, the sperm are expelled. As they pass through the vas deferens and through the lumen of the penis, accessory fluids are added by the vesicular glands and the prostate glands. A gel is added to the semen by the Bulbourethral gland (also called the Cowper's gland).

SUMMARY OF BOAR REPRODUCTIVE ORGANS

1. Testes (testicles)—The organ that produces sperm. The testes are enclosed in the scrotum, a diverticulum of the abdomen. The chief function of the scrotum is thermoregulatory—to maintain the testes at a temperature several degrees lower than that of the body. Sperm-producing cells—spermatogonia—are located in the seminiferous tubules and testosterone is produced by interstitial (Leydig) cells.
2. Epididymis—A passageway for sperm from the seminiferous tubules. The epididymis is also the place where sperm mature, are stored, and concentrated.
3. Vas deferens (ductus deferens)—The duct that leads from the epididymis to the pelvis part of the urethra. Its primary function is to move sperm into the urethra at the time of ejaculation.
4. Vesicular glands—The pair of glands that flank the vas deferens near its point of termination. The vesicular glands are the largest of the accessory glands and are located in the pelvic cavity. They secrete a fluid that provides a medium of transport, energy substrates, and buffers for the spermatozoa.
5. Prostate gland—The gland that is located at the neck of the bladder, surrounding the urethra. The prostate gland contributes fluid and salts (inorganic ions) to semen. Fluid from the prostate gland is basic and acts to neutralize acidic vaginal secretions.
6. Bulbourethral gland (cowper's gland)—The two glands that are located on either side of the urethra in the pelvic region. They produce the gel fraction of boar semen, which forms a plug in the cervix of the sow.
7. Urethra—The long tube that extends from the bladder to the end of the penis. The vas deferens and vesicular glands open to the urethra close to its point of origin. The urethra serves as the passageway of both urine and semen.
8. Penis—The boar's organ of copulation. The penis is composed essentially of fibrous tissue. At the time of erection, cavernous spaces in the penis become engorged with blood, muscles holding the penis relax, and the penis extends beyond its sheath.
9. Glans penis—The tip of the penis. The pig's glans penis is shaped in a counterclockwise direction.

FIGURE 5–3
Drawing of Spermatogenesis in the Boar.

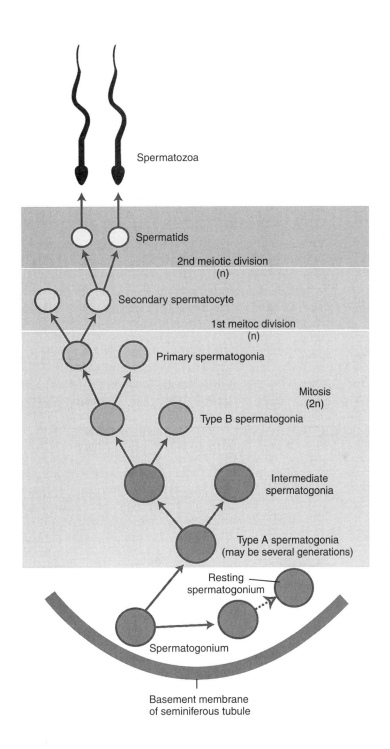

Semen is routinely collected from boars. While fertility of frozen semen is lower for the pig than for cattle, the fertility of fresh semen in pigs can be comparable to that with natural service. Semen is composed of seminal fluids and sperm cells. The boar has two separable fractions of seminal fluids: the liquid and the gel fractions. The pre-sperm fraction which is usually discarded. When semen is collected, a clean filter (such as cheese cloth) is used to separate the gel from the sperm and remaining seminal fluids.

Boars ejaculate 50 to 500 mL of semen; the expected volume is more commonly 200 mL. The number of sperm per ejaculate varies widely among boars and even among ejaculates for the same boar. A fertile boar ejaculate may contain between 10 billion and 200 billion sperm cells. Insemination doses should contain from 2 billion to 5 billion sperm per dose and a volume of 80 to 100 mL. Depending upon the individual boar and the extender to be used, semen can be stored for 3 to 7 d and still result in a fertile insemination. Sperm viability, typically measured as the percentage of motility of "normal" sperm cells, should be 70% to 90% (or higher).

Total sperm production takes about 4 to 6 wk. During periods of heat stress, the boar's immature sperm will be preferentially killed by the heat stress. Thus, the boar may be subfertile and may have signs of dead sperm in its ejaculate 4 to 6 wk after the heat stress episode.

THE GILT OR SOW

A gilt is a young female pig. Opinions vary about when a gilt becomes a sow—some consider her a sow when she is first bred, farrows, or is rebred.

DEVELOPMENT OF GILT OR SOW REPRODUCTIVE ORGANS

The gonad of the female pig is called the ovary. Rather than simply expelling the gamete (sperm or egg) as the male does, the female must provide a site for fertilization (usually the oviduct) and a place to nurture the developing embryos (the uterus).

The ovary of the pig begins to develop in utero just as the male reproductive tract develops. Follicles develop in an ordered manner from dormant through antral follicles

TABLE 5-2
Normal Values for Reproduction in the Pig.

Item	Female	Male
Age at puberty[a]	5–8 mo	5–8 mo
Weight at puberty[a]	81–104 kg (180–230 lb)	81–104 kg (180–230 lb)
Duration of estrus	1–5 d (1–2 d for gilts)	—
Length of estrous cycle	20–21 d	—
Weaning to estrus[b]	4–7 d	—
Time to ovulate	12 h before end of estrus[c]	—
Gonad weight (each)	3–7 g	250–300 g
Gamete production	10–30 ova/estrous cycle	100 billion sperm per ejaculate

[a]Some breeds have a much earlier onset of estrus (ex., Meishan). Housing in confinement without boar exposure greatly increases the age at puberty.
[b]Some sows, especially those weaned early, may have a weaning-to-estrus length of 14 d or more.
[c]Sows may ovulate 36 to 90 h after the onset of estrus. For gilts, ovulation can be 12 to 48 h after the onset of estrus.

FIGURE 5–4
Drawing of a Histological Section of a Pig Ovarian Follicles.

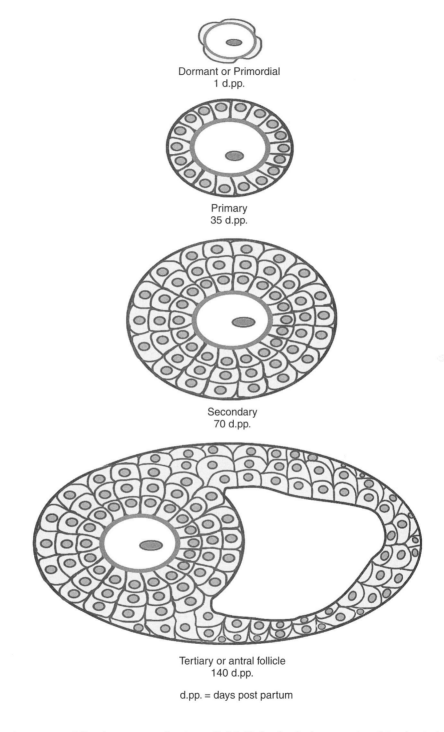

Dormant or Primordial
1 d.pp.

Primary
35 d.pp.

Secondary
70 d.pp.

Tertiary or antral follicle
140 d.pp.

d.pp. = days post partum

(Figure 5–4). The dormant and primordial follicles look the same in a histological section. However, when the dormant follicle is activated, it becomes a primordial follicle. Dormant and primordial follicles are present at birth in the pig. The primary follicle has a single layer of cuboidal cells surrounding the egg. Primary follicles are observed in the gilt at around 35 days after birth. When the cell layer around the ova multiplies (becomes multicellular), the follicle is classified as a secondary follicle. Secondary follicles are observed around d 70 of age. By days 70 to 100, occasional antral or tertiary follicles are observed, but they are more prevalent by d 140 of age.

Beginning at 70 d of age, the follicles are responsive to large, repeated doses of FSH and LH. In the post-pubertal gilt, FSH causes the follicle to grow and LH causes ovulation. Ovulation is an exocrine function (like sperm production) whereby the gamete is released into the lumen of the oviduct. The ovaries of a d 70 and a d 180 gilt look very different.

The eggs are ovulated in a burst. They are caught by the infundibulum of the oviduct and travel down the oviduct where they come in contact with sperm if natural or artificial insemination has occured. Fertilized eggs then travel down the oviduct by a combination of ciliary action and peristaltic contractions of the oviduct. Embryos implant in the uterus.

The uterus is a M-shaped structure that is connected to the oviduct and provides a large area for implantation and fetal growth (see Figure 5–5). At the base of each of the two uterine horns is the uterine body. Caudal to the uterine body is the cervix—a braided structure that separates the uterus from the vagina. The opening of the vagina gives rise to the vulva. By spreading the lips of the vulva, the clitoris is visible, especially when the sow is in estrus.

SUMMARY OF SOW REPRODUCTIVE ORGANS

1. Ovaries—The two irregular-shaped organs that are suspended in the abdominal cavity near the backbone and just in front of the pelvis. The ovaries produce the eggs (ova), estrogen, and the corpus luteum, which secretes progesterone.
2. Oviducts (fallopian tubes)—Small cilia-lined tubes or ducts that lead from the ovaries to the horns of the uterus. The infundibulum is a funnel-shaped structure located at the end of the oviduct near the ovary. At ovulation, the egg passes into the infundibulum of the oviduct, through the oviduct where it is fertilized, and moves into the uterine horn. The ovum may take 3 to 4 d to go from the ovary to the uterine horn.
3. Uterus—The muscular organ where the fertilized eggs attach themselves and develop until parturition. The uterus consists of the two horns, the body, and the neck (cervix). In swine, the fetal membranes that surround the developing embryo are in close contact with the entire lining of the uterus.
4. Cervix—A thick-walled, inelastic structure about 6 in long. The cervix is funnel shaped with ridges that have a configuration that conforms to the end of the boar's penis. The main function of the cervix is to prevent microbial contamination of the uterus. In swine, semen is deposited directly into the cervix during natural mating. When birth occurs, the cervix dialates to allow passage of the piglets.
5. Vagina—The birth canal. At the time of birth, the vagina expands and serves as the final passageway for the fetus. The biochemical and microbiological environment of the vagina also protects the upper reproductive tract from invading microorganisms.
6. Vulva (urogenital sinus)—The external opening of both the urinary and genital tracts. During estrus, the vulva of the sow becomes swollen (one sign of estrus).

FIGURE 5–5
Drawing of Follicular Development.

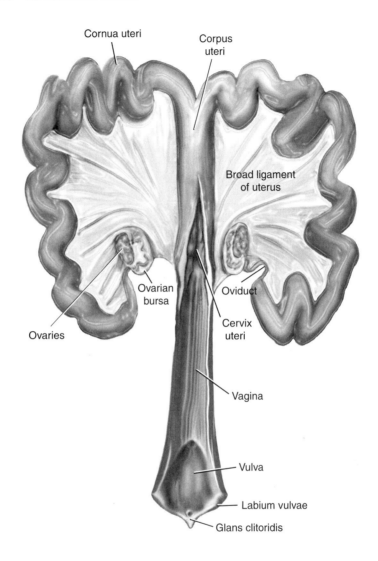

THE ESTROUS CYCLE

The estrous cycle is observed, on average, every 21 d while healthy adult, female pigs are not pregnant. The estrous cycle is characterized by two phases: the follicular phase, which ends with ovulation and the luteal phase. The follicular phase is a period of declining FSH and episodic pulses of LH. Toward the end of the follicular phase, a large spike in LH results in ovulation of the eggs (see Figure 5–6). The spike in LH is proceeded by a rise in blood concentration of estrogen produced by the follicle. Estrogen concentration becomes very low after ovulation. Progesterone concentration, which is very low during most of the follicular phase, rises during the early luteal phase and reaches a plateau about 5 d after the start of the estrous cycle.

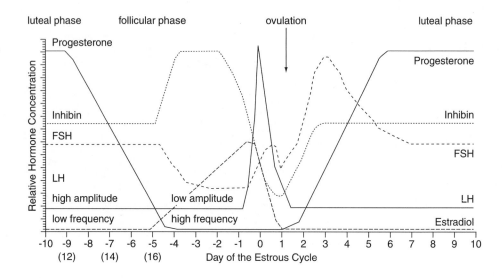

FIGURE 5–6
Hormone Changes during an Estrous Cycle of the Pig. LH will Be Pulsitile with High or Low Pulse Amplitude or Frequency.

Progesterone is synthesized by the corpus luteum (CL), which forms after ovulation. If fertilization takes place, the progesterone is maintained because the developing fetus prevents regression of the CL (luteolysis). If the pig is not pregnant, regression of the CL is caused by prostaglandins produced by the uterus.

Understanding hormonal changes during the estrous cycle is important both from a basic biology perspective and to evaluate and understand reproductive management techniques. For example, prostaglandin products are sold to regress the CL. Prostaglandins are used to abort pregnant sows, synchronize estrus, or synchronize farrowing times.

PREGNANCY

Pregnancy begins with fertilization of the eggs. The embryos and the pituitary work in concert to maintain the CL. Various sources of estrogen serve to maintain the CL during pregnancy.

The embryos migrate into the uterus and distribute themselves more-or-less uniformly in the uterine horns. The capacity of the uterus can limit the number of fetuses, but as a practical matter, most uterine horns are below capacity during pregnancy. While many sows can maintain 16 or more piglets to term, the average number of pigs born is only 12 on well-managed farms.

Pregnancy ends with regression of the CL and initiation of parturition. The CL regression is thought to be caused by glucocorticoids and the prostaglandin $PGF_{2\alpha}$. The resulting luteolysis initiates parturition.

If the sow is not pregnant, the CL secretes signals that cause the uterus to secrete $PGF_{2\alpha}$.

LACTATION AND REBREEDING

Lactation is a period that promotes repair and rejuvenation of the uterus, in addition to providing milk to piglets. Britt (1996) and Britt et al. (1999) describe three phases after parturition:

The first phase is the *hypergonadotropic* phase, in which the CL is regressed and progesterone concentrations are low. FSH and LH are increased and follicular development resumes.

The second phase is the *transition* phase, in which FSH and LH are suppressed and the uterus is involuting. The transition phase is from d 2 to d 14 after farrowing.

The third, or *normalization,* phase is from d 14 to 21 postpartum. From d 14 postpartum and later, FSH and LH levels increase and the uterus becomes involuted. Follicles grow and estrogen levels increase. The ovaries and uterus respond to weaning by ovulating 4 to 5 d later.

BEHAVIORAL BIOLOGY

Among mammals, the domestic pig is unique in both quantity and quality of matings. When boars ejaculate, they produce about 200 mL of semen over about a 10-min ejaculation. The large semen volume and long duration of ejaculation are unique among common domestic animal species. In addition to physical differences, pigs show interesting and often unique mating behaviors. A good understanding of sexual behaviors helps the pig handler become efficient in managing these behaviors.

ESTRUS DETECTION

The adult female pig is receptive to sexual interaction only during a few days when she is said to be in estrus. Full behavioral estrus includes the following behavioral traits:

- Boar seeking
- Change in mucous secretion from "slimy" to "sticky"
- Enlargement of the clitoris (vulva must be opened to observe)
- Lowered feed intake
- Expression of the standing reflex or lordosis behavior:
  - Ears perked
  - Feet firmly placed
  - Swollen, red vulva
- Increase in activity, except when near a boar:
  - Pacing
  - Searching
  - Mounting other sows
  - Standing still while allowing other sows to mount
  - Occasional vocalization

The sow in estrus uses special cues to identify a boar. Sows can clearly discriminate male and female by smell (olfaction). Interesting studies were performed a few decades ago by French scientist Jean-Pierre Signoret. He constructed a T-shaped pen and examined the preference of adult pigs for members of the opposite sex. The pigs were given one stimuli in arm A and another in arm B. If they spent more time near one arm, then they had a preference. His findings were quite revealing about how pigs perceive their world. Sows not in estrus were just as interested in being near a castrated male as being near a boar. But when sows were in estrus, they spent much more time near an intact boar than near a castrated male.

Signoret tested the boar's preference for sows in estrus compared with sows not in estrus. He reported that the boar spent equal time near estrus and nonestrus sows. Recreation of Signoret's work showed that 3 out of 11 boars actually did spend more time near an estrus sow in the T-maze test. Thus, some boars can distinguish an estrus from a nonestrus sow while the majority cannot make such a discrimination. Most boars search for receptive sows in heat by trial and error. They try to mount anything that stands still.

Understanding these aspects of boar behavioral biology explains why boars are relatively easy to train to mount a dummy for semen collection and why some individual boars can develop finer olfactory discrimination abilities. These individuals will find sows in estrus more quickly and breed more efficiently.

MATING AND THE BREEDING AREA

Natural mating for the boar includes the following phases: courtship, mounting, intromission, ejaculation, dismount, and additional courtship behaviors. In artificial insemination (AI), the sow often does not experience courtship and mounting. These components are thought to be necessary for maximum reproductive rates, but not for average or minimal reproductive rates. It is beneficial to maximize AI reproductive rates, so it is important to consider ways to replace the courtship and mounting experiences. Handlers can replace mounting by applying back pressure on the sow, which causes the sow to show the classic standing reflex. Application of back pressure may also improve reproductive performance.

At mating, the sow shows the standing reflex, which at times seems like reduced activity. On the inside, the sow's reproductive tract is very active. The boar's courtship behavior helps stimulate the sow's reproductive tract to increase motility and prepare for fertilized eggs. The sow releases oxytocin, and perhaps other hormones, during courtship behaviors. Release of these hormones is thought to be one physiological signal that prepares the sow's reproductive tract for pregnancy.

North Carolina scientist Billy Flowers gave oxytocin to sows that were to be artificially inseminated to see if he could improve farrowing rate or litter size. He was successful in improving reproductive performance by using 20 IU of oxytocin per insemination. Inexperienced AI technicians who gave oxytocin 2 min before insemination increased the farrowing rate by 12% and increased the number of pigs born alive by 0.8 pigs/litter. This information adds more evidence to support the idea that courtship behavior and the resulting changes in sow physiology are important in obtaining high reproductive rates.

Producers know that either oxytocin or an experienced human can provide a more complete courtship experience to the sow. However, they also need to understand the features of the boar that are stimulating to the sow. Signoret provides some answers.

Signoret performed a now-classic study in 1961 in which he determined the features of the boar that were stimulating to the sow. His findings are summarized in Table 5–3. When Signoret applied back pressure alone, only 48% of the sows showed the standing response. When he played audiotapes of boar sounds, 70% of the sows showed the standing response. By using preputial secretions (containing low concentrations of pheromones), Signoret found 80% of the sows showed the estrus stance (we now know saliva contains greater concentrations of boar pheromones). When he combined the smell and sound of a boar, 90% of the sows stood for insemination. When sows could see, smell, and hear a boar, a full 97% of the sows expressed the lordosis response.

The best results are obtained by having a live boar, but for 100% AI facilities, the smell and sound of the boar are important cues. The Flower's oxytocin treatment (cited earlier) and the boar-like stimuli probably provide similar physiological arousal and each improves reproductive performance.

Sometimes too much of a good thing can hurt reproductive success. Gilts are stimulated by the presence of a boar, and pork producers have known for many years that by placing a boar in the pen next to sows, the sows will show estrus behavior. However, expression of estrus behavior requires energy, and estrus sows get tired. So while boar stimulation is good, over-stimulation can be harmful. Hemsworth (1999) shows data that indicate 87% of gilts were mated in the aisle while only 52% of gilts were found in estrus and bred in their home pen. Hemsworth (1999) suggested housing gilts about 3 ft (1 m) away from boars. In this way, gilts receive some, but not too much, boar contact. He also suggests not breeding in the gilt's home pen. The recommendation is to breed in either a designated mating area or the boar's pen (which would be loaded with boar pheromones).

TABLE 5–3

Percentage of Sows Showing the Classic Lordosis Response when Exposed to Different Features of the Boar.

Stimulus	% Showing Standing Response
Back pressure only	48
Playing boar grunts	70
Odor of prepuberal secretions	80
Smell and sound of boar	90
Smell, sight, and sound of boar	97

Source: Adapted from Signoret and Du Mesnil Du Buisson (1961).

SUMMARY

Efficient reproduction is vital to the economic survival of pork producers. This chapter addresses the biology of the reproductive process. The anatomy and physiology of the male and female reproductive systems were described. Normal values for economically important reproductive traits, such as age at puberty, duration of estrus and the estrous cycle in females, and gamete production in females and males were also discussed, as well as behavioral traits associated with mating and management practices that favor successful reproduction.

QUESTIONS AND ACTIVITIES

1. Any group of growing, developing gilts will show variation in size of the vulva. On one end of the population, some gilts have a very small vulva. Those gilts are not good choices for replacement breeding females because the odds of them having a high reproductive rate are low. List some physiological (particularly endocrinological) measures that might correlate with a small-sized reproductive tract.
2. LH synthesis and secretion are stimulated by hypothalamic gonadotropin-releasing hormone (GnRH). Two general approaches to reduced GnRH concentrations are to kill the GnRH-secreting cells with a specific agent or to immunize animals against GnRH. What would happen to male and female reproduction if GnRH was partially or totally blocked?
3. A pork producer in the South noticed a decreased sperm count from his boars that are used for semen collection. The only change observed was that their diet was changed to reduce costs—the diet now includes cottonseed meal. Is there a link?
4. PG600 is a product sold by Intervet (*http://www.intervet.com/home_1024.html*) as an aid in the induction of estrus in prepuberal gilts and for weaned sows experiencing a delayed return to estrus. PG600 contains 400 IU pregnant mare serum gonadotropin (PMSG) and 200 IU human chorionic gonadotropin (HCG). Why would these hormones from horses and humans induce estrus?
5. At least two companies sell a product that is $PGF_{2\alpha}$ (Upjohn and Am Tech Group, Inc.). These products are used to regress the CL. Why might this be useful in sows and cycling gilts? How might this help breeding?

LITERATURE CITED

Britt, J. 1996. Biology and management of the early weaned sow. Proc. Am. Assoc. Swine Prac., 417–426.

Britt, J. H., G. W. Almond, and W. J. Flowers. 1999. Diseases of the reproductive system. In: B. Straw et al. (eds.) Diseases of Swine. Iowa State University Press, Ames.

Flowers, W. L. 1998. Management of reproduction. In: Progress in pig science. Nottingham University Press. Nottingham. UK. Pages 383–405.

Hemsworth, P. H. 1999. Behavioral problems. In: Straw (ed.) Diseases of Swine. Iowa State University Press, Ames.

ADDITIONAL READINGS

Bearden, H. J. and J. Fuquay. 1997. Applied Animal Reproduction. 4th ed. Prentice Hall, Englewood Cliffs, NJ.

Bronson, F. H. 1990. Mammalian Reproductive Biology. University of Chicago Press, Chicago.

Cupps, P. T. 1991. Reproduction in Domestic Animals. 4th ed. Academic Press, San Diego, CA.

Gordon, I. 1997. Controlled Reproduction in Farm Animals Series (1: Cattle and Buffaloes; 2: Sheep and Goats; 3: Pigs; 4: Horses, Deer, and Camelids). CAB International, Oxon, UK.

Hafez, E. S. E. 1993. Reproduction in Farm Animals 6th ed. Lea & Febiger, Philadelphia.

King, G. J. 1993. Reproduction in Domesticated Animals. Elsevier, New York.

Knobil, E. and J. D. Neill. 1998. Encyclopedia of Reproduction. Academic Press, San Diego, CA.

Lamming, G. E. 1984. Marshall's Physiology of Reproduction. Reproductive Cycles of Vertebrates. Churchill Livingstone, Philadelphia.

McDonald, L. E. 1989. Veterinary Endocrinology and Reproduction 4th ed. Lea and Febiger, Philadelphia.

Yen, S. S. C., R. B. Jaffe, and R. L. Barbieri. 1999. Reproductive Endocrinology 4th ed. W. B. Saunders, Philadelphia.

INTERNET RESOURCES

At the University of Wyoming, Dr. William J. Murdoch has a course at:
http://www.uwyo.edu/ag/anisci/wjm/repro/homepg.htm.

At Southern Illinois University, Dr. Todd A. Winters has a course at:
http://www.siu.edu/~tw3a/431.htm.

At the University of Guelph, Dr. Mary Buhr has a course at:
http://www.aps.uoguelph.ca/teaching/10-412.html.

A variety of pig reproduction fact sheets:
http://www.ianr.unl.edu/pubs/Swine/

Pig Reproduction, especially placentation:
http://www.ansi.okstate.edu/resource-room/reprod/all/animations/placentation_pig.htm

6

PIG GENETICS

INTRODUCTION

Pig improvement through genetic selection relies on tools and knowledge developed over centuries of effort. This chapter describes the biological basis on which genetic changes are made and introduces methods of selection for various traits of economic importance. Emerging tools involving molecular genetics and other new technologies are explored and described.

THE CHANGING PIG

For centuries, animal breeders have capitalized on the impressive variation in the genome of the domestic pig. Imagine two pigs standing side-by-side: a European wild boar and a modern meat-type pig. It is hard to imagine that one derives from the other. But, in fact, the recombination of genes from the wild ancestor results in the current pig found on farms around the world—from the fattest to the leanest and the most to the least prolific.

This chapter reviews procedures for the gradual improvement of biological traits through genetic selection. More modern techniques that will undoubtedly dominate the industry in the near future include tools of molecular biology that will spread and hasten genetic change at rates that far exceed the older techniques. The following describes the progression of genetic selection over the centuries.

The Progression of Genetic Selection	
• 11,000 yr ago through about mid 1900s	Breed like to like; propagate same
• Mid 1900s through 1980s	Breed for desired genotypes using carefully measured phenotypes
• 1980s through 1990s	As above, plus use of sophisticated Best Linear Unbiased Predictor (BLUP) computer models for selection
• 1990s	As above, plus marker-assisted selection and concentration of certain genes with known effects
• Future	Transgenic pigs with specialized traits or features of interest to food production

The effects of genetic selection have not changed from prehistoric times. What has changed, as new tools for selection are incorporated, is the rate of genetic change. Using mathematical procedures like best linear unbiased prediction (BLUP), animal breeders can effect change in the pig genotype in much less time than with previous methods. Applying modern tools of marker-assisted selection will speed genotype improvement even further. The international effort called the Pig Genome Project will help to elucidate important genes as the pig genome is mapped. New genotypes created in research laboratories such as transgenic technologies will be applied to farm animal production in ways similar to the application of animal models to biomedical problems.

THE U.S. GENE MAPPING PROGRAM CAN BE VIEWED ON THE WORLD WIDE WEB AT:
http://www.public.iastate.edu/~pigmap/pigmap.html

ADDITIVE VERSUS NONADDITIVE GENE ACTION

MENDELIAN GENETICS

Gregor Mendel, the Czechoslovakian monk, bred plants in the late 1800s and carefully recorded the outcome. Mendel's work was lost for decades, then rediscovered during the birth of the field of quantitative genetics. Quantitative genetics resulted from the assignment of traits to genes transmitted through selective breeding.

Mendelian genetics refers to the simple quantitative inheritance of genes that express themselves in defined and predictable phenotypes. Animals of *AA* genotype, bred to *AA*, will yield all *AA* offspring. This is the outcome of breeding like to like. If an animal with the genes *AA* is bred with an animal bearing *aa* alleles, *all* offspring will be *Aa*. However, when two animals, both with the genotype *Aa* are bred (to the wonder of early animal breeders), breeding "like to like" does not yield like offspring.

If pigs with the genotype *Aa* are bred with a mate bearing the *Aa* genotype, the offspring genotypes fall in the predictable, Mendelian distribution of:

Genotype	Genotype Distribution	Phenotype
AA	1/4	A-like
Aa	1/2	A-like
aa	1/4	a-like

The phenotype distribution falls in only two categories: those like A and those like a. Certain aspects of coat color in pigs have this general genetic mechanism. If a purebred white sow is bred to a boar of any other color, the offspring are all white. Thus, the purebred could be said to have the *AA* genotype and a white-coat-color phenotype. All offspring would have a white coat color and would be either *AA* or *Aa*, regardless of the boar's coat color in the boar's genes.

In practice, some traits are controlled by one or just a few genes and are said to follow Mendelian genetics. Still other traits are controlled by many genes, and the distribution of individuals follow a normal distribution. These traits are polygenic and, thus, may have several genes that code for a trait. Most traits of economic importance to the pig industry are polygenic.

When more and more genes code for a particular trait, the distribution changes from just three phenotypes to more and more genotypes depending on the number of genes. The distribution of genotypes and phenotypes approaches a normal distribution with just three pairs of genes, segregating at random in a population. Common, economically important traits in commercial pork production have a normal distribution and, thus, many genes are involved, each contributing an incremental effect on a given trait. From time to time, a major gene is discovered that contributes a much greater effect on the trait.

The objective in genetic improvement is to move the mean of the population in a desired direction (to increase or decrease). For some traits, like lean gain, producers seek to increase the value. For other traits, like backfat thickness (or % body fat), producers seek to decrease the mean of the herd (although some modern lines may be too lean). The objective for the animal breeder is to shift the population either left or right.

Figure 6–1 shows an example herd profile adapted from PIC USA, Inc. To affect genetic change, sires and dams must be superior to the present herd average. By eliminating the poor-performing animals through culling (discarding lower-valued animals) and by adding genetically superior animals, the herd average improves.

Some traits have more than two possible genes at a given locus. For example, instead of just *A* and *a* genes there may be *A*1, *A*2, *A*3, or *A*4 genes. Such a polygenic trait would be more difficult to select. It is quite possible that traits of economic importance are controlled by many genes.

A highly inbred genetic line is one in which relatives have been mated to create uniform animals that are homozygous (AA or aa) for many traits at many loci. Purebred livestock were selected over many generations to be uniform in certain traits and only using certain families of animals to start with. Thus, although pure breeds of livestock are more homozygous than crosses, breeds are not as homozygous as inbred lines.

FIGURE 6–1
Use of Selection to Improve the Population of Pigs.

Source: Adapted from PIC USA, Inc.

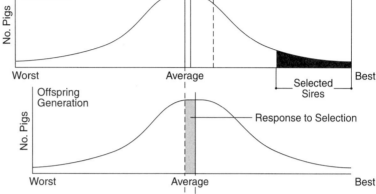

Certain phenotypic traits change more consistently in incremental steps from several genes. In an oversimplified example of gene effects on average daily gain (ADG), the unit of growth for animals containing these genes might be:

Genotype	ADG, lb/d	ADG, kg/d
GG	1.6	0.72
Gg	1.5	0.68
gg	1.4	0.64

Currently, there is no known, single gene that would have such an effect on ADG, but it could be discovered. Simple Mendelian genetics can improve ADG. As pigs with faster ADG are selected, producers are more and more likely to be selecting for one or two copies of the G gene. This type of incremental gene action is called *additive gene action*—the small contribution of many genes to a trait. Traits that respond well to the effects of additive gene action are those associated with growth and body composition (especially fat content). For traits that respond through additive gene action, the offspring trait is the average trait of the parents. Heritability estimates are measures of additive gene action. Heritability refers to the proportion of the variation that is attributed to genetics relative to the total trait variation (due to genetics and environment). Table 6–1 gives heritability estimates for common traits.

HETEROSIS

Heterosis has its greatest effect on traits that tend to be low in their heritability estimates. Thus, reproductive performance and health tend to be thought of as traits that respond with greater heterosis. Table 6–2 shows measures of heterosis that represent the views of many authors.

SECTION II BIOLOGY OF THE PIG

TABLE 6–1
Heritabilities and Genetic Correlations Estimated from the NPPC National Genetic Evaluation Program. The Heritability Estimates Are Given on the Diagonal and the Genetic Correlations Are below the Diagonal.

	BF	ADG	LMA	IMF	Min	pH	WHC	Inst
Backfat	.46							
Average daily gain	.14	.50						
Loin muscle area	−.61	−.13	.48					
Intramuscular fat	.30	.06	—	.25	.47			
Muscle color (Minolta)	.08	.11	.02	.11	.25			
Ultimate pH	.03	−.11	−.11	0	−.49	.38		
Water-holding capacity	−.05	.07	.13	−.02	.52	−.92	.19	
Tenderness (Instron)	−.17	−.07	.15	−.17	.18	−.42	.22	.20

Source: National Pork Producers Council, Genetic Evaluation Program (1995).

TABLE 6–2
Average Estimates of the Effects of Crossbreeding on Performance Traits in Pigs.

Trait	Heterosis (%)
Maternal heterosis	
Number of embryos	7
Litter size	
At birth	7–10
At weaning	20
Litter weight	
At birth	8
At weaning	20
Paternal heterosis	
Testicle weight	20
Total sperm	30
Improved conception	10–14
Individual traits	
Average daily gain	5
Feed efficiency	5

Some traits do not respond as a simple addition of individual gene effects. Traits that express a greater effect for one or more genes than the sum of the individual gene effects are influenced by nonadditive gene action. When the average of the offspring differs from the average of the parents, the trait is said to express heterosis, or hybrid vigor.

There are several types of heterosis:

a. *Individual heterosis*—The improvement in performance traits in a pig such as growth rate, feed efficiency, survival, etc., due to its crossbreeding relative to that of its pureline parents.
b. *Maternal heterosis*—The improvement in performance in sows and their offspring from using a crossbred dam (e.g., improved pre- and post-natal environment, larger litter size, greater rebreeding rate, etc.).
c. *Paternal heterosis*—The improvement in performance in boars and their offspring from using a crossbred sire (e.g., improved libido, persistent breeding, longevity, etc.).

HERITABILITY ESTIMATES

Estimates for heritability for economically important growth and carcass traits are given in Table 6–1. These estimates are adapted from the National Pork Producers Council 1995 Genetic Evaluation Program (NGEP). All traits have a heritability estimate that is greater than zero. With these moderate heritability estimates, producers could make genetic progress in these traits through selection.

The data for meat quality traits such as tenderness and water-holding capacity have lower heritability estimates and, thus, slower genetic improvement is possible for these traits. Still another problem with measures of tenderness is the need to kill animals to test for tenderness. This is not necessary for measures such as ADG and body fat percentage. Because slaughter eliminates the ability of high-quality animals to reproduce, genetic improvement in meat quality may be slower. Identification of genetic markers or genes that control pork quality will greatly hasten the search for improved pork quality. Progeny tests are conducted to evaluate performance traits of breeding lines.

Breeds of pigs vary in meat quality, especially intramuscular (within the muscle) fat percentage (Table 6–3). Packers who seek quality pork products have encouraged use of the Duroc and Berkshire breeds to add flavor and tenderness as well as intramuscular fat.

TABLE 6–3
Least Squares Means for Muscle Quality Traits Estimated from NGEP Data.

SIRE LINE	INTRAMUSCULAR FAT %	MINOLTA REFLECTANCE	ULTIMATE pH
Berkshire	2.41^{bc}	22.6^a	5.91^a
Danbred HD	2.33^c	23.0^a	5.75^{cd}
Duroc	3.03^a	23.2^a	5.85^{ab}
Hampshire	2.57^b	25.3^b	5.70^d
NGT Large White	2.15^c	23.4^a	5.84^{ab}
Nebraska SPF Duroc	2.71^{ab}	23.1^a	5.88^{ab}
Newsham Hybrid	2.25^c	22.7^a	5.82^{bc}
Spotted	2.35^c	23.3^a	5.83^{bc}
Yorkshire	2.33^c	23.0^a	5.84^{ab}

[a,b,c,d]Means with different superscripts are statistically different ($P < .05$).
Source: National Pork Producers Council Genetic Evaluation Program (1995).

GENETIC IMPROVEMENT

Seedstock producers and commercial producers, who make their own replacement seedstock, seek constant genetic improvement. The rate of genetic improvement (ΔG) is determined by three factors:

- Heritability estimate (h^2) of the trait
- Selection differential (SD)—the difference of the selected animals from the herd average, calculated as:
  (average of selected animals − herd average)
- Generation interval (GI)—the average age of breeding stock when replacements enter the herd

The formula for genetic improvement is:

$$\Delta G = \frac{h^2 \times SD}{GI} = \text{rate of genetic improvement}$$

For example, assume a producer wishes to lower the trait for backfat thickness. With replacement seedstock averaging 0.7 in and the herd average at 1.0 in of backfat, an h^2 of 0.46 (see Table 6–1) and a generation interval of 2.0 yr, the formula works out like this:

$$\Delta G = \frac{0.46 \times (0.7 - 1.0)}{2.0} = -0.069 \text{ in/yr}$$

At the end of one year, the producer expects the herd to lower its backfat thickness by 0.069 in (starting from 1.0 in). The year-end backfat thickness would be 0.931 in (not quite a 0.10-in improvement in backfat thickness). The h^2 is fixed and the generation interval, for a given herd, is difficult to change. The easiest way to increase the rate of genetic progress is to have a greater SD.

Many animal breeding projects have quantified genetic improvement over time. Figures 6–2 and 6–3 show data from herds that have been selected for changes in economically important traits. The USDA scientists at Beltsville, Maryland selected genetic lines to produce boars and gilts with increasing or decreasing backfat thickness. Over the 12 generations, the lean line had 50% less backfat than the fat line. Fat pigs tended to be shorter in body length and had a shorter height. Fat pigs were more round-shaped with a larger circumference. Breeding stock companies use the same process to reduce backfat thickness of its commercial seedstock. In addition to growth and body composition, reproductive traits like ovulation rate can be improved with selection, even though the h^2 is lower and genetic progress is slower.

CROSSBREEDING SYSTEMS

Crossbreeding systems are used to capitalize on traits controlled by both additive and nonadditive gene action. If the maternal line is selected for the desired body composition and meat quality, using crossbred females, advantages in health, reproduction, and mothering ability (especially milk production) may be captured by the heterosis (Table 6–3). The amount of heterosis obtained by crossbreeding systems is outlined in

FIGURE 6–2

USDA Scientists Hetzer and Miller (1972) Reported these Responses to Different Measures for Duroc and Yorkshire Pigs Selected for Low and High Backfat Thickness. By the 13th Generation, the "Fat" Line had About 6 cm (2.4 in) of Backfat while the "Lean" Line had 2.5 to 3.0 CM (about 1 in) of Backfat Thickness. Note the Correlated Changes in Body Length and Shape.

Source: Adapted from J. Anim. Sci. 35:746.

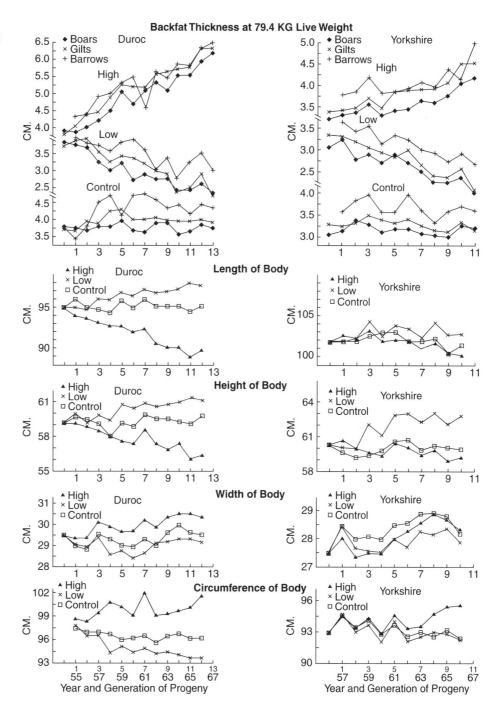

FIGURE 6–3
Response of Selection for Reduced Backfat Thickness (P_2) by a Commercial Company, PIC USA, Inc., That Uses the Biological Principles Discovered Earlier.

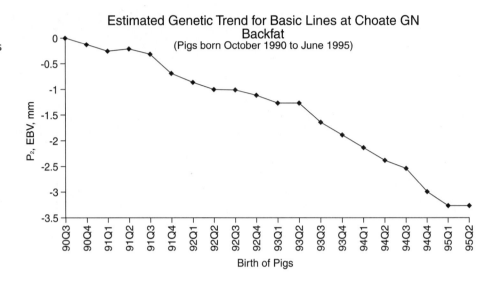

TABLE 6–4
An Example of the Computation of the Expected Heterosis in a Crossbreeding System.

Sire Line	Dam Line Composition	Mating	Percent Heterosis	Progeny Line Composition
A	B	A × B	100	1/2A + 1/2B
A	1/2A + 1/2B	A × 1/2A; A × 1/2B	50	3/4A + 1/4B
B	3/4A + 1/4B	B × 3/4A; B × 1/4B	75	3/8A + 5/8B
C	3/8A + 5/8B	C × 3/8A; C × 5/8B	100	3/16A + 5/16B + 8/16C
A	3/16A + 5/16B + 8/16C	A × 3/16A; A × 5/16B A × 8/16C	81	19/32A + 5/32B + 8/32C

Source: Adapted from Farmers Hybrid literature by H. I. Sellers (1994).

Table 6–4. A similar logic can be used to determine the amount of heterosis for any matings of combinations of pure lines.

The sire line has traditionally been a purebred animal because producers felt there was little advantage to use of crossbred boars. This view is largely changed today. Producers now favor use of crossbred boars, which are thought to have better health, more sperm production, and better libido. These are reasons enough to use crossbred boars, even if the benefits are limited.

Crossbreeding systems seek to obtain as much of the available heterosis as possible. For a simple two-way cross such as A × B, all the offspring have 100% of available heterosis, while the parents have 0% of the possible heterosis. Use of rotational crossing systems has the advantage that only boars (or semen) must be brought onto the farm, thus reducing biosecurity concerns. The disadvantage of rotational systems is that less that 100% of the available heterosis is obtained in the progeny. With terminal crossing systems, 100% of the available heterosis is obtained. The biosecurity risk is increased because new females and males must enter the farm.

The modern business of pork production must use the best genetics and animal breeding systems that are affordable. Nearly all of the large-scale, commercial producers today use a terminal breeding system. To handle their biosecurity concerns, large farms use multiplication units that receive grandparent or great-grandparent stock, thus minimizing the number of animals that enter the herd (and thus reducing risk of disease entry). Biosecurity is improved because of the low number of animals entering the herd and the ability to test isolated breeding stock in the multiplication unit before possible contamination of the core herd. Thus, genetic progress and biosecurity represent a balance that must be struck for attainment of the herd goals—healthy animals of good market value.

> **BALANCE BETWEEN GENETIC PROGRESS AND BIOSECURITY**
> Genetic progress and biosecurity must be weighed carefully.
> **Genetic progress**   **Biosecurity**
> △

TWO-WAY CROSSES—TERMINAL AND ROTATIONAL

The simplest crossbred animal is a two-way terminal cross. Crossing a maternal line with a meat line makes a great deal of sense—for example, a Landrace sow with a Hampshire boar. The resulting $Y \times H$ pigs would be sold as (mostly white) crossbred market animals. None of the offspring would be kept as replacements. Replacement maternal lines and sire lines would either have to be created from internal purebred herds or purchased. Parents have 0% of the available heterosis while offspring have 100% of the available heterosis.

The two-breed rotational cross requires that the producer purchase only replacement purebred boars of two genotypes. The sows (line A, for example) are first mated to boars (line B, for example). The next generation is selected from among the offspring and would be mated to line-A boars. The third generation is mated to line-B boars, and the cycle continues. Sows are always crossbred and boars are always purebred in this scheme. The boars have 0% of the available heterosis; the sows and piglets have an average of 68% of the available heterosis because the dam has some of the sire's breed in her genes.

THREE-BREED CROSSES—TERMINAL AND ROTATIONAL

Crossbreeding programs that use three breeds have the advantage that the maternal line is a cross, so females will exhibit heterosis in their health, reproduction, and maternal traits. The three-breed terminal cross captures 100% of the possible heterosis in the market hogs, while the three-breed rotational cross market hogs capture less than 100% (an average of about 86%).

Although some heterosis is lost by use of rotational crosses, they are still used. One major advantage of the rotational cross is that replacement gilts do not need to be purchased. Aside from their expense, bringing replacement gilts onto the farm poses a biosecurity risk to herd health (see Figure 6–4).

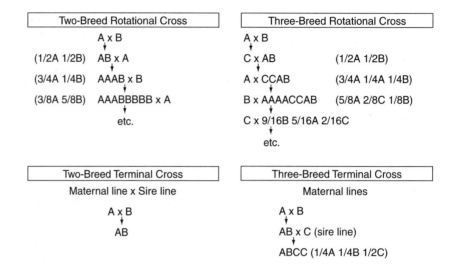

FIGURE 6–4
Four Crossbreeding Systems Used in Commercial Pork Production. A, B, and C Represent Three Distinct Breeds or Lines.

OTHER CROSSES

Three other crossbreeding programs are the rotaterminal, the four-breed rotational, and the four-breed terminal cross systems. The three- or four-breed rotaterminal uses a rotational system for the maternal side but uses a terminal sire. The four-breed terminal system uses a two-way cross sow (it could even use a three-way cross sow) and a two-way cross boar. This is the first time in the examples of breeding systems that a crossbred boar is used. The crossbred boar may have advantages in sexual drive, sperm production, longevity, and health.

Commercial lines of pigs resemble a five-way cross in many cases. The sow lines are most commonly composed of three breeds: Yorkshire, Landrace, and Duroc. The sire lines may be Hampshire, Pietrain, Duroc, or other breeds.

From a meat quality effect, the Hampshire breed has the disadvantage of adding a factor for poor meat quality, known as the "Hampshire effect." Note in Table 6–3 that meat from the Hampshire breed has the lowest ultimate pH and the highest Minolta color score (indications that the pork may be pale and lack water-holding capacity). These values indicate a greater tendency toward poor meat quality. The added benefits of the Hampshire for lean gain may offset some of the meat quality problems.

Many crossbreeding systems use Duroc in the maternal and/or the sire lines. The Duroc adds extra intramuscular fat—a desirable effect for eating quality of pork—and fewer meat quality problems. The Berkshire breed has been associated with greater pork quality by some pork buyers as well, but the data in Table 6–3 do not support use of the Berkshire over the Duroc for meat quality traits.

MOLECULAR GENETICS

Several recent examples highlight the potential applications of molecular genetics in commercial pork production. The industry will continue making progress in applying technology in molecular genetics to the improvement of commercial lines of pigs.

MARKER-ASSISTED SELECTION

Traditional selection relies on measuring the phenotype as a predictor of the animal's genotype. Environmental effects are often highly significant determinants of the observed phenotype, so more precise selection technologies that improve the odds of actually selecting genetically superior animals are always desired.

Short of having identified the actual causative gene, the pig breeder would like to have a gene that is closely linked to the desired trait. This selection technique is called marker-assisted selection by use of quantitative trait loci (QTL). QTL are known genes that are either (1) physically close on the chromosome to the desired genes or (2) will be shown later to be directly involved in the expression of the desired gene.

For example, assume that a gene that serves as a marker for lean growth was found. This gene would not, by itself, be the sole determinant of fast, lean growth. Rather, this gene would be closely linked to some of the genes that are positive (or negative) for lean growth. When animals with this gene are retained, lean growth increases. Thus, this marker will add to the phenotypic information available to the animal breeder. The selection, therefore, can be more rapid.

THE *HAL* GENE

Early scientists anesthetized pigs with the gas anesthetic halothane and found some pigs developed a rapid increase in body temperature and often died. This condition, known as Porcine Stress Syndrome, or PSS, is identical to the human condition known as malignant hyperthermia. PSS pigs often have poor meat quality (known as PSE pork—pale, soft, and exudative or watery).

When Canadian scientists first located the gene for PSS/malignant hyperthermia, they called it *HAL*-1843 (HAL-1843 is a registered trademark of Innovations Foundation, Toronto, Canada). The trait is a recessive gene that is considered a mutation. *Nn* pigs are normal and show few signs of stress susceptability. *Nn* pigs are called monomutants or carriers because they have one copy of the gene. These pigs are leaner, but have some meat quality problems. Pigs with the *nn* genotype are called dimutants because they have two copies of the gene. Dimutants are at risk when they experience even mild stress. Pigs of *nn* genotype can have significant meat quality problems.

Most geneticists and industry leaders feel the mutated *HAL* gene should be eliminated from the pig herds of the world. Consumers do not like the meat from HAL^{mm} pigs, and these pigs are more stress susceptible and therefore are at risk from an animal welfare point of view. The *HAL* gene is not needed to get lean growth and lean pork.

Pig breeders who use the *HAL* gene do so in a controlled manner. The sow herd would be negative for the gene (*HAL*− or *NN*). The boar lines would be either *Nn* or *nn*. If the boars are *Nn*, one-half of their offspring will be of the *Nn* genotype. If the boars are *nn*, 100% of their offspring will be *Nn*.

THE *ESR* GENE

Iowa State University scientists, in collaboration with PIC USA, Inc., discovered a gene that influences ovulation rate in pigs. The gene is a variation of the estrogen sulfate receptor (*ESR*) gene. Pigs can have one or two copies of this gene and each copy is reported to add about one pig born alive/litter. The gene is a registered trademark of PIC USA, Inc. under the name *PIC*lit$^+$. PIC USA, Inc. is reportedly multiplying this gene in its seedstock.

THE K88 ADHERENCE GENE

Pigs, like other animals, are susceptible to various strains of *E. coli* that cause intestinal problems, including diarrhea. The *E. coli* attach to an intestinal receptor by a protein called K88. Some pigs lack the K88 antigen and, therefore, *E. coli* do not attach to the intestinal wall and colonize. Pigs that lack the K88 antigen have less problems with scours from *E. coli*. The relationship between *E. coli* adherence and disease susceptibility is not perfect. The genetic mechanisms of this condition are not fully worked out, but this is a good example of how genetics can play a significant role in disease resistance.

THE PIG SKIN COLOR GENE

After knowing for centuries that white coat color is dominant in pigs, molecular biologists identified the gene that codes for white skin in pigs. The gene is located on chromosome 8 and codes for white only. Other genes code for other skin colors. Swedish scientists identified the gene and the test for the color gene is licensed outside Sweden exclusively to PIC. The trademark is called PICment. Sows with PICment have all-white offspring no matter the color of the sire. Because packers have a slight preference for white-colored pigs, tracking this trait should have an economic impact.

SUMMARY

This chapter described genetic selection, the biological basis on which genetic changes are made, and the methods of selection for various traits of economic importance. Mendelian genetics, the simple quantitative inheritance of genes, were explained. Some traits that are controlled by one or a few genes follow Mendelian genetics. However, other traits are controlled by many genes and follow a normal distribution; these are polygenic traits. Most traits of economic importance to the pig industry are polygenic. Some traits, particularly reproductive and health show hybrid vigor or heterosis.

The objective of genetic improvement is to move the mean of the population in a desired direction. Producers seek genetic improvement in traits such as percent of body fat, average daily gain, feed efficiency, conception rates, and litter size and weight.

The chapter explained heterosis, heritability estimates, and genetic improvement. Crossbreeding systems, including two-way crosses, three-way crosses, and others, were also discussed.

Molecular genetics has great potential in improving commercial lines of pigs. The industry will continue making progress in applying technology to pig production.

QUESTIONS AND ACTIVITIES

1. Assume a pork producer wants to make his herd more lean. He finds that his herd is too fat to capture the best price on his pork processor's buying grid (based on weight and backfat thickness), so he sets about to lower his herd backfat thickness. As a start, he measures backfat thickness at the last rib and finds that his pigs average about 1.2 in of backfat. His herd boars presently have a backfat thickness of 1.0 in and his sows (at market weight) have 1.4 in of backfat. If he now selects boars and replacement gilts that average 0.8 in of backfat, what will his backfat thickness be after one year? Assume a generation interval of 1.5 yr. State the heritability estimate to be used.

2. Boars are leaner than castrated males (barrows). From most to least fat, the pig genders rank barrow, gilt, and boar. In one paper, boars had 30% less backfat than gilts and 40% less backfat than barrows with the same time on feed (Nold et al., 1997). If boars are assumed to be 35% leaner than barrows and gilts (on average), then what must the replacement boars and gilts average backfat be in question 1? What level of backfat thickness do you recommend the pork producer purchase in his replacement breeding stock?

3. In Chapter 1, the Iberian pig was discussed. This pig is fatter, has a different fatty acid profile, and has the perception of better taste than conventional pork. In addition, pork from Iberian pigs commands a premium in some European markets. This trait's heritability estimate is unknown. However, if the heritability of the desired fatty acid profile is 0.10, 0.25, or 0.50, how quickly can a "pork flavor" sensory score be improved starting with an average of 3.0 units and seeking to increase the herd average to 4.0 (on an 8-point scale with 1 being the worst flavor)? (See Perez-Enciso et al., 2000).

4. Perez-Encisco et al. (2000) reported a QTL regarding the flavor of pork from Iberian pigs. If the QTL was used in a marker-assisted selection program, how many generations would it take to concentrate two copies of the desired gene through successive backcrosses with an American breed of pig? Show the results by diagram.

5. If you had the funds, which pig traits would you spend money on in a search for major genes? Put them in order of economic importance. Some examples include litter size, coat color, pork tenderness, boar taint odor, ease of handling, aggressive behaviors, and others.

LITERATURE CITED

Hetzer, H. O. and R. H. Miller. 1972. Rate of growth as influenced by selection for high and low fatness in swine. J. Anim. Sci. 35:730–746.

Nold, R. A., J. R. Romans, W. J. Costello, J. A. Henson, and G. W. Libal. 1997. Sensory characteristics and carcass traits of boars, barrows, and gilts fed high- or adequate-protein diets and slaughtered at 100 or 110 kilograms. J. Anim. Sci. 75:2641–2651.

National Pork Producers Council. 1995. Genetic Evaluation Program. National Pork Board. Des Moines, IA.

Perez-Enciso, M., A. Clop, J. L. Noguera, C. Ovilo, A. Coll, J. M. Folch, D. Babot, J. Estay, M. A. Oliver, I. Diaz, and A. Sanchez. 2000. A QTL on pig chromosome 4 affects fatty acid metabolism: Evidence from an Iberian by Landrace intercross. J. Anim. Sci. 78:2525–2531.

Zimmerman, D. W. and P. J. Cunningham. 1975. Selection for ovulation rate in swine: population, procedures and ovulation response.

INTERNET RESOURCES

Pig Genome references:
http://www.ri.bbsrc.ac.uk/pigmap/pigmap.html
http://www.genome.iastate.edu/community/links.html

Pig Genetics:
http://www.ipg.nl/
http://mark.asci.ncsu.edu/genehp.htm

7

MODERN BREEDS OF PIGS

INTRODUCTION

A breed is a group of animals of a given species that has been selected by people (in contrast to natural selection) to possess a uniform, heritable, and distinctive appearance (definition adapted from Porter, 1993). In addition to a given appearance, pigs were bred to emphasize certain production traits. Early in human history, the emphasis was on either meat or lard production. Other traits of interest were hardiness, temperament, prolificacy, milk production, mothering ability, and meat quality.

A century ago, the Berkshire was a lard-type pig and the Tamworth was a meat-type pig. Some breeds were developed for other reasons, not related directly to productive traits. Heritable traits such as skin and hair color and body size, and even traits useful to a geographic region, could be selectively bred. For example, the Irish Greyhound was bred with long legs to allow it to jump over the sheep fences of Ireland. Owners could call the foraging Irish Greyhounds from their house and the pigs could run and jump over fences to get to the table scraps.

Modern pig breeds have been improved while maintaining standard body types, coat colors, and ear carriage. Pigs can be white, gray (different shades), black, red (different shades), or multicolored. Multicolored patterns include red or black spots and white belts (white over the shoulders). Among coat colors, white is dominant to other colors; black is dominant to red. The white belt tends to be dominant over black or red. In many cases, the inheritance is incomplete, meaning the color inheritances just mentioned are characteristic of the majority of breedings. For example, a black pig bred to a white pig (if the white pig is a true purebred) will yield a majority of pigs that are all white. However, some pigs may have black spots, or what is called blue spots (black and white hairs in a patch that looks gray or blue).

Ear carriage is another distinguishing trait for domestic pigs. Pigs can have ears that are erect (pricked) or floppy (drooped). The extent of the ear droop is variable. Some Landrace lines have very droopy ears, for example. It was thought years ago that droopy ears caused the pigs to be more calm because they could not see very well. This theory has been neither confirmed nor denied in the scientific literature.

Pure breeds of pigs are an essential component of successful, worldwide pork production. However, the percentage of the U.S. pig herd that is purebred is very small. Table 7–1 lists the numbers of registered litters from the eight major American breeds of pigs. The total number of purebred litters registered in the United States in 1990 was 85,236. Thus, registered, pure breeds of pigs represent about 1% of the pigs in the United States. Two important points about purebreds are:

- Pure breeds and lines of pigs are necessary to make genetic progress and to maintain genetic diversity.
- Pure breeds of pigs represent a needed, but very minor, part of the total economic value of the commercial pig industry in the United States.

Today, several single-owner farms have over 100,000 sows. Each of these farms uses a type of crossbred sow and each farm has more pigs of a given crossbreed than all the purebred litters in the United States.

Eight major breeds of hogs have been used for breeding in North America. In general, the five dark breeds—Berkshire, Duroc, Hampshire, Poland China, and Spot—are historically known and used for their siring ability and potential to pass along their durability, leanness, and meatiness to offspring. The three white breeds—Chester White, Landrace, and Yorkshire—are sought for their reproductive and mothering abilities.

TABLE 7–1

The Numbers of Pure Breeds of Pigs Born in 1990 in the United States. In Comparison, the Breeding Stock Companies Have a Much Larger Percentage of the U.S. Sow Herd.

Breed	Origin	Number of Litters in 1990	Percentage of the U.S. Sow Herd
Yorkshire	England	23,861	0.34%
Duroc	New Jersey and New York	22,179	0.32%
Hampshire	Colonial United States	18,925	0.27%
Chester White	Ohio	5,544	0.08%
Spot	Ohio	6,443	0.09%
Landrace	Denmark	4,365	0.06%
Berkshire	England	2,071	0.03%
Poland	Ohio	1,848	0.03%
Total Pure Breeds		85,236	1.22%
PIC USA, Inc.	England	Millions	
DeKalb Swine	Texas[a]	Millions	

[a]DeKalb Swine breeders started by purchasing the breeding herd from Lubbock Swine Breeders of Lubbock, Texas.

Generally, purebred hogs are raised to be sold to commercial producers as seedstock for crossbreeding programs. The purpose of crossbreeding is to mix the best traits of several breeds to produce a superior offspring with the sought-after characteristics of leanness, meatiness, good feed efficiency, fast growth, and durability. Each breed's characteristics are closely considered when breeding stock is selected by the nation's hog producers to balance their crossbreeding programs.

Breeds and lines of the future are expected to possess traits beyond just quantity of muscle. Important traits for lines of pigs in the future likely will include:

- Quantity and quality of lean gain
- Meat-eating quality, including tenderness, juiciness, and flavor
- Prolificacy—high conception rate and litter size
- Disease resistance
- Enhanced feed intake, while increasing lean gain
- Increased sexual drive
- Lower aggressive behaviors
- Fewer "abnormal" behaviors (tail biting, navel sucking, etc.)
- Desirable behaviors while handling

MAJOR U.S. BREEDS

Yorkshire: The most sought-after breed, Yorkshires are good mothers and produce large litters. They exhibit a long, big frame and are white with erect ears. The Large White breed is very much like the Yorkshire in appearance, but it is a distinct breed.

American Yorkshire Club
P.O. Box 2417
West Lafayette, IN 47906
Phone: (765) 463-3593

Landrace: Landrace breeds in different countries may be different (e.g., Swedish Landrace versus UK Landrace). This breed is known for the sow's mothering ability. These hogs have very large, floppy ears, are long-bodied, and have the highest weaned average of any breed, as well as the highest average post-weaning survival rate.

Landrace Association, Inc.
P.O. Box 2340
West Lafayette, IN 47906
Phone: (765) 497-3718

Duroc: These hogs, noted for their fast growth and good feed efficiency, are a light to dark red color and have droopy ears. On the average, this breed needs less feed to make a pound of muscle than the other breeds.

United Duroc Swine Registry
P.O. Box 2397
West Lafayette, IN 47906
Phone: (765) 497-4628

Hampshire: These hogs are black with a white belt that extends from one front leg, over the shoulder, and down the other front leg. They have erect ears and are popular for their lean, meaty carcasses.

> Hampshire Swine Registry
> P.O. Box 2807
> West Lafayette, IN 47906-0807
> Phone: (765) 497-4628

LESS COMMON U.S. BREEDS

Chester White: These pigs are solid white and have medium-sized, droopy ears. They usually have large litters and are sought for their mothering ability. Boars of this breed are usually aggressive.

> Chester White Swine Record Association
> P.O. Box 9758
> Peoria, IL 61615
> Phone: (309) 691-0151

Berkshire: These hogs are black with six white points (nose, tail, and legs), and have erect ears and a short, dished snout. They work well in enclosed facilities and are noted for their siring ability. Of late, the Berkshire is prized for the superior flavor of its meat. The Berkshire Gold program markets meat from Berkshire pigs.

> American Berkshire Association
> P.O. Box 2436
> West Lafayette, IN 47906-2436
> Phone: (765) 497-3618

Spot: This breed is white with black spots and has medium-sized, droopy ears (like the Poland China). These hogs are known for producing pigs with a high growth rate.

> National Spotted Swine Record
> P.O. Box 9758
> Peoria, IL 61612-9758
> Phone: (309) 693-1804

Poland China: Like the Berkshire, this breed has six white points on a black body. These hogs have medium-sized, droopy ears and produce meaty carcasses with large loin eyes.

> Poland China Record Association
> P.O. Box 9758
> Peoria, IL 61612-9758
> Phone: (309) 691-6301

UNCOMMON OR SPECIALIZED U.S. BREEDS

Tamworth: Tamworth pigs are lean and known for their ability to survive outdoors on rough terrain. They are red with erect ears.

Meishan: Meishan pigs were imported into the United States and Europe in the past few decades. They are known for their prolificacy, having three-plus more pigs per litter than the better U.S. breeds. Meishans are fat, small-bodied, swayback, pot bellied, and they have extensive wrinkles on their face and body.

Pietrain: Pietrain is a heavily muscled breed imported into the United States from Germany and Belgium. The Pietrain is lean, and while the original breed was positive for the Halothane gene, some lines are now Halothane negative.

Hereford: This breed of pig resembles Hereford cattle. Hereford hogs have a white face and a red body color.

COMMERCIAL LINES

Commercial lines are the most common types of pigs found in commercial production in North America and Western Europe. Commercial seedstock companies do not generally sell purebred lines of sows or boars. They create hybrid animals that are technically crossbred animals—they are derived from genetic lines, not breeds. Commercial companies are not as concerned with maintaining breed integrity for traits like coat color and ear posture as they are with maintaining line integrity, uniformity, productivity, and continued genetic improvement.

PIC (FORMERLY PIG IMPROVEMENT COMPANY)

PIC USA is the largest seedstock supplier in the United States and the world. PIC is known for its Camborough female, which is composed of Yorkshire/Large White and Landrace. This Yorkshire x Landrace (YL) sow is bred to a Duroc boar to make the standard Camborough-15 female, which was replaced by the Camborough-22 (C-22), the next generation of maternal sow. The C-22 sow is bred to a terminal, crossbred boar to make terminal market animals. Future PIC USA market barrows and gilts are expected to be all white.

PIC maintains experimental lines that have Meishan, Pietrain, and the other common pure breeds of pigs.

DEKALB SWINE BREEDERS/MONSANTO CHOICE GENETICS

DeKalb Swine Breeders started with a nucleus herd obtained from Lubbock Swine Breeders (Lubbock, Texas), which was developed by T. Euel Liner and his son-in-law, Roy Poage. In the late 1970s, DeKalb sold more commercial gilts than any other company. PIC USA now holds that position. DeKalb sells a white-line maternal line and several lines of dark-colored white boars. DeKalb also maintained experimental lines that have Meishan, Pietrain, and the other common pure breeds of pigs. Recently, DeKalb Swine Breeders was purchased by Monsanto, which scaled back traditional breeding and placed more effort in advanced technologies.

OTHER BRITISH LINES

Other important British seedstocks include NPD (recently purchased by PIC), Cotswold, and Newsham (among others). Newsham also has an U.S. subsidiary in Colorado.

OTHER EUROPEAN LINES

Other important lines include Seghers and NPD from the United Kingdom and Danbred and Dalland from Denmark, as well as many others.

SUMMARY

A breed of pigs is a group selected to process uniform, heritable, and distinctive appearance and characteristics. Specific characteristics of a particular breed may include body composition (high or low fatness), prolificacy, milk production, meat quality, and other economically important traits. This chapter lists and describes current breeds of pigs in the United States. The hundreds of breeds existing around the world are described by Porter (1993).

QUESTIONS AND ACTIVITIES

1. Write a history of any single breed of domestic pig.
2. By researching the Internet and the library, list as many breeds of pigs that are identified (hint: there are hundreds).
3. How did breeds develop in colonial days? Compare the history of pork production in the United States (see Chapter 1) with the site of origin of American breeds.
4. Is color important as a breed trait? Why?
5. For what trait was the Wessex Saddleback bred? Where is it found and in what production system?
6. Where was the Hereford hog developed and what does it look like?

LITERATURE CITED

Porter, Valerie. 1993. Pigs: A Handbook to the Breeds of the World. Cornell University Press, Ithaca, NY.

INTERNET RESOURCES

For rare breeds, contact:
American Livestock Breeds Conservancy
P.O. Box 477
Pittsboro, NC 27312
http://www.albc-usa.org/
Telephone: 919-542-5704

For excellent information on pig breeds, see Oklahoma State University's Web page at:
http://www.ansi.okstate.edu/BREEDS/SWINE/

The Canadian Swine Breeders Association is found at:
http://www.canswine.ca/about.html

Information on rare pig breeds can be found at:
http://www.pigparadise.com/
http://www.albc-usa.org/

8
GROWTH, DEVELOPMENT, AND SURVIVAL

INTRODUCTION

Any discussion of pig biology must include an overview of growth because cellular and tissue growth is fundamental to all of agriculture, including pig production. The term "growth," as applied to animal production, is often considered to be simply an increase in body weight. The increase in body weight associated with growth consists of systematic increases in the main chemical constituents of the body—fat, protein, and mineral elements. Differences in the rate of increase for each component result in changes in body composition as pig growth proceeds. In the commercial pig industry, an increase in lean growth has a significant economic value. To capture this enhanced economic value, pig producers must understand and manage the biology of growth. This chapter addresses the broader definition of growth and describes the processes of growth and development in the pig from conception to maturity.

THE NATURE OF GROWTH

Growth was defined by Brody (1945) as "the constructive or assimilatory synthesis of one substance at the expense of another (nutrient) which undergoes dissimilation." This definition implies that changes in chemical composition of the body occur during the process of increasing in size. Such changes do indeed occur, as illustrated by Reeds et al. (1993) in Table 8–1.

TABLE 8–1
Changes in Body Composition of Pigs During Growth.

BODY WEIGHT		BODY PROTEIN		BODY FAT		PROTEIN TO FAT RATIO
(KG)	(LB)	(KG)	(LB)	(KG)	(LB)	(LB)
1.3	2.9	0.15	0.33	0.06	0.13	2.5
5.9	13.0	0.83	1.83	1.28	2.82	0.6
22	48.4	3.9	8.6	3.3	7.26	1.2
38	83.6	6.2	13.6	7.5	16.5	0.8
65	143	10.6	23.3	15.4	33.9	0.7
85	187	13.0	28.6	27.0	59.4	0.5
110	242	16.6	36.5	45.0	99.0	0.3

Source: Adapted from Reeds et al. (1993), p. 7.

Growth, as indicated by increase in body weight, can be expressed as absolute gain in a specified period of time or as relative gain (or percentage gain). The absolute growth rate:

$$\frac{larger\ weight\ -\ smaller\ weight}{time\ in\ d}$$

can be misleading because the average value may not represent the actual value at any single point in time. For example, a pig whose average daily weight gain from birth to slaughter at 160 d of age was 0.68 kg (1.5 lb):

$$\frac{243\ lb\ -\ 3\ lb}{160\ d} = 1.5\ lb/d \qquad \frac{110\ kg\ -\ 1.36\ kg}{160\ d} = .68\ kg/d$$

might never have gained exactly 1.5 lb (0.68 kg) on any one day (see S-shaped growth curve in Figure 8–1).

FIGURE 8–1
S-shaped Growth Curve Typical of all Biological Systems.

Source: Adapted from Brody (1945).

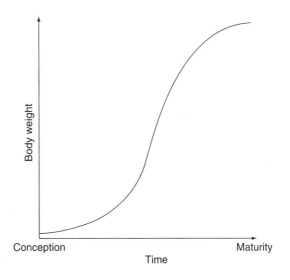

The absolute gain on the day after conception might have been a few micrograms, followed by a gain of 0.1 lb (0.045 kg) on the day after birth, and of 2.5 lb (1.14 kg) on the day before slaughter. Nevertheless, on a practical basis the absolute average daily gain to slaughter is an important measure of growth as used in the broad sense. By contrast, relative growth rate:

$$\frac{(larger\ weight - smaller\ weight)}{(smaller\ weight)}$$

declines steadily as the pig approaches maturity (again, see Figure 8–1). Baxter (1987) provided regression equations that predicted the changes in body size as the pig grows. Box 8–1 depicts some body dimensions of the pig and Box 8–2 shows the relationship of foot dimensions to body weight for pigs up to 10 kg (22 lb) body weight. These measures are very useful when designing pens, feeders, or other handling equipment. Growth of the portions of the pig follows allometric equations. Examples of body growth curves for body measures include: (a = length in mm; w = body weight in kg):

The general formula is $y = a + w^{.33}$, where y = dimension (ex width)
Width at shoulders: a = 61, w = .33 power
Width at ham: a = 59, w = .33 power
Body length: a = 300, w = .33 power
Body height (floor to top of back): a = 159, w = .32 power
Foot width: a = 14.3, w = .33 power

All equations provided by Baxter (1987) follow an exponential curve with varying coefficients as illustrated in Box 8–2.

Box 8–1 Some Body Dimensions of the Pig

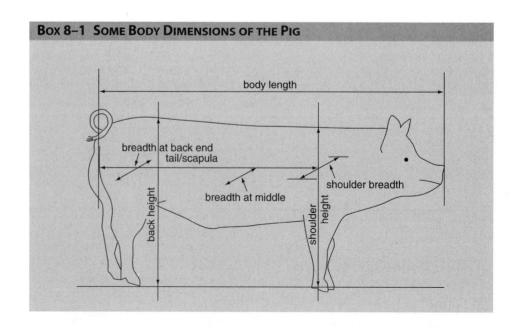

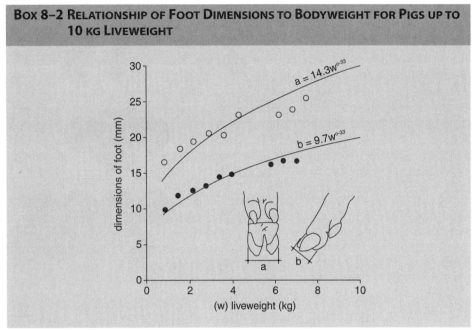

Source: Adapted from Baxter (1984).

The growth rates of individual parts of the pig are not identical; for example, the head and shoulders reach mature size before the posterior parts of the body. In early life, the head and shoulders make up a higher proportion of the total body weight than they do later. Indeed, individual organs and tissues grow at different rates, and even individual cells of different types within a single organ grow at different rates.

HYPERPLASIA AND HYPERTROPHY

Growth of tissues and organs and of the whole pig occurs in two phases: increase in number of cells (hyperplasia) and increase in size of cells (hypertrophy). Immediately after conception, most growth is by hyperplasia. During prenatal and early postnatal life, the two processes occur concurrently (hyperplasia-hypertrophy). Finally, at some point in postnatal life, hyperplasia ceases in some tissues (for example, nervous tissue) and adult size of specific organs is attained. The two processes of growth are illustrated diagrammatically in Figure 8–2. Cells in some tissues, such as the epithelium of skin and the digestive tract and blood cells, have continuous turnover throughout life (life of the red blood cell of the pig is about 70 d, that of intestinal mucosal cells is a few d).

In the case of tissues that reach mature size by proceeding through the three phases (hyperplasia, hyperplasia-hypertrophy, and hypertrophy), the ultimate size is presumably determined genetically (i.e., the upper limit imposed by the genes cannot be exceeded, although environmental factors may limit ultimate tissue or body size). Strong evidence suggests that some nutrient deficiencies imposed early in life, even prenatally in some cases, affect the subsequent growth and development in pigs. The exact nature of this stunting effect and its relationship to impaired

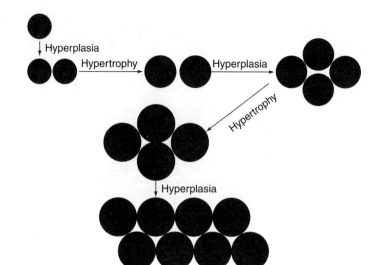

FIGURE 8–2
Diagram of Growth by Hyperplasia and Hypertrophy.

function is unknown, but it is probably associated with such factors as reduced cell number or altered metabolic function of cells.

MUSCLE AND FAT ACCRETION

The time sequence of cell proliferation and cell growth in individual tissues of the pig is not completely mapped, but for muscle and fat, it appears that most cell division is complete at birth or shortly thereafter. Most postnatal increases in muscle and fat mass are associated with an increase in size rather than the number of muscle and fat cells. Muscle and fat accretion are determined by a complex of hormones and other metabolites under endocrine control (pituitary, thyroid, pancreas, liver, and other organs are involved) in the pig, as in other animals.

PRENATAL DEVELOPMENT

Prenatal development is a critical phase of the life cycle because events of this period influence the subsequent growth, development, and productive life of the pig. As indicated previously, cell multiplication and differentiation occur early in prenatal life to form tissues and organs of specialized structure and function. Detailed embryology of the pig is well beyond the scope of this discussion, but the subject is thoroughly covered in Patten (1948), Marrable (1971), Hafez (1975), Harrison (1971), and in symposia (Book and Bustad, 1974; Brinster, 1974; Tumbleson, 1986; Tumbleson and Schook, 1996). Walker (1964) described the dissection of the fetal pig. The present discussion is confined to the events of embryonic and fetal development considered to be important in pig production.

FERTILIZATION

Fertilization, defined as the union of the male and female pronuclei to form a zygote, involves the penetration of the surface of the ovum (zona pellucida) by several sperm, but entry by only one. Fertilization normally occurs in the oviduct a few hours after mating or insemination. Cell division begins soon after fertilization. The fertilized ova pass down the oviduct and enter the uterus by the third d after mating. Cell specialization and rearrangement begin by d 6.

ATTACHMENT (IMPLANTATION)

By d 11, embryos show signs of attachment to the endometrium of the uterus, where a slight indentation marks the site of attachment of each embryo. True union of the embryo with the uterine lining at the site of attachment to form the placenta is not complete until d 18 (embryo length is 5 to 10 mm at this stage).

DIFFERENTIATION OF ORGANS

By 20 to 25 d, the fetus is about 35 mm long (Figure 8–3) and many of the major organs can be seen, including the five brain regions, the heart, pharynx, trachea, esophagus, stomach, lung, liver, and the four limb buds. The elongation of the intestines has begun. The urinary system is composed of a large network of tubules and represents a great mass of tissue in relation to total body size. The body cavity has not yet divided into the pericardial, pleural, and peritoneal cavities, but what will later be the diaphragm, separating the pleural and peritoneal cavities, is beginning to form. By d 28, nostrils, eyes, testes, ovaries, blood-forming centers, hair follicles, early stages of the mammary glands, growth plates of long bones, and digits have appeared. By d 35, long bones and vertebrae ossify and the brain cerebellum differentiates. By d 50, most endocrine glands are secreting hormones, blood serum proteins are being synthesized, secondary sex organs of males and females are differentiated, and by d 95, testes enter the inguinal canal in males and move to the scrotum before birth, which occurs at about 114 d.

PLACENTA

The placenta of the pig is termed *epitheliochorial*. The extra-embryonic membranes can be separated from the uterus with no damage to the uterus and, unlike the case in some other species, the blood must pass through the chorion epithelium to reach the fetus. The following six tissues separate the maternal and fetal blood in the pig:

Maternal tissues: (1) uterine endothelium, (2) uterine connective tissue, (3) uterine epithelium;

Fetal chorion composed of: (4) trophoblasts, (5) connective tissue layer, (6) endothelium.

FIGURE 8–3
Stages of Development of Pig Embryo; (a) 10–12 mm Embryo (12–15 d); (b) 35 mm Embryo (20–25 d).

Source: Adapted from Harrison (1971).

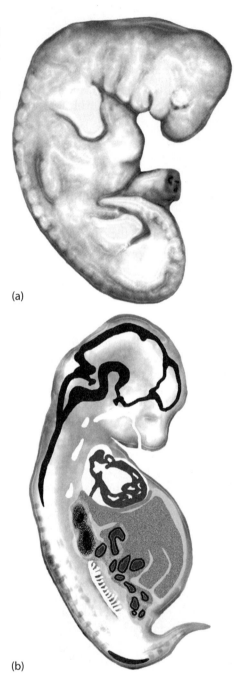

(a)

(b)

The transport of large organic molecules across the placenta is limited in the pig. Therefore, there is almost no passive immunity provided by the maternal blood available to the fetus. The blood of the newborn pig is therefore almost devoid of immune antibodies (immunoglobulins, IGs).

FETAL MEMBRANES

Each embryo has its own covering membranes (except in the case of identical twins, which is rare in pigs). After attachment, each fetus lives within its own protective membranes and depends on the maternal blood circulation flowing across the six-layered epitheliochorial placenta for nutrients. Continued growth and development of the fetus depends, therefore, on the integrity and growth of the placenta.

The fetal membranes and their functions include (for a more detailed description, see Pond et al., 1991): (1) Yolk sac—this serves mainly in very early embryonic life by carrying nutrients to the embryo. As the allantois develops, the yolk sac becomes a shriveled vestige. (2) Amnion—this single-layered membrane surrounds the embryo and, when completed, is filled with fluid that engulfs the fetus and protects it from mechanical injury and from adhesions between the membrane and fetal body surface. Amnion formation starts at about 1 wk and is complete by the time of implantation at d 18. The amnion is attached to the body ventrally at the umbilical ring, through which nutrients and metabolites pass via the umbilical artery (flow from fetus to dam) and umbilical vein (flow from dam to fetus). (3) Serosa (trophoderm or somatopleure)—the serosa encloses the other extra-embryonic cavities with their respective surrounding membranes and, together with the splanchnopleure of the allantois, represents the chorion, the three-layered fetal part of the placenta. (4) Allantois—this single-layered membrane arises from the hindgut and rapidly enlarges and dilates, except at the point where it leaves the body cavity (bell stalk), where it becomes a constricted connection with the hindgut. The allantois has an abundant blood supply and its rapid growth and vascularity are associated with a rapid increase in its importance as a vehicle for supplying nutrients and oxygen to the fetus. It supplants the yolk sac in this function and eventually occupies a large space within the cavity of the serosa and becomes fused with it to form the fetal part of the placenta (Figure 8–4).

Most of the growth of the placenta is completed by d 70 of gestation, whereas the fetus grows most rapidly after that time (Figure 8–5). Fetal weight increases faster than length, particularly near the end of gestation. Bazer et al. (2001) described the relationship between placental and fetal growth and the negative association of litter size in early gestation with placental development. Low birth weight, in turn, reduces subsequent fetal and neonatal pig survival.

GROWTH OF THE UTERUS

The uterus increases in weight throughout pregnancy from about 0.5 to 1.5 kg (1.0 to 3.3 lb) immediately after conception to 3.0 to 4.0 kg (6.6 to 8.8 lb) at term. The amount of uterine growth depends on the number and size of the fetuses. The empty uterine weight represents only the uterine tissue itself and does not include the weight of the

FIGURE 8–4

Semidiagrammatic View of Fetal Membranes and Placenta.

Source: Harrison (1971).

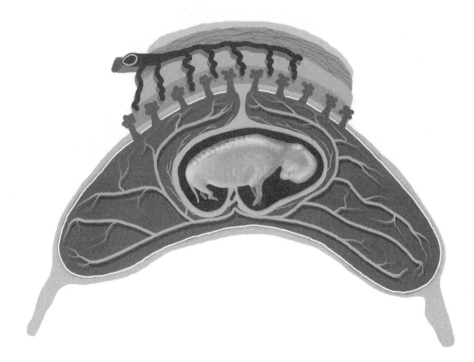

FIGURE 8–5

Length (from end of snout to base of tail) and Weight of Fetuses and Weight of Membranes During Pregnancy. X Indicates American Results, ●, ○, and △ Indicate Danish Results.

Source: Moustgaard (1962).

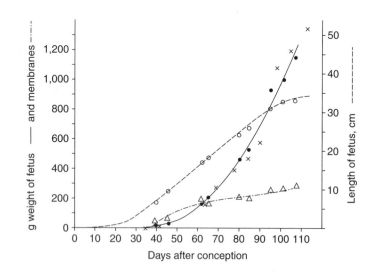

placenta. The weight of the placenta has a rather constant relationship to fetal weight; the placenta weighs about 10% of the fetus (a ratio of placental weight to fetal weight at parturition is about 1:10).

GROWTH OF THE FETUS

The ultimate size of the individual fetus at term is inversely related to the number of fetuses in the litter. Weight of the individual fetus at birth is directly related to weight of the placental (chorionic) membranes associated with it, which, in turn, is related partly to the amount of vascular connection from the placenta to the fetus and the amounts of nutrients reaching the fetus. However, the greater weight of fetuses at the ends of the uterine horn rather than in the middle, and at the upper (ovarian) end rather than at the lower (cervical) end, cannot be explained on the basis of the vascular structure of the uterus.

Fetal growth is under control of hormones and growth factors; for example, growth hormone (somatotropin), insulin, insulin-like growth factors (IGFs), and other endocrine products are involved. The complex interrelationships among and between these factors are discussed in Campion et al. (1989), Hollis (1993), Lawrence and Fowler (1997), and Ford et al. (2001). The size of the fetus increases dramatically during the final half of gestation. Bazer et al. (2001) adapted the extensive data of Knight et al. (1977) to produce the growth curves of the fetus and the placenta from d 20 through d 100 of gestation (comparable to the data shown in Figure 8–5). Viable and strong newborns have been known to weigh as little as 400 g (0.9 lb) and more than 2,000 g (4.4 lb) at birth after a normal pregnancy. The amounts of protein, energy, and mineral elements deposited in the fetus increase by more than 3%/d, so the absolute amounts deposited/d in late gestation are extremely large. The changes in gross chemical composition of the pig fetus at various stages throughout prenatal life illustrate the shift that occurs in the proportions of water, fat, protein, and ash between d 30 and term—112 to 115 d. Water content, as a percentage of fresh weight, decreases from about 95% to 80% from d 30 to term, whereas fat, protein, and ash increase by twofold to threefold during the same period. The relatively wide range in chemical composition at birth reflects, in part, the variation in body size at birth. Gestation length, which ranges from 111 to 117 d (114 d on average), appears to have little effect on birth weight or body composition of newborn pigs taken naturally to term.

PRENATAL MORTALITY

Prenatal mortality in pigs represents a huge loss in potential reproductive efficiency. Bazer et al. (2001) estimated that early embryonic mortality is 20% to 45%, most of which occurs before d 25 of pregnancy. Contributing factors include chromosomal abnormalities, heat stress, excessively low and high planes of nutrition, infection, and adverse biochemical interactions between the developing embryo/fetus and the maternal endometrium. Some of these factors can be controlled by management (e.g., heat stress, infectious diseases, plane of nutrition), but others are not well understood. Uterine crowding in large litters appears to be responsible for some fetal deaths after 30 d of gestation, but clearly the largest proportion of prenatal mortality occurs earlier.

Mummified Fetuses

Fetal death occurring after the beginning of ossification of the skeleton (later than 30 d) results in necrosis and shriveling of the fetus and the birth of "mummified" pigs. The vestiges containing the atrophied fetal membrane and enclosed partially resorbed fetus are retained in the uterus to term and expelled at parturition with the full-term viable littermates. Mummified fetuses are distinct from stillborn pigs in that the latter survive up to the time parturition begins.

ADAPTATIONS AT BIRTH

Change in Oxygen Supply

The severance of the umbilical cord at birth requires that the lungs begin to function to provide oxygen formerly supplied from the maternal blood. This sudden need is met by the presence of two structures in the circulatory system (foramen ovale and ductus arteriosus) that remain anatomically open at birth. Both entities function during prenatal life in directing blood flow (oxygen from maternal blood) within the fetus in the absence of lung function, but they are no longer needed when the lungs begin to function as the source of oxygen at birth. The foramen ovale, an opening in the wall between the two atria of the heart, ceases to function a few d after birth although it may not actually close until later. The anatomical closure of the ductus arteriosus (fetal channel between pulmonary artery and aorta) is not complete until several wk after birth, although it normally becomes functionally closed at birth.

Danger of Blood Loss

When the umbilical cord is severed at birth, the umbilical artery and vein have no further function. Excessive blood loss due to hemorrhaging at the site of the severed umbilical cord can be controlled by pinching the cord with blunt scissors or by tying a knot in the cord close to the umbilicus. Typically umbilical blood clots from normal stretching and trauma of birth.

Stillbirths

Several h may elapse between the birth of the first and last pig in a litter. This delay can result in suffocation if the umbilical cord breaks before birth. The result is one or more stillborn pigs. Other factors, including an array of infectious reproductive diseases and genetic differences, also contribute to the incidence of stillborn pigs. The frequency of stillbirths in herds free of infectious reproductive diseases varies from herd to herd and from season to season, but may approach 5% or higher on occasion, even in well-managed herds. Stillborn pigs are smaller, on average, than live littermates. Stillborn pigs can be differentiated from piglets that died shortly after birth by the fact that the lungs of a stillborn pig will sink when placed in water, whereas those of a pig that died after birth contain trapped air and will float in water. Floating the lungs can be a useful procedure to troubleshoot herds with a high incidence of dead newborn pigs as a means of confirming the cause of death.

POSTNATAL DEVELOPMENT

The growth and development of the pig from birth to maturity is a continuation of the prenatal period, but is characterized by a greater rate of daily gain, of which a major part is muscle and fat, deposited mostly as a result of increased muscle and fat cell size (hypertrophy). Several of the physiological parameters in the growth and development of the young pig are discussed here. Changes in rate of weight gain, body composition, and patterns of protein and fat deposition from weaning to market weight are discussed later in this chapter. Each has important economic implications in commercial pig production.

BIRTH WEIGHT AND SURVIVAL

A close relationship exists between birth weight and survival. An increase in litter size (number of live pigs) at birth reduces the chance of survival due to the inverse relationship between litter size and individual piglet birth weight. Likely factors explaining the lower survival rate are: (1) low body energy stores; (2) differences in the placental supply of nutrients; (3) uterine crowding, in turn, related to endocrine or spatial constraints on development; and (4) relative physiological immaturity.

Body fat stores are limited (only about 1% body fat in newborn pigs compared with 20% or more in newborn humans) and glucose synthesis from other substances is limited, forcing the neonate to depend largely on carbohydrate stores (glycogen) and immediate dietary nutrients for survival. Records from more than 17,000 piglets, summarized by Speer (1970), showed that piglets with a low birth weight have a greater risk of preweaning death (Table 8–2).

The survival rate of small pigs can be improved greatly (increased from 49% to 74%) by supplementing them by stomach tube individually once or twice daily with 15 cc of fortified milk diets (Moody et al., 1966) (Table 8–3).

TABLE 8–2
Relationship of Birth Weight and Survival.

WEIGHT RANGE		NUMBER OF PIGS	WEIGHT DISTRIBUTION OF POPULATION %	SURVIVAL %
KG	LB			
<0.9	<2.0	1,035	6	42
0.9–1.1	2.0–2.4	2,367	13	68
1.1–1.3	2.5–2.9	4,197	24	75
1.4–1.5	3.0–3.4	5,012	28	82
1.6–1.8	3.5–3.9	3,268	19	86
>4.0	>1.8	1,734	10	88
Total		17,613	100	(Mean) 77

Source: Adapted from Speer (1970), Iowa State University Swine Nutrition Herd Records (unpublished data).

TABLE 8–3
Survival of Light Weight Suckling Pigs Supplemented Daily with Liquid Sow Milk Replacer.

Item	Control (Nursing Sow)	Supplemented (Nursing Sow Plus Supplement)[a]
Number of pigs	69	69
Number survived	34	51
Percent survival	49	74

[a] Pigs weighing less than 2.0 lb at birth within each litter received by stomach tube 15 cc of reconstituted sow milk replacer once or twice daily from day of birth to 7 d old.
Source: Adapted from Moody et al. (1966).

PATTERNS OF MORTALITY IN YOUNG PIGLETS

The highest rate of mortality is during the first d after birth (10% to 13%) with smaller losses during the second and third d (3%) and even less during the fourth d (2%). This is followed by a gradual decline to 0.02% during the period from 28 d to market weight. Deaths are due to a variety of causes, including starvation (largest cause), trauma, gastrointestinal diseases, respiratory and other infectious diseases, and malformations. Total preweaning death losses in the U.S. pig industry are typically 10% to 20%.

IMPORTANT PHYSIOLOGICAL FACTORS THAT AFFECT SURVIVAL AND DEVELOPMENT

Because of its small energy reserve, due to limited liver and muscle glycogen and body fat, the pig must nurse shortly after birth or the blood glucose level falls dramatically. The normal value of 100 mg of glucose /100 ml blood at birth may drop to 10 mg/100 ml or less within the first two d in the fasted newborn pig. This hypoglycemia leads to lethargy, coma, and death unless corrected by nursing or by oral or intravenous glucose. Pigs not rescued from coma by glucose administration within about 1 h usually do not survive. Fasting also results in reduced body temperature, hastened by low environmental temperature, during the first 2 or 3 d. The two stressors, operating together, exacerbate the problem.

Hypoglycemic coma can occur even in piglets allowed to nurse. These cases are usually associated with a cold or drafty environment in combination with an inadequate milk supply. Immediate glucose administration and removal of the piglet to a warm environment are required to save such pigs. Body fat and glycogen reserves increase steadily during the first few weeks of life, so susceptibility to hypoglycemia is progressively less as age progresses.

BODY TEMPERATURE CONTROL AND ENVIRONMENTAL TEMPERATURE

The inefficient body temperature control of the baby pig is related to its relatively immature physiological state at birth, including limited body energy reserves, limited subcutaneous fat, sparse hair coat, and large body surface area relative to body weight, especially in small piglets. The newborns of wild pigs (increasing in numbers in many rural areas of the United States) are more cold resistant than the newborn of domestic pigs during the first 2 wk after birth. This is due partly to a higher oxygen consumption of wild pigs during exposure to cold and partly to their more dense hair cover. Heat loss in a cold environment is reduced by skin vasoconstriction and by the fact that individual pigs in a litter huddle together, reducing their effective surface area, and therefore their rate of heat loss.

Maintenance of body temperature when environmental temperature is low requires a higher level of heat production by the animal (higher metabolic rate). At an environmental temperature below the critical temperature (the environmental temperature below which an increase in metabolic rate is necessary to maintain body temperature), the newborn pig is able to increase heat production to a surprisingly efficient extent while maintaining body temperature constant. Normal body temperature at birth is 38°C or 100°F; at 1 d of age normal body temperature is 39°C or 102.2°F. However, extremes in environmental temperature that exceed the newborn pig's ability to respond with an increase in heat production to maintain body temperature result in chilling, lethargy, coma, and death. Following exposure to near-freezing temperatures for several hours, the baby pig can rebound quickly after warming and feeding, even when the rectal temperature drops below 91°F and the pig is comatose.

DIGESTION AND ABSORPTION OF FOOD

Colostrum and Immune Antibodies

The colostrum of the sow is very high in gamma globulin and its associated antibodies. Immediate passive immunity is provided as soon as the piglet nurses. The immunoglobulin content of the colostrum gradually declines, but considerable amounts persist for several days. During the first 24 to 48 h, the intestinal mucosa lining is permeable to almost any protein and other large molecules, including infectious pathogenic agents. Following ingestion of colostrum, the pig's ability to absorb intact proteins is lost. The phenomenon is called "gut closure."

Colostrum-acquired antibodies (passive immunity) persist in the blood serum for 6 wk or longer after birth. Active antibody production by the piglet begins at about 3 wk of age. Most parameters of immunity in the pig reach adult values at about 1 mo of age (Blecha, 2001). Three wk of age is a critical immunological period for the pig, because colostrum-acquired antibodies (passive immunity) are low and active antibody production (active immunity) has not attained full development. In addition to immunoglobulins, sow colostrum also contains factors, including lactoferrin, insulin, and insulin-like growth factors (IGFs), that stimulate intestinal growth and development. Current trends toward weaning at less than 3 wk of age require intensive management and husbandry practices because of the vulnerability of these very young pigs to infectious agents (see Figure 8–6).

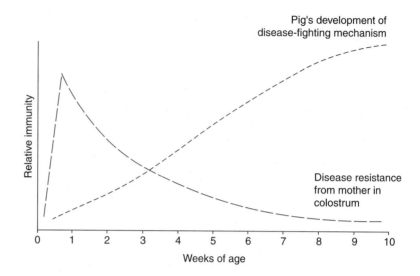

FIGURE 8–6
The Development of the Disease-Fighting Mechanism.

DIGESTIVE SYSTEM DEVELOPMENT

Rearing of Newborn Piglets Away from the Sow

The newborn pig adapts well, after colostrum ingestion, to a liquid milk replacer diet for the first few d after birth, and can be trained to consume entirely dry feed (with free access to fresh drinking water) beginning at about 4 to 5 d of age. Liquid milk replacer or cow's milk can be supplied either from a bottle and nipple or from a bowl at frequent intervals. A daily feeding schedule of a limited amount of liquid feed at 0800, 1230, 1700, and 2200 h has been used successfully to raise piglets for the first few d or up to 2 or 3 wk of age. Limited feeding at these relatively infrequent intervals (piglets nurse about once each h over a 24-h day when left with the sow) is needed to control diarrhea, which is prevalent when pigs overeat at a single meal to compensate from a fast of several hours between meals. Transition to a dry diet is successful by progressively mixing more dry feed with the liquid diet each meal over a period of 1 to 2 d starting at 4 d of age.

Piglets being reared away from the sow should be caged separately for at least the first wk to avoid injury and continued agitation caused by persistent suckling of ears and nuzzling of other pigs when kept together. Automated feeding devices for baby pigs are commercially available.

Digestive Enzyme System Development

1. Carbohydrates—The newborn pig readily absorbs glucose, but cannot digest sucrose or polysaccharides efficiently. Lactose is used efficiently by the newborn pig, but becomes less used in the older pig because of the decline in secretion of lactase, which splits the disaccharide—lactose—into its component monosaccharides—glucose and galactose. Lactase is an adaptive enzyme. Therefore, when lactose is fed

to older pigs, most can adapt with increased lactase production and digest most of the lactose. The young pig is unable to utilize xylose or fructose. Utilization of the complex carbohydrates (polysaccharides) increases with age. Pigs weaned after 3 wk of age are commonly fed cereal grains, corn, and other sources of complex carbohydrates as their major source of energy.

Unlike the ruminant, the pig cannot break down cellulose, the complex carbohydrate common in forages. Pigs, like other higher animals, do not produce cellulase, the enzyme that breaks down cellulose to release glucose, whose linkage in starch differs from its linkage in cellulose. Cellulase is produced by microbes residing in the rumen and in the large intestine of the pig and other nonruminants. Therefore, to the extent that intestinal microbes in the pig can break down cellulose, some of this energy is available to the pig. Mature sows and boars, because of the microbial activity in their large intestine and cecum, are able to use high-fiber diets to meet 20% to 30% of their maintenance energy requirements.

2. Fat—The pig is able to hydrolyze and absorb large amounts of fat very efficiently from birth. Most of the fat is hydrolyzed in the small intestine under the influence of pancreatic lipase. Apparent digestibility of fat is generally more than 80% in pigs. In addition to the action of lipases, dietary fats are prepared for absorption by emulsification and reduction in particle size through the action of bile acids secreted via the bile duct into the small intestinal lumen.

3. Protein—The hydrolysis of protein to its constituent amino acids is limited during the first 36 to 48 h by the presence of a trypsin inhibitor in colostrum. As described previously, gut closure subsequently occurs and proteins must be hydrolyzed before they can be used. Apparent protein digestibility by the young pig is high (92% to 95%). Secretion of pepsin from the stomach and other proteases from the pancreas and small intestine are low at birth and gradually increase during the first 8 wk. Milk proteins are utilized more efficiently than plant proteins during the first few wk, but by 5 wk of age, plant and animal proteins are utilized equally well.

DENTITION

The dental pattern of the pig is illustrated in Figure 8–7. At birth, the pig has eight teeth—four incisors and four canines. The full complement of temporary teeth is 28, and consists of 12 incisors (three pairs on upper and lower jaws), four canines (one pair on upper and lower jaws), and 12 molars (three pairs on upper and lower jaws). The permanent teeth erupt at varying intervals from 3 to 4 mo for the first molar, and 18 to 20 mo for the third molar. During the preceding time, the dentition shows a mixture of deciduous and permanent teeth. The mature pattern is three pairs of incisors, one pair of canines, four pairs of premolars, and three pairs of molars—upper and lower—for a total of 44 teeth present by about 24 to 27 mo of age.

The eight teeth present at birth are very sharp and have been descriptively termed "needle teeth." In the past, it was common to clip these temporary teeth shortly after birth to avoid injury to other pigs and to the nipples of the dam during nursing. This practice is not uniformly practiced in the commercial industry in the United States today.

The adult canines grow very long and protrude out of the mouth as tusks in the male. It is common practice to trim them to avoid injury to humans and other pigs.

FIGURE 8–7
Dental Pattern of the Pig.

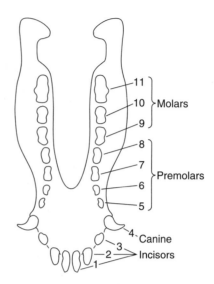

GROWTH AND BODY COMPOSITION FROM WEANING TO MARKET WEIGHT

Most of the weight gain and changes in body composition in pigs occur between weaning and market weight [approximately 240 to 280 lb (109 to 127 kg) in the United States]. Growth and body composition after weaning are influenced by nutrition, growth-modifying agents added to the feed, genetics, environment (e.g., ambient temperature, available pen space, pig density), and sex (intact or castrated males, females). The main factors affecting growth rate and body composition during the post-weaning period were described by Mitchell et al. (2001). Feed additives and injected hormones having effects on feed intake, growth rate, and carcass leanness in pigs are summarized here.

Beta-adrenergic Agonists

This class of compounds structurally resembles the naturally occurring hormones epinephrine and norepinephrine. The beta agonists—clenbuterol, cimaterol, raptopamine, and others—reduce fat deposition, increase carcass lean content, increase daily rate of weight gain, and improve feed utilization by 5% to 25% when added in small amounts to the diet of finishing pigs. Only one of these compounds, raptopamine, has been approved by the U.S. Food and Drug Administration (FDA) for addition to swine feeds. The mode of action of raptopamine and other beta agonists in producing these improvements in performance appears to be related to increased muscle protein synthesis and decreased fat synthesis. The improved performance is obtained by trace amounts (20 mg or less/kg of diet) of raptopamine. This and other non-nutritive feed additives yet to be introduced for use by the swine industry promise to continue to improve efficiency of utilization of feed resources. It must be recognized that such new technologies may influence nutrient requirements. In the case of raptopamine, dietary protein requirement is increased 10% to 15%, due to the greater protein deposition in pigs that are fed the compound. Thus, the economic benefits of increased efficiency of lean growth brought about by any new technology must be weighed against the extra diet cost of the higher protein content. The economic and biological advantage of dietary raptopamine for finishing pigs has been clearly established.

Growth Hormone (Somatotropin)

It was shown more than 70 yr ago that growth hormone injection into animals could be used to stimulate growth. The cost of supplying a purified source was prohibitive (the hormone had to be extracted from pituitary glands of slaughtered pigs and the supply required for commercial application in the livestock industry was grossly inadequate). Evock et al. (1988) showed that porcine growth hormone from recombinant DNA technology and derived from large-scale production, was equal to growth hormone obtained from slaughtered animals in terms of its growth-promoting effects. In the 1990s, many research groups showed the efficacy of daily injection of recombinant DNA-derived growth hormone (pST) in improving growth rate and carcass leanness of pigs. To date, pST injection on a continuous and frequent (once or twice daily) schedule appears to be the only effective delivery system to administer the hormone. Addition of pST to the feed is ineffective. Until a means of feeding pST is developed, it appears that the physiological benefits of the hormone on pig performance will not be economically feasible. The widely used practice of daily injections of growth hormone in dairy cattle to increase milk production is economically favorable because individual handling of cows is a daily practice; not so with pigs.

CHANGES IN RATE OF WEIGHT GAIN

Average daily gain gradually increases from weaning at 2 to 3 wk of age to midway through the growing-finishing period when it reaches a peak and then gradually declines to market weight at about 240 to 280 lb (109 to 127 kg). The general shape of the growth curve is depicted in Figure 8–8. Compare this curve with the sigmoidal (S-shaped) growth curve of all biological systems in Figure 8–1. The extreme right side of the curves in Figures 8–1 and 8–8 represent the stage of growth when daily weight gain plateaus and mature body weight is reached (average daily gain = 0).

CHANGES IN BODY COMPOSITION

Changes in body composition between weaning and slaughter weight were summarized by de Lange et al. (2000), based on a thorough review of research in the United States and many other countries. Genetic background has the greatest impact on body composition, although many other factors are involved. Body composition can be described either by physical or chemical measurements. Physical body composition includes measures of muscle, visceral organs, fat deposits, bone, and skin weights, whereas chemical composition generally refers to the amounts of the four main classes of nutrients: water, protein, lipids, and ash. Both physical and chemical composition measurements are based on empty body weight (i.e., the total body

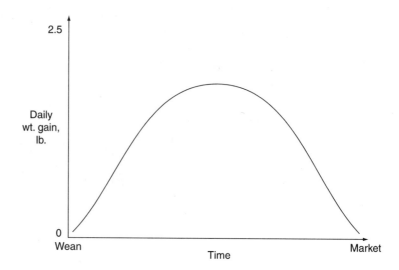

FIGURE 8–8

Changes in Average Daily Weight Gain from Weaning to Market Weight.

weight after all digestive tract contents have been removed). Carcass value is affected by both physical and chemical composition, although physical measurements such as backfat depth, loin eye area, carcass length, and percentage of trimmed lean cuts (Boston butt, picnic, loin, and ham) are normally used in evaluation and pricing by meat packers, as discussed in Chapter 9.

Chemical composition changes dramatically between 10 and 24 wk of age. Boars are leaner than gilts at any given weight during growth, and both sexes are leaner than castrated males (barrows). Intact males continue to increase in daily protein accretion beyond 5 mo of age while females and barrows tend to plateau. Daily fat accretion continues to slaughter weight in boars, gilts, and barrows. Tess (1981) compared the fat content of contemporary (in 1981) pigs at 10, 17, and 24 wk of age with that of pigs selected for several generations for high or low backfat. His data illustrate vividly the dramatic increase in fat content of the carcass during the growing/finishing period and the striking effects of genetic background on body fatness at each age. The results are summarized in Table 8–4.

TABLE 8–4

Changes in Carcass Fat as a Percentage of Empty Body Weight at 10, 17, and 24 wk of Age in Three Genotypes of Pigs.

GENOTYPE	CARCASS FAT, %		
	10 WK	17 WK	24 WK
Contemporary[a]	7.4	17.1	24.4
Lean[b]	7.6	15.7	26.2
Obese[c]	13.0	27.8	40.5

[a]Typical modern-type pigs of 20 yr ago.
[b]Pigs selected for many generations for low backfat thickness (Hetzer and Harvey, 1967).
[c]Pigs selected for many generations for high backfat thickness (Hetzer and Harvey, 1967).
Source: Adapted from Tess (1981).

CHAPTER 8   GROWTH, DEVELOPMENT, AND SURVIVAL

TABLE 8–5
Chemical Composition (%) of the Empty Body of Pigs at Birth, 15 lb, 50 lb, and 242 lb.[a]

CONSTITUENT	AGE (BIRTH) OR WEIGHT (LB)				
	BIRTH	6.8 KG (15 LB)	25 KG (55 LB)	110 KG (242 LB) LEAN PIGS	110 KG (242 LB) FAT PIGS
Ash	3	3	3	3	3
Lipid	2	15	12	15	35
Protein	18	16	16	18	14
Water	77	66	69	64	48

[a]Note the inverse relationship between lipid and water concentration and the general increase in lipid and decrease in water as body weight increases.
Source: Adapted from de Lange et al. (2000), p. 66.

The chemical composition of the empty body of pigs from birth to approximately 242 lb (110 kg) was estimated by de Lange et al. (2000) from data derived from research done in several countries over several decades. Table 8–5 depicts the changes in ash, lipid, protein, and water percentages as affected by age and weight. A rough estimate of the mass of each chemical constituent in the body at each body weight and in lean and fat pigs at 220 lb (100 kg) can be determined by multiplying the percentage of each constituent by the corresponding body weight. Note that modern pigs are much heavier at slaughter 240 to 280 lb [(109 to 127 kg) or more] and these modern pigs are considerably leaner, even at these heavier weights, than those of the past.

SUMMARY

This chapter describes the nature of growth and addresses changes in body composition as the pig increases in size during prenatal and postnatal life. The processes of growth from fertilization through prenatal and postnatal life to maturity are described. The differentiation of organs and the growth and structure of the placenta, fetal membranes, uterus, and the fetus itself are described. Physiological factors affecting survival and development during prenatal and postnatal life are discussed. Such factors include low energy reserves, immature body temperature regulation at birth, and limited digestion and absorption of nutrients related to ongoing development of the digestive system.

QUESTIONS AND ACTIVITIES

1. In a spreadsheet program create a graph based on a table that has x-axis data in body weight in kg (from 3 to 300 kg) and the width of the pig's shoulder for the y-axis data. Generate y-axis data using the regression equation for body width at the shoulders ($a = 61 * w^{.33}$, where a is width in mm and w is body weight in kg). Based on this graph and the raw data, how wide should feeder openings be (assuming the shoulder is the widest part of the pig) for pigs at the following weights? 40 lb, 150 lb, 200 lb, 250 lb, 600 lb.

2. Would you rather have a QTL for a gene that induces muscle hyperplasia or muscle hypertrophy? Why?
3. What is the feed efficiency for a chicken during the period from hatch to 6 wk of age? What is the feed efficiency of a modern pig from weaning at 14 d and for the next 6 wk? Are chickens inherently more efficient than pigs or is it a function of our production systems?
4. Why is a newborn pig more susceptible to death than a 4-mo old pig?
5. The newborn pig nurses to obtain colostral antibodies by passive diffusion. The piglet, born with very little blood antibody, will increase blood concentrations of immunoglobulin 30-fold after nursing on the first day of life. Do you think birth order influences blood immunoglobulin concentrations? Why?
6. Piglets nurse more often on average than once per hour. As they develop, nursing bouts get less frequent, but the meals are large on average. When pigs are weaned they fast from dry feed for a period that is inversely proportional to their weaning age. Why do you think this is? At three weeks of lactation age, do you think the piglet gets enough milk from the sow to satiate it?
7. Compare the growth and development of the forest-dwelling European wild boar and today's modern pig. In which major ways have the pigs changed?

LITERATURE CITED

Baxter, S. 1984. Intensive Pig Production. Granada Publishing Ltd., London.

Bazer, F. W., J. J. Ford, and R. S. Kensinger. 2001. Reproductive Physiology. In: W. G. Pond, and H. J. Mersmann (eds.), Biology of the Domestic Pig. pp. 150–224. Cornell University Press, Ithaca, NY.

Blecha, F. 2001. Immunology. In: W. G. Pond, and H. J. Mersmann (eds.) Biology of the Domestic Pig. pp. 688–711. Cornell University Press, Ithaca, NY.

Book, S. A. and L. K. Bustad. 1974. The fetal and neonatal pig in biomedical research. J. Anim. Sci. 38:997–1002.

Brinster, R. L. 1974. Embryo development. J. Anim. Sci. 38:1003–1012.

Brody, S. 1945. Bioenergetics and Growth. Hafner, New York.

Campion, D. R., G. J. Hausman, and R. J. Martin. 1989. Animal Growth Regulation. Plenum Press, New York.

de Lange, C. F. M., S. H. Birkett, and P. C. H. Morel. 2000. Protein, fat, and bone tissue growth in swine. In: A. J. Lewis and L. L. Southern (eds.) Swine Nutrition. pp. 65–81. 2nd ed. CRC Press, Boca Raton, FL.

Evock, C. M., T. D. Etherton, C. S. Chung, and R. E. Ivy. 1988. Pituitary growth hormone (pGH) and a recombinant pGH analog stimulate pig growth performance in a similar manner. J. Anim. Sci. 66: 1928–1941.

Ford, J. J., J. Klindt, and T. H. Wise. 2001. Endocrinology. In: W. G. Pond, and H. J. Mersmann (eds.) Biology of the Domestic Pig. pp. 653–687. Cornell University Press, Ithaca, NY.

Hafez, E. S. E. 1975. The Mammalian Fetus. Thomas Publishers, Springfield, IL.

Harrison, B. M. 1971. Embryology of the Chick and Pig. Wm. C. Brown, Dubuque, IA.

Hetzer, H. O. and W. R. Harvey. 1967. Selection for low or high fatness in swine. J. Anim. Sci. 26:1244–1251.

Hollis, G. R. (Ed.). 1993. Growth of the Pig. CAB International, Wallingford, Oxon, UK.

Knight, J. W., F. W. Bazer, W. W. Thatcher, D. E. Franke, and H. D. Wallace. 1977. Conceptus development in intact and unilaterally hysterectomized-ovariectomized gilts: Interactions among hormonal status, placental development, fetal fluids, and fetal growth. J. Anim. Sci. 44:620–637.

Lawrence, T. L. J. and V. R. Fowler. 1997. Growth of Farm Animals. CAB International, Wallingford, Oxon, UK.

Marrable, A. L. 1971. The Embryonic Pig: A Chronological Account. Pitman, New York.

Mitchell, A. D., A. M. Scholz, and H. J. Mersmann. 2001. In: W. G. Pond and H. J. Mersmann (eds.) Biology of the Domestic Pig. pp. 225–308. Cornell University Press, Ithaca, NY.

Moody, N. W., V. C. Speer, and V. W. Hays. 1966. Effects of supplemental milk on growth and survival of suckling pigs. J. Anim. Sci. 25:1250. (Abstract).

Moustgaard, J. 1962. Growth of fetus and membranes. In: J. T. Morgan and D. Lewis (eds.). Nurtition of Pigs and Poultry. Butterworths, London, UK.

Patten, B. M. 1948. Embryology of the Pig. 3rd ed. McGraw-Hill, New York.

Pond, W. G., J. H. Maner, and D. L. Harris. 1991. Pork Production Systems. Van Nostrand-Reinhold, New York.

Reeds, P. J., D. G. Burrin, T. A. Davis, M. A. Fiorotto, H. J. Mersmann, and W. G. Pond. 1993. Growth regulation with particular reference to the pig. In: G. R. Hollis (ed.) Growth of the Pig, Chapter 1, pp. 1–33. CAB International, Wallingford, Oxon, UK.

Speer, V. C. 1970. Relationship of birth weight and survival. Iowa State University Swine Nutrition Records (unpublished mimeograph).

Tess, M. W. 1981. Simulated effects of genetic change upon life-cycle production efficiency of swine. Ph.D. thesis, University of Nebraska, Lincoln.

Tumbleson, M. E. (ed.) 1986. Swine in Biomedical Research (vols. 1, 2, and 3). Plenum Press. New York.

Tumbleson, M. E. and L. B. Schook (eds.). 1996. Advances in Swine in Biomedical Research (vols. 1 and 2). Plenum Press, New York.

Walker, W. F., Jr. 1964. Dissection of the Fetal Pig. W.H. Freeman, San Francisco.

INTERNET RESOURCES

Tri-State Swine Nutrition Guide, Bulletin 869-98, Pig Growth and Development:
http://www.ag.ohio-state.edu/,ohioline/b869/b869_16.html

Growth- Growth Modeling:
http://www.ansc.purdue.edu/swine/porkpage/growth/model.htm

Growth and Development of Meat Animals:
http://meat.tamu.edu/growth.html

Introduction to Swine Biology:
http://cal.vet.upenn.edu/swine/bio/hm.html

Swine Modeling Project:
http://www.bae.uky.edu/,ltumer/Reswine.htm

9
PORK COMPOSITION AND QUALITY

INTRODUCTION

Pork is an important food source in all parts of the world. Total meat consumption continues to increase and per capita consumption of pork is increasing in most of the developing countries. Pork accounts for about 40% of total meat consumption and ranks first among all meat sources in total consumption (followed in descending order by beef, poultry, lamb, and goat meat). The high palatability and nutritional value of pork and the variety of processing and cooking methods available in many cultures contribute to its nearly worldwide acceptance. The consumer ultimately determines the amount and kind of pork produced. This chapter examines the composition and quality of pork and the factors that affect them. Continuous adjustment by producers to changing consumer preferences and demands will always play an important role in the pig industry in an increasingly global market.

NUTRIENT COMPOSITION OF PORK

When taken from the same muscle and prepared in a similar way, pork, beef, lamb, and chicken contain comparable amounts of important nutrients. Pork from modern, muscular pigs, unlike that of earlier generations, is lean and contains about the same amount of cholesterol as meat from broiler chickens, beef, and lamb (Table 9–1).

Today's fresh pork is, on average, 31% lower in fat, 14% lower in calories, and 10% lower in cholesterol than pork produced in 1983 (National Pork Producers Council,

TABLE 9–1
Cholesterol Content of Some Common Animal Products.

Animal Product	Cholesterol (mg/100g)
Egg yolk, chicken	1,602
Shrimp, canned	150
Cheese, cheddar	106
Turkey, roasted	105
Chicken, roasted	90
Lamb	95
Beef	95
Pork, variety of cuts	89
Tuna, canned	55
Liquid cow's milk, 2% fat	8

Source: Adapted from Ensminger et al. (1986).

1999). The reduction in fat, calories, and cholesterol is the result of a combination of changes in pork production practices, including improvements in genetics, nutrition, and husbandry. New information on the chemical composition of pork is available on the Internet at http:/www.nal.usda.gov/fnic/cgi-bin/list_nut.pl. The data are assembled from the USDA Database for Standard Reference. Composition data from a broiled, fresh composite of the separable lean of trimmed retail cuts are shown in Table 9–2 (USDA, 2001). Similar tables are available from the USDA for other fresh and processed pork products. It is noteworthy that cooking decreases the water content of pork and increases the concentrations of fat, protein, minerals, and fat-soluble vitamins. The content of some water-soluble vitamins is decreased following cooking, due to partial destruction by heat.

FAT AND CHOLESTEROL

The fat, calorie, and cholesterol contents of common, fresh, broiled pork cuts are shown in Table 9–2. The fatty acid composition of the fat varies directly in response to diet, as discussed in Chapter 10. Digestibility of pork fat is high and similar to that of plants. In general, pork fat is more unsaturated (softer) than beef or lamb fat and becomes progressively less saturated when pigs are fed diets high in polyunsaturated fatty acids, such as fish meal or most plant oils. The presence of a major recessive gene (*imf*) responsible for increasing the rate of intramuscular fat deposition has been reported from The Netherlands (see Houde, 1999). The dominant *IMF* allele is responsible for the normal rate of fat deposition within the muscle (marbling). Individual pigs homozygous for the recessive *imf* allele have more than twice as much intramuscular fat as those homozygous for the dominant *IMF* allele (3.9% versus 1.8%). Efforts are underway to develop a screening test for use in genetic selection for increased marbling in pork. The feeding of fish and plant oils containing the highly heralded omega-3 and omega-6 polyunsaturated fatty acids to pigs homozygous for the recessive *imf* allele offers the potential for producing pork with high levels of these unique fatty acids. Such pork would be expected to fill a niche market for consumers concerned about the fat composition of animal products.

TABLE 9–2
Pork, Fresh, Composite of Trimmed Retail Cuts (leg, loin, and shoulder), Separable Lean Only, Cooked.

Nutrient	Units	Value per 100 Grams of Edible Portion
Proximates		
Water	g	60.31
Energy	kcal	212
Energy	kj	887
Protein	g	29.27
Total lipid (fat)	g	9.66
Carbohydrate, by difference	g	0.00
Fiber, total dietary	g	0.0
Ash	g	1.18
Minerals		
Calcium, Ca	mg	21
Iron, Fe	mg	1.10
Magnesium, Mg	mg	26
Phosphorus, P	mg	237
Potassium, K	mg	375
Sodium, Na	mg	59
Zinc, Zn	mg	2.97
Copper, Cu	mg	0.061
Manganese, Mn	mg	0.018
Selenium, Se	mcg	45.0
Vitamins		
Vitamin C, total ascorbic acid	mg	0.3
Thiamin	mg	0.846
Riboflavin	mg	0.345
Niacin	mg	5.172
Pantothenic acid	mg	0.684
Vitamin B-6	mg	0.434
Folate, total	mcg	6
Folic acid	mcg	0
Folate, food	mcg	6
Folate, DFE	mcg_DFE	6
Vitamin B-12	mcg	0.75
Vitamin A, IU	IU	7
Vitamin A, RE	mcg_RE	2
Vitamin E	mg_ATE	0.260
Lipids		
Fatty acids, total saturated	g	3.410
4:0	g	0.000
6:0	g	0.000

Nutrient	Units	Value per 100 Grams of Edible Portion
Lipids		
8:0	g	0.000
10:0	g	0.010
12:0	g	0.010
14:0	g	0.120
16:0	g	2.110
18:0	g	1.090
Fatty acids, total monounsaturated	g	4.350
16:1 undifferentiated	g	0.310
18:1 undifferentiated	g	3.910
20:1	g	0.090
22:1 undifferentiated	g	0.000
Fatty acids, total polyunsaturated	g	0.750
18:2 undifferentiated	g	0.650
18:3 undifferentiated	g	0.020
18:4	g	0.000
20:4 undifferentiated	g	0.050
20:5 n-3	g	0.000
22:5 n-3	g	0.000
22:6 n-3	g	0.000
Cholesterol	mg	86
Amino Acids		
Tryptophan	g	0.372
Threonine	g	1.337
Isoleucine	g	1.371
Leucine	g	2.348
Lysine	g	2.632
Methionine	g	0.775
Cystine	g	0.373
Phenylalanine	g	1.168
Tyrosine	g	1.020
Valine	g	1.588
Arginine	g	1.819
Histidine	g	1.169
Alanine	g	1.705
Aspartic acid	g	2.715
Glutamic acid	g	4.582
Glycine	g	1.390
Proline	g	1.176
Serine	g	1.209

Pork, fresh, composite of trimmed retail cuts (legs, loin, and shoulder), separable lean only, cooked.
Source: USDA Nutrient Database for Standard Reference (Released 14 July 2001).

PROTEIN AND AMINO ACIDS

Pork is not only high in protein content, but also contains a better balance of amino acids than most plant proteins and its amino acids are utilized more efficiently by the human digestive system. The amino acid composition of fresh pork loin is shown in Table 9–2.

VITAMINS

Pork is not particularly high in fat-soluble vitamins, but is an excellent source of water-soluble vitamins. Pork is especially high in thiamin (B_1). A serving of 100 g of pork provides nearly three-fourths of the daily thiamin requirement for adult humans. Pork, as well as all animal proteins, is an excellent source of vitamin B_{12}. Pork is similar to milk, beef, and lamb in riboflavin (B_2) and niacin and is much higher than plants in these two vitamins. Furthermore, the niacin of some cereal grains and their by-products is biologically unavailable for use by humans because it is bound in the plant in an indigestible form. The vitamin content of fresh loin is shown in Table 9–2.

MINERAL ELEMENTS

Pork is an excellent source of some minerals, but is a poor source of others. It is a satisfactory source of phosphorus (P), but a poor source of calcium (Ca). Pork is high in potassium (K), but low in sodium (Na). It is high in iron (Fe), zinc (Zn), manganese (Mn), and magnesium (Mg). Mineral bioavailability is high in pork compared to plant sources. This favorable property of pork is often overlooked in comparisons of animal and plant sources. The mineral content of fresh pork loin is shown in Table 9–2.

CARCASS MEASUREMENTS AND COMPOSITION

The body weight and carcass measurements of a typical modern market hog are summarized in Table 9–3 (NPPC, 1999). NPPC (1999) also provided a chart showing the carcass breakdown by weight from a 184.2-lb (83.6 kg) carcass (Figure 9–1). The lean carcasses representing contemporary U.S. market hogs are the result of genetic selection for leanness and associated traits. Pork leanness can be changed dramatically in a few gen-

TABLE 9–3
Carcass Measurements of a Typical Modern Market Hog.

MEASUREMENT	UNIT	AVERAGE VALUE
Live body weight	kg (lb)	113.5 (250)
Carcass weight	kg (lb)	83.5 (184)
Back fat at 10th rib	cm (in)	2.25 (0.9)
Loin eye area	cm² (in²)	32.5 (5.2)
Total lean meat	kg (in)	40.2 (88.6)

Source: Adapted from NPPC (1999).

erations of selection, as illustrated by the contrasting body shape and fatness of lines of Duroc and Yorkshire pigs selected by USDA scientists in the 1950s and 1960s.

Pork leanness is also affected by body weight at slaughter—body fatness increases steadily as animals approach slaughter weight. This is depicted in Figure 9–2 (Mitchell et al., 2001).

FIGURE 9–1
Carcass Characteristics and Weights of Wholesale Cuts of the Modern Market Hog.

Source: Pork Facts (1998/99), National Pork Producers Council.

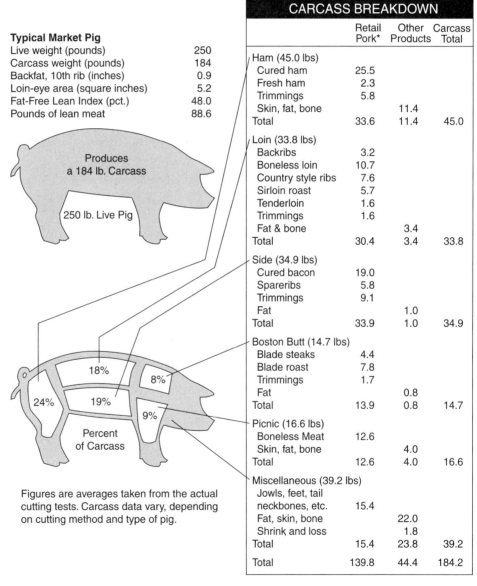

Typical Market Pig
Live weight (pounds)	250
Carcass weight (pounds)	184
Backfat, 10th rib (inches)	0.9
Loin-eye area (square inches)	5.2
Fat-Free Lean Index (pct.)	48.0
Pounds of lean meat	88.6

Produces a 184 lb. Carcass

250 lb. Live Pig

24% 18% 8%
19% 9%
Percent of Carcass

Figures are averages taken from the actual cutting tests. Carcass data vary, depending on cutting method and type of pig.

CARCASS BREAKDOWN			
	Retail Pork*	Other Products	Carcass Total
Ham (45.0 lbs)			
Cured ham	25.5		
Fresh ham	2.3		
Trimmings	5.8		
Skin, fat, bone		11.4	
Total	33.6	11.4	45.0
Loin (33.8 lbs)			
Backribs	3.2		
Boneless loin	10.7		
Country style ribs	7.6		
Sirloin roast	5.7		
Tenderloin	1.6		
Trimmings	1.6		
Fat & bone		3.4	
Total	30.4	3.4	33.8
Side (34.9 lbs)			
Cured bacon	19.0		
Spareribs	5.8		
Trimmings	9.1		
Fat		1.0	
Total	33.9	1.0	34.9
Boston Butt (14.7 lbs)			
Blade steaks	4.4		
Blade roast	7.8		
Trimmings	1.7		
Fat		0.8	
Total	13.9	0.8	14.7
Picnic (16.6 lbs)			
Boneless Meat	12.6		
Skin, fat, bone		4.0	
Total	12.6	4.0	16.6
Miscellaneous (39.2 lbs)			
Jowls, feet, tail neckbones, etc.	15.4		
Fat, skin, bone		22.0	
Shrink and loss		1.8	
Total	15.4	23.8	39.2
Total	139.8	44.4	184.2

* Retail cuts on semi-boneless basis. Fully boneless would show lower retail weights.
Source: NPCC, Purdue University and Texas A & M University, 1994.

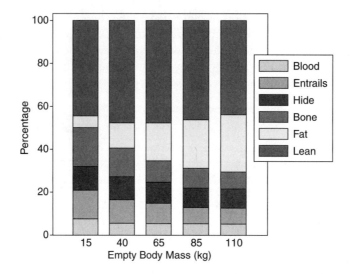

FIGURE 9–2
Changes in the Proportions of Different Body Fractions of the Pig Relative to Empty Body Mass During Growth from 16 to 112 kg. Data adapted from Susenbeth and Keitel (1988).

Source: Mitchell et al. (2001).

The notable increase in slaughter weight of contemporary pigs in the United States compared with their weight 10 yr ago (without an increase in percentage of body fat) is largely the result of genetic selection for rapid growth and carcass leanness. This higher slaughter weight is a benefit to the meat packer because more lean pork can be processed in a given packing plant without increasing slaughter capacity (number of pigs killed/d).

PORK QUALITY

The acceptability of pork to the consumer is determined by nutritional content, wholesomeness, and palatability and by factors, real or perceived, related to safety. Both palatability and safety are critical to the consumer in making food choices and are therefore in the vital interest of pork producers. Pork acceptability and safety are among an array of societal issues discussed in Chapter 3.

PALATABILITY

Palatability factors include tenderness, color, juiciness, and flavor. Variations among these traits are less with pork than with beef or lamb, so differences in breed, age, and feed have a relatively minor effect on pork palatability. Method and temperature of cooking pork can greatly affect flavor, juiciness, and tenderness. High cooking temperature decreases cooking yield (Table 9–4).

Three serious problems of pork palatability face the pig industry: PSE and DFD pork, boar odor, and the Rendement-Napole gene effect.

PSE Pork (Halothane gene) and DFD Pork

Extremes in color, firmness, and water-holding capacity range from pale, soft, and exudative (PSE) to dark, firm, and dry (DFD). Both PSE and DFD traits detract from

TABLE 9-4
Effect of Internal Cooking Temperature of Grilled Pork Chops on Cooking Yield and Palatability.[a]

PARAMETER	INTERNAL TEMPERATURE		
	RARE (140°F)	MEDIUM (158°F)	WELL-DONE (176°F)
Cooking Yield, %	73.8	67.0	58.3
Flavor[a]	9.8	10.1	10.2
Juiciness	11.4	9.4	6.1
Tenderness	11.3	10.8	9.1

[a] The higher the score, the more desirable.
Source: Adapted from Simmons et al. (1985).

pork's appearance and marketability. Pig producers are well advised to eliminate these traits from their herds. PSE pork results mainly from an accelerated rate of postmortem glycolysis (breakdown of muscle glycogen to produce lactic acid), resulting in low muscle pH and protein denaturation while the carcass temperature remains high after slaughter. The subject has been reviewed by Bowker et al. (2000). PSE pork is produced from mildly stressed pigs that carry one or both copies of the Halothane gene (the genetics of these pork quality traits were discussed in Chapter 6). (See Genetiporc, http://www.genetiporc.USA.com/Corner/index.htm for photographs of pork from a carrier and a noncarrier of the Halothane gene.)

Boar Odor

Meat from boars has an objectionable odor and flavor when cooked. The flavor is produced by a testosterone derivative and by skatole, produced by intestinal microbial breakdown of the amino acid tryptophan. The presence of the testosterone derivative, 5-alpha-androstenone, in pork is detectable by taste panel studies in boars weighing greater than 200 lb and occasionally in gilts. Skatole is detected in pork from boars, barrows, and gilts. Because boars grow faster and are leaner than castrated males, it would be an economic advantage to the producer if a method could be developed to reduce boar odor without castration. Research is underway toward that end.

The Rendement-Napole Gene

The Rendement-Napole (RN^-) gene was discovered in France in 1985 (Houde, 1999). The trait is predominant in the Hampshire breed and its crosses. The acid-meat condition and decreased water-holding capacity of pork caused by the RN^- gene has been found only in Hampshire populations in the United States (Miller et al., 2000). The gene exists in the form of two alleles: the dominant RN^- and the recessive rn^+ (Houde, 1999). When the RN^- allele of the gene is present, the meat has less water-holding capacity and is exudative and, thus, has a large amount of drip loss in the meat showcase. (See http://www.genetiporcusa.com/Corner/index.htm for photographs of pork from a carrier and a noncarrier of the RN^- gene.)

PORK SAFETY

The four overriding issues regarding safety of pork are: (1) the threat to humans of contracting the parasitic disease trichinosis by eating *Trichinella*-infected pork; (2) the threat of eating pork contaminated with pathogenic strains of organisms, such as *Salmonella, E.coli,* and *Campylobacter;* (3) the threat of eating pork containing antibiotic residues from pigs injected illegally with or fed antibiotics before slaughter; and (4) antibiotic-resistant strains of bacteria. Each of these concerns is addressed in Chapter 3, and in more detail in Chapter 20.

TRICHINOSIS

The parasite *Trichinella spiralis* encysts in the muscle of the host. If present in pork, it can infect humans consuming the pork. Worldwide, the problem is persistent because pigs and other hosts (rodents and other mammals) are frequently infected, making the parasite difficult to eradicate. The number of cases of trichinosis in the United States has declined to only a few cases annually. The perception of a threat of infection by eating pork persists disproportionately to its actual importance. Cooking pork to an internal temperature of 77°C (170°F), freezing it for 20 d at −15°C (5°F), or irradiating it by gamma radiation (recently approved for use in the United States) destroys the parasite. Effective drugs are now routinely administered to infected humans. However, proper preparation of pork ensures protection from *Trichinella* infection.

PATHOGENIC STRAINS OF *SALMONELLA, E. COLI,* AND *CAMPYLOBACTER*

All of these serious pathogens, because they are common in the environment, can contaminate pork and other food products. Destruction of each organism occurs by proper cooking before consumption. Contamination can occur at any stage in the processing and marketing cycle, all the way from the farm to the kitchen table. Federal, state, and local health regulations are in place to safeguard the pork supply, but the household consumer is ultimately responsible to minimize the danger of illness by appropriate handling and cooking of pork. Pork producers must do their share in curtailing the spread of pathogens by following accepted sanitation and husbandry standards.

TISSUE RESIDUES OF ANTIBIOTICS

For many years, the Food and Drug Administration has maintained and enforced regulations and procedures to assure the sale of only antibiotic-free pork. Although many pigs in the United States are fed subtherapeutic levels of antibiotics to improve growth, feed utilization, and reproductive traits, continuous surveillance of the pork supply ensures that violators are identified and prosecuted. Preslaughter withdrawal times are determined for each approved antibiotic and the constraints are strictly enforced. Approval of new antibiotics for use in pig diets must meet strict conditions at great financial investment by the company applying for marketing rights. There is ongoing debate over this controversial practice, and the welfare of the consumer is always the guiding force.

ANTIBIOTIC RESISTANCE

Antibiotics have been added to pig diets at subtherapeutic levels for more than 50 yr. The improved performance resulting from this practice has remained at 3% to 10% over the entire period (NRC, 1998). Yet, challenges to ban the use of antibiotics continue because of the documented development of strains of pathogenic organisms resistant to certain antibiotics, some of which are found in nature. Such resistance poses a potential threat to human health. Clear evidence of a causal relationship between antibiotic resistance and human health remains unavailable. Several countries prohibit the use of antibiotics in pig diets because of this potential danger. The ultimate action will undoubtedly be based on human health perceptions, documented evidence of deleterious effects of the practice on human health, and economics. If imposed, the ban on antibiotics in pig diets in the United States would probably increase the overall cost of pork production. The impact of a ban would be small or nil in some production systems, but significant in others.

SUMMARY

Pork is an important food in all parts of the world, ranking first among all meat sources, and accounting for nearly 40% of total meat consumption. This chapter describes the fat, protein, vitamin, and mineral element compositions of pork, carcass measurements and composition, pork quality, and pork safety. Consumer acceptance of pork is determined by palatability, safety, and price. Acceptance factors such as pork appearance and taste are affected by genetics and the presence of boar odor. The safety of eating pork is an issue to consumers, who may fear the presence of the *Trichinella* parasite, pathogenic bacteria, and antibiotic residues resulting from illegal administration prior to slaughter. The development of bacterial resistance to antibiotics has aroused consumer concern for a causal relationship between antibiotic resistance and human health. The actions taken by the pork industry, producers, and the government to address these issues were briefly described. The global importance of pork requires continued efforts by all segments of the industry to ensure the safety and acceptability of pork.

QUESTIONS AND ACTIVITIES

1. Let us assume we could genetically engineer pigs to have a different pork composition. What would you change first? What would be the consumer reaction to a genetically-modified improved pork product in your view?
2. One could possibly genetically engineer a pig with zero cholesterol. Would those embryos be expected to survive?
3. Pigs have an interesting metabolism that allows them to adopt a fatty acid profile in their body fat that is similar to their diet. If a farmer were to feed his pigs (a) peanuts or (b) fish oil, how might the pork be different than if they fed no added fat using a corn-soybean meal diet?
4. Food safety has become of increased public concern in recent years. What might be done on the farm to prevent or reduce the risk of food-borne pathogens being delivered to consumers on pork products?

5. What would happen to the shelf life of pork if higher levels of antioxidant nutrients, like vitamin E were fed to the pigs prior to processing? Can you find scientific literature to support your view?
6. One company is now certifying its pork is free of Trichinae. What are some positive and negative results that may arise from this practice. On balance is the result of such certifications positive or negative?

LITERATURE CITED

Bowker, B. C., A. L. Grant, J. C. Forrest, and D. E. Gerrard. 2000. Muscle metabolism and PSE pork. Available at: http://www.asas.org/2000 proc.htm. Accessed April 17, 2001.

Ensminger, A. H., M. E. Ensminger, J. E. Konlande, and J. R. K. Robson. 1986. Food for Health: A Nutrition Encyclopedia. p. 194. Pegus Press, Clovis, CA.

Genetiporc USA. 1999. Meat Quality. Available at: http://www.genetiporc.usa.com/Corner/index.htm. Accessed April 17, 2001.

Houde, A. 1999. Major genes and meat quality. Genetiporc USA. Available at: http://www.genetiporcusa.com/Corner/index.htm. Accessed December 22, 1999.

Miller, K. D., Ellis, J. C. Forrest, and D. E. Gerrard. 2000. Frequency of the rendiment-Napole RN^- allele in a population of American Hampshire pigs. J. Anim. Sci. 78:1811–1815.

Mitchell, A. D., A. M. Scholz, and H. J. Mersmann. 2001. Growth and body composition. In: W. G. Pond and H. J. Mersmann (eds.) Biology of the Domestic Pig. pp. 223–308. Cornell University Press, Ithaca, NY.

NPPC. 1999. Pork Facts. National Pork Producers Council, Des Moines, IA.

NRC. 1998. Nutrient Requirements of Swine. 10th ed. National Academy Press, Washington, DC.

Simmons, S. L., T. R. Carr, and F. K. McKeith. 1985. Effects of internal temperature and thickness on palatability of pork loin chops. J. Food Sci. 50:313–315.

USDA. 2001. Nutrient Database for Standard Reference, Release 14. Available at: http://www.nal.usda.gov/fnic/cgi-bin/list_nut.pl. Accessed March 20, 2002.

INTERNET RESOURCES

USDA/ARS Nutrient Database for Standard Reference, Release 14:
http://www.nal.usda.gov/fnic/foodcomp/Data/SR14/sr14.html

Pork Quality Research-Swine Genetics Page:
http://www.anr.ces.purdue.edu/anr/anr/swine/backup/genetic/quality.htm

Evaluating Pork Carcasses for Quality:
http://mark.asci.ncsu.edu/nsif/95proc/evaluating.htm

Ontario Pork Carcass Appraisal Project:
http://cgil.uoguelph.ca/pub/swine/opcap2.html

SECTION III

NUTRITION AND FEEDING

10

NUTRIENTS, NUTRITION, AND LIFE-CYCLE FEEDING

INTRODUCTION

Feed represents 50% to 85% of the cost of pork production. Therefore, providing adequate nutrition during all stages of pig growth, reproduction, and lactation deserves high priority. This chapter describes the nutrients required by pigs, outlines the principles of nutrition, and discusses life-cycle feeding. Although most pork producers depend heavily on commercial feed companies or specialists to supply well-balanced diets for each stage of the life cycle, it is important that producers have a general knowledge of adequate nutrition to recognize nutritional problems and address them appropriately. The topics covered are addressed in greater detail in several recent books and monographs (Miller et al., 1991; Pond et al., 1991, 1995; Close and Cole, 2000; Cole et al., 2000; Burrin, 2001; Garnsworthy and Wiseman, 2001; Kellems and Church, 2001; Lewis and Southern, 2001; Yen, 2001) and in Nutrient Requirements of Swine, 10th revised ed. (NRC, 1998).

NUTRIENTS

In the context of animal nutrition, any dietary substance required for normal life processes is defined as a nutrient. The six broad classes of nutrients are water, carbohydrates, lipids, proteins, inorganic mineral elements, and vitamins.

WATER

Without water, plant and animal life cannot exist. Water accounts for up to 90% of the body weight of newborn animals and about 70% of that of adults. It is important in several aspects of normal metabolism. Water is required: (1) in body temperature regulation; (2) in all biochemical reactions in the body; and (3) as a solvent for a wide variety of substances serving as a medium for transport of dissolved substances in blood and other fluids, tissues and cells, and in excretions such as urine and sweat.

Water is available to body tissues from drinking water, water contained in feed, and water liberated from normal metabolic reactions and from fat and protein breakdown, such as during starvation or illness. Losses of water occur via feces, urine, and vaporization from the lungs and skin. Losses from sweat are minor in swine because they have few sweat glands.

CARBOHYDRATES

Together with lipids, carbohydrates represent the main source of energy for swine. Carbohydrates are the major components of plants; they comprise up to 85% of the dry matter in corn and cereal grains and up to 70% in forages. In contrast, animal tissues contain less than 1% carbohydrates. Therefore, the conversion of plants to animal tissue involves major metabolic transformations within the animal body.

The simple sugar glucose is the basic unit from which more complex carbohydrates are built. The structure of glucose is shown in Figure 10–1. In plants, glucose is formed from carbon dioxide (CO_2) and water (H_2O) by photosynthesis, using ultraviolet light from the sun. Glucose so produced is stored in the plant as starch, hemicellulose, cellulose, and lignin. Table 10–1 shows the classification of carbohydrates into mono-, di-, tri- , poly-, and oligosaccharides.

Polysaccharides make up the largest fraction of total carbohydrates ingested by swine under most feeding systems. Starch, the most soluble and easily digested polysaccharide, is present in plant seeds, whereas the polysaccharide cellulose, found in plant stems and leaves, is not hydrolyzed by digestive enzymes of the host animal, but is broken down by fermentation by microflora in the digestive tract. Oligosaccharides are present in a variety of plants, but are not efficiently used by pigs. Simple carbohydrates (mono- and disaccharides) are readily digestible by pigs. The polysaccharide,

FIGURE 10–1
Structure of Glucose.

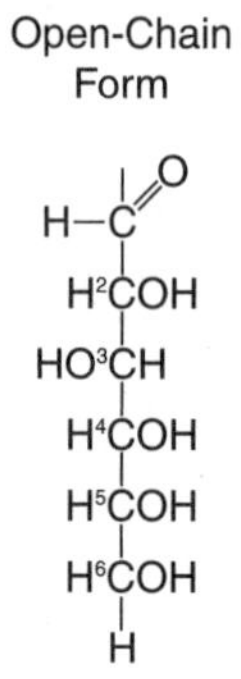

Open-Chain Form

$$\begin{array}{c} H-\overset{1}{C}=O \\ H-\overset{2}{C}-OH \\ HO-\overset{3}{C}-H \\ H-\overset{4}{C}-OH \\ H-\overset{5}{C}-OH \\ H-\overset{6}{C}-OH \\ H \end{array}$$

TABLE 10–1
Classification of Carbohydrates.

COMPOUND	CONSTITUENTS	MONOSACCHARIDES	WHERE FOUND
Monosaccharides			
	Pentoses (5-C sugars)		
	Arabinose		Pectin; polysaccharides
	Xylose		Corn cobs, wood, polysaccharides
	Ribose		Nucleic acids
	Hexoses (6-C sugars)		
	Glucose		Disaccharides, polysaccharides
	Fructose		Disaccharides (sucrose)
	Galactose		Milk (lactose)
	Mannose		Polysaccharides
Disaccharides			
	Sucrose	Glucose-fructose	Sugar cane, sugar beets
	Maltose	Glucose-glucose	Starchy plants and tubers
	Lactose	Glucose-galactose	Milk
Trisaccharides			
	Raffinose	Glucose-fructose-galactose	Cottonseed, sugar beets
Polysaccharides			
	Pentosans		
	Araban	Arabinose	Pectins (e.g., apples)
	Xylan	Xylose	Corn cobs, wood
	Hexosans		
	Starch[a]	Glucose	Grains, seeds, tubers
	Cellulose[b]	Glucose	Cell wall of plants
	Glycogen	Glucose	Liver and muscle of animals
	Insulin	Fructose	Potatoes, tubers
Mixed Polysaccharides			
	Hemicellulose	Mixtures of pentoses and hexoses	Fibrous plants
	Pectins	Pentoses and hexoses mixed with salts of complex acids	Apples, citrus fruits
	Gums	Pentoses and hexoses	Acacia trees, specific plants
Oligosaccharides[c]			
	Fructo- and galacto-oligosaccharides, (e.g., raffinose, stachylose)	Monosaccharide units, 3 to 9 similar or different monosaccharides	Soybeans, other legumes

[a]The most common carbohydrate fed to swine; readily digestible by animals.
[b]A major cell wall constituent of seed coats and of plant stems and leaves; indigestible by animals; broken down by microbes in the large intestine and cecum of pigs to volatile fatty acids that the pig can use for energy.
[c]They may be linear or branched chains in various linkages. They are not hydrolyzed by mammalian enzymes, but are fermented by microflora in the digestive tract of the pig.

Source: Adapted from Pond et al. (1995), Basic Animal Nutrition and Feeding, John Wiley & Sons, Inc. NY, p. 80.

starch, is by far the most important source of energy used in pig feeding. Starches, which are highly digestible by pigs, differ only slightly in chemical structure from cellulose, yet the latter is totally undigested by the pig because, like other mammals and birds, it lacks the digestive enzyme cellulase. Cellulase is produced by microbes that inhabit the intestinal tract of the pig, enabling the breakdown of cellulose to volatile fatty acids (VFAs) in the large intestine and cecum. Some of these VFAs are absorbed by the pig to provide energy. In older growing pigs and mature sows, up to 25% or 30% of the maintenance energy requirement can be met by high-fiber diets. Cellulose and hemicellulose are often present in plants in combination with lignin, a highly insoluble and indigestible mixture of polymers of phenolic acid. Lignification increases with plant maturity and lignin is the chief structural constituent in mature trees. Cereal grains and corn are low in lignin, whereas grasses and legumes are higher in lignin.

LIPIDS

Like carbohydrates, lipids are organic compounds containing mainly carbon (C), hydrogen (H), and oxygen (O). They are relatively insoluble in water. Commonly used fats and oils (e.g., lard, tallow, butter, corn oil, soybean oil) are made up largely of triglycerides consisting of an ester of glycerol (a 3-C alcohol; see Figure 10–2a) and three fatty acids. The structure of an ester and the linkage between glycerol and fatty acids in glycerides are illustrated in Figure 10–2b. The structures of monoglycerides, diglycerides, and triglycerides (esters of glycerol and one, two, and three fatty acids, respectively) are depicted in Figure 10–2c, where R, R', and R" represent three different fatty acids. Fatty acids (FAs) are chains of 2 to 24 carbon atoms. A fatty acid is saturated if it contains only single bonds between carbon atoms. Fatty acids have one or more double bonds between carbon atoms.

The structure of linoleic acid, an 18-C fatty acid with two double bonds, is shown in Figure 10–3. Linoleic acid is required in the diet of pigs because they cannot synthesize it from other fatty acids. The longer the chain and the greater the number of double bonds in the molecule, the more likely it is for a triglyceride to

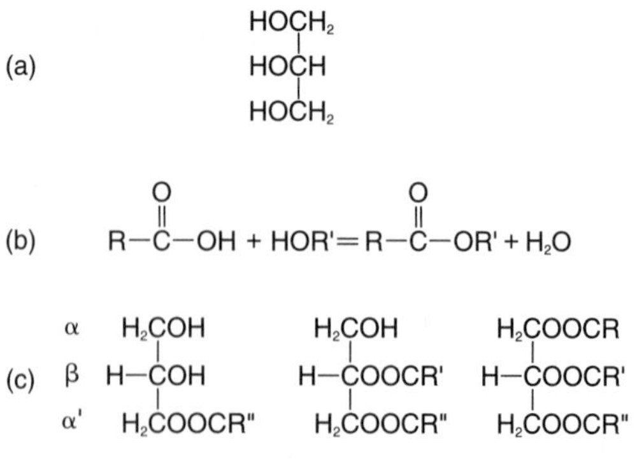

FIGURE 10–2
Structure of (a) Glycerol, (b) Ester Linkage between Glycerol and Fatty Acids (HOR' can represent any alcohol; in this case HOR' may be present in any monoglyceride), and (c) Monoglyceride, Diglycerides, and Triglycerides (R, R', R", respectively, represent three different fatty acids).

FIGURE 10–3
Structure of Linoleic Acid.

be liquid at room temperature. Examples of saturated fats are butter and coconut fat; examples of unsaturated fats are corn oil and soybean oil. The fatty acid composition of lard tends to mimic that of the fat in the diet of the pig. Pigs fed typical corn-soybean meal diets have more unsaturated (softer) fat in the carcass than pigs fed a barley-soybean meal diet because of the higher unsaturated fatty acid content of corn oil than of fat from barley. Thus, the composition of the dietary fat has a direct effect on the fat composition of the pork produced.

In addition to their contribution to the energy needs of the pig, lipids have the following general functions:

1. They are a source of essential FAs (linoleic and linolenic acid).
2. They are a carrier of the fat-soluble vitamins (A, D, E, and K).
3. They are an integral component of cell membranes (largely as phospholipids).
4. Steroids, another type of lipid molecule, are of importance in reproduction and growth, including hormones (e.g., estrogens, testosterone, progesterone), cholesterol, and precursors of vitamin D.
5. They are partners in the transport of a variety of metabolically important proteins throughout the body (lipoproteins involved in cholesterol trafficking contain fatty acids, triglycerides, and other lipids).

Lipids represent the primary energy storage depot in animals. Triglycerides, synthesized in the body from carbohydrates ingested by the pig, are deposited in storage depots when energy intake exceeds energy requirement for maintenance and other productive functions. The body fat content of the pig increases from about 1% of body weight at birth to 25% to 30% or higher in adulthood. As growth and development proceed, body fat concentration increases and body water decreases, while body protein and mineral concentrations remain relatively constant.

PROTEINS AND AMINO ACIDS

All cells synthesize protein as an obligatory part of life processes. The building blocks of protein are amino acids. More than 200 amino acids exist in nature, but only about 20 are present in significant amounts in most proteins. Amino acids are analogous to letters of the alphabet and proteins are analogous to words. Each protein has a specific amino acid composition with amino acids arranged in a specific order, just as letters of a word appear in a set order. Therefore, thousands of proteins exist in nature, just as thousands of words compose a language.

The fundamental structure of amino acids is illustrated by glycine, the simplest amino acid (see Figure 10–4). The essential constituents are a carboxyl group ($-COOH$) and an amino group ($-NH_2$) attached to the carbon atom adjacent to the carboxyl group. This $-NH_2$ group is designated the alpha-amino group.

FIGURE 10–4
Structure of Glycine.

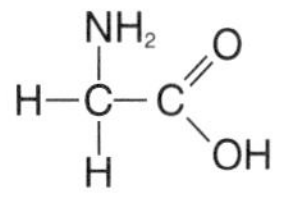

Protein is synthesized by joining the carboxyl group of one amino acid with the alpha-amino group of another to form long chains. This linkage of one amino acid with another is called a *peptide bond*. The reaction is depicted in Figure 10–5a and b. All naturally occurring amino acids are in the L-configuration, which is, with few exceptions, the most biologically useful form. Synthetic amino acids are usually found as a mixture of the L and D isomers. Most proteins in the pig body are present as constituents of a variety of tissue types; for example, muscle, skin, hair, blood, nerves, fat depots, and vital organs. Muscle protein is characterized by its high digestibility and favorable balance of amino acids, making it a nutritious animal protein, whereas hair is noted for its resistance to digestion and its unfavorable balance of amino acids, making it a very poor dietary protein source. Each body tissue has its own characteristic amino acid composition and configuration.

The pig requires a dietary source of 10 amino acids, as listed in Table 10–2. These amino acids are termed *essential* (indispensable) because the pig is unable to synthesize them from other substances. All other amino acids are termed *nonessential* (dispensable) because they can be synthesized in the body from other sources. These nonessential amino acids are incorporated into body protein along with essential amino acids. A portion of the nonessential amino acids, along with any excesses of essential amino acids, are catabolized and excreted in the urine and feces, posing an environmental nitrogen challenge. A dietary deficiency of a single essential amino acid in the diet interrupts protein synthesis and thereby reduces pig growth rate, reproduction, and health and results in excretion of the unused nitrogen, further contributing to potential environmental nitrogen pollution.

Proteins function in many ways: (1) they are the structural constituents of virtually every cell in the body; (2) they are the constituents of blood, hormones, enzymes, and other metabolic compounds in the body; and (3) they protect the animal from infectious agents with specific proteins called immune function antibodies.

Protein metabolism consists of two general processes: synthesis (anabolism) and breakdown (catabolism). Ingested proteins are broken down to their constituent amino

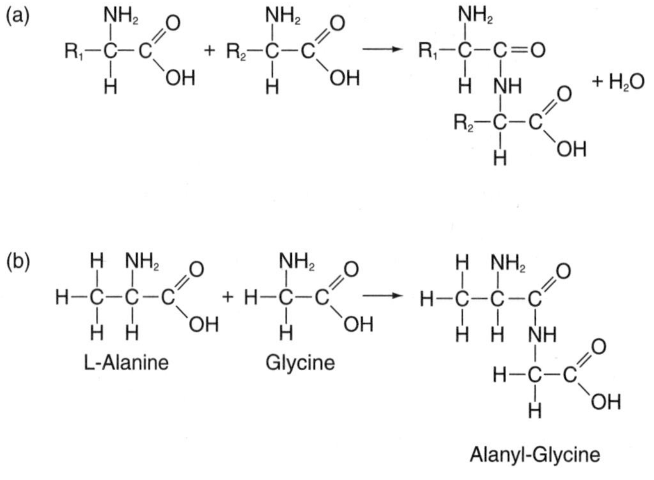

FIGURE 10–5
Structural Linkage of Amino Acids in the Synthesis of Protein (a) Peptide Bond (b) Linkage of Alanine with Glycine to Form the Dipeptide Alanyl-Glycine.

TABLE 10–2
Essential (indispensable) Dietary Amino Acids for Swine.

Arginine (required only for growth, not for maintenance)
Histidine
Isoleucine
Leucine
Lysine
Methionine (one-half of the requirement can be met by cystine)
Phenylalanine (one-half of the requirement can be met by tyrosine)
Threonine
Tryptophan
Valine

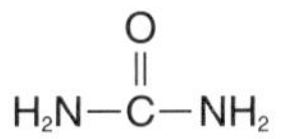

FIGURE 10–6 Structure of Urea.

$$H_2N-\overset{\overset{\displaystyle O}{\|}}{C}-NH_2$$

acids in the digestive tract. These amino acids are absorbed into the body where they are either used for protein synthesis or degraded. Processes of protein synthesis and breakdown occur throughout the body. The main end-product of protein and amino acid breakdown in swine is urea. The chemical structure of urea is depicted in Figure 10–6.

Urea and other nitrogenous products of metabolism are excreted in the urine. Dietary proteins and amino acids not absorbed from the digestive tract, coupled with proteins, amino acids, and other nitrogenous compounds already used by the body (sloughed intestinal cells and secretions from organs of digestion [e.g., digestive enzymes]), are lost in the feces. Small amounts of nitrogen are also lost from sloughing of skin and hair.

Animals that consume the same amount of nitrogen that they excrete via these routes are said to be in nitrogen balance. Growing pigs are in positive nitrogen balance, whereas underfed animals are in negative nitrogen balance. The nutritive value of dietary proteins is evaluated in several ways, including digestibility and nitrogen balance. Protein utilization in swine is conventionally estimated by determining the nitrogen content of the feed and excreta (urine plus feces) and multiplying by 6.25 to convert to protein equivalent (most proteins contain about 16% N; $6.25 \times 16 = 100$). To determine the apparent protein digestibility of a feed, the difference between the nitrogen intake of the animal during a specific number of d and the nitrogen excreted in the feces during the same time period equals the nitrogen digestibility. For example, if a pig consumes 5 lb of a dry diet containing 20% protein daily ($5 \times 0.20 = 1.0$ lb) and excretes 1 lb of dry feces containing 20% protein daily ($1 \times 0.20 = 0.20$ lb), the apparent digestibility of the protein is 1.00 minus $0.20 = 0.80$, or 80%. To calculate nitrogen balance, one would collect both feces and urine and, using the same procedures, subtract the nitrogen in the feces plus urine from the nitrogen intake, and express nitrogen balance as the amount of nitrogen (usually measured in g) retained by the animal/d [N balance = g of N intake/d minus the sum of g of N in feces/d (20% of N intake) and g of N in urine daily = N retained/d].

Modern pigs have a high capacity for lean growth. Therefore, palatable diets containing proteins of high nutritive value are critical in achieving performance that matches the genetic potential. Pure sources of economically competitive amino acids produced by biotechnology can be added to swine diets so as to reduce the total protein content of the diet without sacrificing rate and efficiency of body weight gain. This reduction in total dietary nitrogen conserves nitrogen and reduces the potential for nitrogen pollution of streams and water supplies.

VITAMINS

Vitamins are distinct from carbohydrates, lipids, and proteins in that they are required in the diet in minute amounts. Vitamins function not as structural components of the body, but as organic catalysts for many enzyme reactions and in other aspects of metabolism. Some are soluble in lipids (designated as fat-soluble), whereas most are water-soluble. The vitamins required in the diet of swine are listed in Table 10–3. Several vitamins are normally present in large enough amounts in typical feedstuffs to meet the needs of growth and reproduction, whereas most must be added to the diet in pure form to meet metabolic needs.

Fat-soluble Vitamins

Each of the fat-soluble vitamins has its own metabolic function and signs of deficiency and toxicity.

Vitamin A. Vitamin A is required for normal night vision, for growth of normal epithelial cells (cells lining the respiratory, digestive, urinary, and reproductive tracts and skin), and for normal bone growth. Vitamin A can be provided as retinol, retinal, or retinoic acid, or as its precursor, carotene. The most abundant form of carotene is beta-carotene. One molecule of beta-carotene produces two molecules of retinol. Retinol is

TABLE 10–3
Vitamins Required in the Diet of Swine.

FAT-SOLUBLE	WATER-SOLUBLE
Vitamin A[a]	Biotin[b]
Vitamin D[a]	Choline[a]
Vitamin E[a]	Folic acid[b]
Vitamin K[b]	Niacin[a]
	Pantothenic acid[a]
	Riboflavin (vitamin B_2)[a]
	Thiamin (vitamin B_1)
	Vitamin B_6 (pyridoxine)
	Vitamin B_{12}[a]
	Vitamin C (ascorbic acid)[d]
	Inositol[c]
	Para-aminobenzoic acid[c]

[a]Routinely added to swine diets.
[b]Metabolic needs are usually met by microbial synthesis in the digestive tract; generally added to sow diets; should be included in wheat-based diets.
[c]Controversial regarding their dietary essentiality.
[d]May be required in the diet during periods of stress, such as after early weaning.

considered to have 100% biopotency. Beta-carotene is poorly absorbed by the pig, and its efficiency of conversion to vitamin A is low (approximately 16%). Thus, even though yellow corn and many forages are high in carotenoids, the biologically available vitamin A from carotenes for the pig is limited. Vitamin A deficiency in the pig causes loss of coordination, posterior paralysis, blindness, abortion, malformed fetuses, decreased liver vitamin A, and increased cerebrospinal fluid pressure. Signs of toxicity are rough skin, hyperirritability, sensitivity to touch, bloody urine and feces, loss of mobility, decreased bone strength, skeletal defects, and eyelessness in newborn.

Vitamin D. Vitamin D activity is present in the steroids, irradiated ergosterol, or ergocalciferol (vitamin D_2, found mostly in plants), and irradiated 7-dehydrocholesterol or cholecalciferol (vitamin D_3, found in animal tissues). Both forms are converted in the liver and kidney to the hormones, 1,25-dihydroxy-D3 or 24,25-dihydroxy-D3. These hormones, in concert with other hormones (parathyroid hormone and calcitonin), maintain blood plasma calcium and phosphorus homeostasis. Pigs can use vitamins D_2 and D_3 with similar efficiency. Sunlight converts each of the precursors to the active form. Thus, exposure of harvested green forage to sunlight converts plant sterols to D_2, and exposure of pigs to the sunlight converts skin sterols to D_3. Exposure of pigs to sunlight for a few minutes daily minimizes or eliminates the need for dietary vitamin D. Vitamin D deficiency causes inadequate bone calcification (rickets in growing pigs, osteoporosis in sows and boars). Signs of toxicity include abnormal deposition of calcium in soft tissues, including kidneys, aorta, and lungs.

Vitamin E. Vitamin E consists of a mixture of tocopherol compounds, the most biologically active of which is alpha-tocopherol. Less active forms include beta- and gamma-tocopherol. Vitamin E is very unstable and can be oxidized by the presence of unsaturated FAs or mineral salts. It acts as an antioxidant in animal tissues and protects pigs from mulberry heart disease (cardiac muscle degeneration) and nutritional muscular dystrophy. Vitamin E is closely linked with the trace mineral element, selenium (Se), in protecting pigs from muscular dystrophy and liver necrosis. Newborn pigs are especially susceptible to vitamin E-selenium deficiency, probably because of inefficient placental transfer of vitamin E from dam to fetus. In sows, vitamin E deficiency results in reproductive failure. Vitamin E is stored in liver, muscle, fat depots, and in heart and other vital organs. Vitamin E toxicity is not known to occur in the pig.

Vitamin K. Vitamin K is required for synthesis of prothrombin, a factor in normal blood clotting. Deficiency results in prolonged clotting time, generalized hemorrhages, and even death. The vitamin exists in two forms in nature: phylloquinone (K_1), found in green vegetables, and menaquinone, (K_2), produced by microbes in the lower digestive tract of animals, including pigs and humans. These microflora normally synthesize enough vitamin K to meet the metabolic needs of swine. Synthetic vitamin K, menadione (K_3), is a widely used feed additive. Vitamin K can be absorbed directly from the intestinal tract or by coprophagy (feces-eating), which serves as an important source of nutrients in many animals, including the pig. Feed additives that alter intestinal microflora, or the use of wire or slatted floor pens, which reduce access to feces, may increase the need for dietary vitamin K supplementation. A well-known antagonist of vitamin K, dicoumerol, is used in a commercial rat poison (Warfarin). It causes death by massive hemorrhages due to an inability of blood to clot.

Water-soluble Vitamins

The water-soluble vitamins vary widely in their specific functions and chemical structure. Unlike the fat-soluble vitamins, they generally are not stored in appreciable amounts in the body, so they must be supplied in the diet on a daily basis. Each of the water-soluble vitamins has its own metabolic functions and signs of deficiency. Toxicity of water-soluble vitamins is highly unlikely, except in the rare case of accidental ingestion of a massive dose.

Thiamin (B_1). Thiamin is required for normal carbohydrate and fat metabolism. It is converted in the liver to form two coenzymes—thiamin pyrophosphate and lipothiamide. Both are critical components of energy metabolism. Thiamin-deficient pigs have elevated blood pyruvic acid and lactic acid levels and reduced appetite. Deficiency signs include enlarged heart, muscle weakness, unsteady gait, and edema. Thiamin is present in most common feedstuffs, so it is not likely to be deficient in most swine diets. Unlike other water-soluble vitamins (except B_{12}), thiamin is stored to some extent in swine tissues. Therefore, pork is a better source of thiamin than is meat from other animals.

Riboflavin (B_2). Riboflavin is a yellow, fluorescent pigment that functions as a constituent of the coenzymes flavin adenine dinucleotide (FAD) and flavin mononucleotide (FMN), which are present in virtually every animal cell. Deficiency signs in pigs include reduced growth (as in the case of all water-soluble vitamin deficiencies), cataracts, conjunctivitis, and lesions of the lips and mouth.

Niacin (nicotinic acid). Niacin functions as a constituent of the coenzymes nicotinamide adenine dinucleotide (NAD) and nicotine adenine dinucleotide phosphate (NADP), both of which are vital in energy metabolism. Signs of niacin deficiency in pigs include diarrhea, vomiting, ulcerated intestine, and dermatitis. Although grains contain relatively high amounts of niacin, it is present in a form unavailable for absorption by the pig. In pigs deprived of niacin, the metabolic requirement can be partially met by supplying a slight excess of the amino acid tryptophan in the diet. Tryptophan can be converted in the body to niacin, but the cost of supplemental niacin is far less than that of tryptophan, so there is no incentive to increase tryptophan intake above the requirement.

Pantothenic Acid. Pantothenic acid is a constituent of an important coenzyme (coenzyme A) involved in the formation of two-carbon fragments in metabolism of carbohydrates, fats, and amino acids. Therefore, a deficiency of pantothenic acid precludes the synthesis of coenzyme A in the body and results in a variety of symptoms in animals. In pigs, the classical sign of deficiency is "goose-stepping," a peculiar abnormal gait resulting from nerve degeneration.

Pyridoxine (B_6). Pyridoxine functions as a coenzyme for a wide array of enzyme systems involved in protein and amino acid metabolism. It occurs in nature as pyridoxine, pyridoxal, and pyridoxamine, all of which are readily absorbed from the intestine and converted to the active form of coenzyme, pyridoxal phosphate. The dietary requirement increases as the protein content of the diet increases. Deficiency signs include convulsions, demyelination of peripheral nerves, reduced antibody response to various antigens, and abnormal lipid metabolism (fatty liver, increased carcass fat, increased blood lipids).

Vitamin B_{12}. Vitamin B_{12} was first known as the animal protein factor (APF) because animals that did not consume a diet containing animal protein developed anemia and nervous signs later shown to be prevented or cured by vitamin B_{12}. Unlike most water-soluble vitamins, it is stored in appreciable amounts in animal tissues but is not synthesized by animals. The only known primary source of vitamin B_{12} is microorganisms. Vitamin B_{12} functions as a co-enzyme in several important enzymes in energy metabolism. It is closely linked with pantothenic acid in fatty acid (FA) oxidation and with the vitamin folacin in metabolism of sulfur-containing amino acids (methionine and cystine) in animal cells. Absorption of vitamin B_{12} from the intestine requires the presence of an enzyme termed "intrinsic factor," which is secreted by cells lining the stomach and upper small intestine. If intrinsic factor is absent, vitamin B_{12} is not absorbed and deficiency results. Vitamin B_{12} deficiency signs in pigs include anemia (reduced red blood cells), impaired thyroid function, rough hair coat, and secondary folacin deficiency. Because the vitamin B_{12} molecule contains the trace mineral element cobalt (Co), a dietary deficiency of cobalt leads to vitamin B_{12} deficiency in ruminants, in which rumen microorganisms synthesize vitamin B_{12}. Microbial synthesis in the digestive tract of swine is not sufficient to meet the pig's needs for vitamin B_{12}, even in the presence of high dietary cobalt.

Folacin. Folacin is the generic term for several derivatives of folic acid, each of which has biological activity. Folacin functions with vitamin B_{12} in several metabolic reactions. The prominent sign of folacin deficiency is anemia that is indistinguishable from that of vitamin B_{12} deficiency—described as macrocytic (large cells), hyperchromic (high amount of red pigment) anemia. Folacin content of most feedstuffs in swine diets is ample, so a deficiency is unlikely.

Biotin. Biotin is required by the pig in extremely small amounts and is synthesized by intestinal microflora. It is a component of several enzyme systems involved with carbohydrate, lipid, and amino acid metabolism. Deficiency signs include scaly dermatitis, alopecia (loss of hair), and reproductive failure. Because of its synthesis in the digestive tract of swine, biotin deficiency is unlikely, but it has been reported in pigs kept on slatted floors and not allowed to practice coprophagy.

Vitamin C (ascorbic acid). Vitamin C is involved in several enzyme systems that drive oxidation and reduction reactions. Therefore, it has become recognized as a powerful antioxidant in animal tissues. Most animals synthesize vitamin C from glucose. The pig apparently synthesizes enough vitamin C to meet its metabolic requirements under most conditions, but improved growth and immune function have been reported in newly weaned pigs fed vitamin C under stressful conditions. Deficiency signs include diarrhea, structural defects in bone and connective tissue, and reduced immune function.

Choline. Unlike other water-soluble vitamins, choline appears not to be a cofactor in enzyme reactions. It is widely distributed in animal tissues as free choline, acetylcholine, and as a constituent of phospholipids, where it serves as a structural component of cell membranes. Choline is also involved as acetylcholine in nerve impulse transmission and it supplies methyl groups in certain metabolic pathways. Choline deficiency signs include fatty liver, kidney hemorrhages, changes in cell membrane structure of lipoproteins needed for lipid transport, and interference with nerve impulse transmission, resulting in an abnormal gait in pigs.

INORGANIC MINERAL ELEMENTS

Table 10–4 lists the mineral elements known to be required in the diet of swine. Mineral elements required in relatively large amounts are termed *macrominerals*, whereas those required in much smaller amounts are termed *trace minerals*. Macrominerals include calcium (Ca), phosphorus (P), sodium (Na), chlorine (Cl), potassium (K), magnesium (Mg), and sulfur (S). Calcium, phosphorus, and magnesium are structural components of the skeleton, whereas sodium, chlorine, and potassium are electrolytes that function in acid-base and water balance. Sulfur is a constituent of several organic compounds that are structural components of soft tissue and skeleton and function in metabolism. Trace elements likely to be deficient in swine diets are iron (Fe), copper (Cu), iodine (I), manganese (Mn), selenium (Se), and zinc (Zn). Other trace elements required by swine are normally present in feedstuffs at levels sufficient to meet the pig's requirement, such as boron (B), cobalt (Co), chromium (Cr), fluorine (F), molybdenum (Mo), and silicon (Si).

Virtually all mineral elements are toxic when ingested in excessive amounts. The National Research Council (1980) published a list of the maximum tolerable levels of dietary mineral elements for swine and other farm animals.

Macrominerals

Calcium (Ca). About 99% of body calcium is stored in bones and teeth. Its functions include: (1) structural component of the skeleton (bone is in a dynamic state of continuous remodeling); (2) required in controlling the excitability of nerve and mus-

TABLE 10–4
Mineral Elements Required in the Diet of Swine.[a]

MAJOR MINERALS	TRACE MINERALS
Calcium (Ca)	Cobalt (Co) (as a constituent of vitamin B_{12})
Phosphorus (P)	Copper (Cu)
Sodium (Na)	Iodine (I)
Chlorine (Cl)	Iron (Fe)
Magnesium (Mg)	Manganese (Mn)
Potassium (K)	Selenium (Se)
Sulfur (S) (as a constituent of many organic compounds)	Zinc (Zn)

[a]Others, including molybdenum, chromium, and fluorine, are present as constituents of enzymes and other metabolites, but are required in such minute amounts that they are not of concern in swine diet formulation. They are present at low concentrations in most feedstuffs.

cle; and, (3) required for normal blood clotting.

Calcium deficiency results in poorly mineralized bone (rickets in growing pigs and osteoporosis in adults). Severe calcium deficiency causes hypocalcemia (low blood calcium), resulting in convulsions and tetany. The concentration of blood calcium is under tight homeostatic control by hormones that regulate accretion and removal of calcium from bone. Thus, blood levels low enough to cause tetany are uncommon in swine and usually occur only when bone density is severely decreased by long-term dietary calcium deficiency. Osteoporosis in sows is associated with bone fractures and the "downer sow syndrome."

Phosphorus (P). About 80% of body phosphorus is stored in bones and teeth, in a Ca:P ratio of about 2:1. The other 20% is distributed in blood plasma (both organic and inorganic forms) and in muscle, fat, nervous tissue, and vital organs, mostly in organic form. Its functions include: (1) structural component of the skeleton; (2) constituent of phospholipids (involved in lipid metabolism and transport and in cell membrane structure); (3) constituent of DNA (deoxyribonucleic acid) and RNA (ribonucleic acid); and (4) constituent of several enzyme systems (FAD, FMN, NADP).

Signs of phosphorus deficiency include rickets and osteoporosis (as in calcium deficiency), lameness, and bone fractures.

Magnesium (Mg). About 50% of magnesium is in bone and 50% is in soft tissues, mostly within cells. Its functions include: (1) structural component of the skeleton; (2) required in oxidative phosphorylation in cells; and (3) required for activation of several enzyme systems. Signs of magnesium deficiency include hyperirritability, muscular twitching, loss of equilibrium, tetany, and death. The magnesium requirement of pigs is met by most of the common feedstuffs, so field cases of magnesium deficiency are unlikely.

Potassium (K), Sodium (S), and Chlorine (Cl). These three elements function together in maintaining electrolyte balance and water movement in the body. Normal ratios among electrolytes in body tissues are remarkably constant. Potassium is located mainly within cells (90% is intracellular) and is exchangeable with sodium, which is located mainly in the extracellular fluid (90%), with only 10% intracellular. Chloride is located almost exclusively in the extracellular fluid. It acts with bicarbonate to balance electrically the extracellular sodium. When excess sodium is excreted by the kidney, it is accompanied by chlorine excretion. The cations, K and Na, and the anions, HCO_3 and Cl, act in synchrony to keep the body in electrolyte balance.

The functions of potassium include: (1) required to maintain acid-base balance; (2) required for normal protein synthesis; (3) required for integrity of heart and kidney muscle; and (4) required for normal heart rhythm (normal electrocardiogram). Signs of potassium deficiency include abnormal electrocardiogram, overall muscle weakness, and emaciation. Most feedstuffs for swine contain adequate levels of potassium to meet the pig's dietary requirement, so deficiency is unlikely.

The functions of sodium include: (1) required to maintain acid-base balance; (2) serves as an energy-dependent sodium "pump" in energy metabolism; and (3) required in nerve impulse transmission associated with its separation from potassium by the cell membrane. Signs of sodium deficiency include reduced growth and milk production and a craving for salt (NaCl).

The functions of chlorine include: (1) required to maintain acid-base balance and regulate osmotic pressure; and (2) secreted as the chief anion in gastric juice in the stomach where it unites with hydrogen to form hydrochloric acid (HCl) to activate the digestive enzyme pepsin. Signs of chlorine deficiency are the same as those for sodium deficiency: reduced growth and milk production and salt craving.

Toxicity of potassium, sodium, and chlorine is unlikely because the kidney increases their excretion when intakes are excessive and when adequate water is provided. Severe water restriction in the presence of excessive salt results in convulsions and death.

Sulfur (S). Sulfur is a constituent of several organic compounds, including amino acids (methionine, cystine) and vitamins (biotin and thiamin). Therefore, sulfur is evenly distributed throughout the body and makes up about 0.15% of body weight. Inorganic sulfur is not required in the diet of swine because its only known functions are related to its presence in amino acids, vitamins, and other organic compounds. Sulfur toxicity is not likely because inorganic sulfur is poorly absorbed from the digestive tract and largely excreted in the feces.

Trace Mineral Elements

Swine have a metabolic requirement for boron, chromium, cobalt, copper, fluoride iron, iodine, manganese, molybdenum, selenium, silicon, and zinc. Only iron, copper, iodine, manganese, selenium, and zinc are included in this discussion because they are likely to be deficient in most common feedstuffs for swine, whereas the others are normally present at levels above the dietary requirement.

Iron (Fe). Iron is a constituent of the red pigments hemoglobin in blood and myoglobin in muscle, which gives blood and meat its red color. Hemoglobin is responsible for carrying oxygen to all cells in the body. Iron deficiency results in anemia, which is characterized by smaller than normal red blood cells containing lower than normal hemoglobin concentration (termed "microcytic hypochromic anemia"). Iron-deficiency anemia is common in baby pigs, whose body stores of iron are low at birth due to inefficient fetal transfer, coupled with the low level of iron in milk.

For piglets raised indoors during the nursing period, iron is routinely supplied by intramuscular injection of an organic iron complex such as iron-dextran at 1 to 3 d of age followed by a second injection at 2 wk of age. Piglets raised outdoors with access to soil ingest sufficient iron from the environment to meet their needs, so iron injection is not needed. Iron-deficiency anemia is associated with increased susceptibility to infections and labored breathing (called "thumps") due to restricted oxygen supply to tissues.

Iron is normally added to the diet after weaning because its bioavailability is low in most natural feedstuffs even though the total iron present may be adequate if it is absorbed efficiently. Iron toxicity is unlikely, but massive doses of iron cause toxic accumulation in tissues. The upper tolerance level for iron is decreased in pigs that are marginally deficient in selenium and vitamin E, resulting in iron toxicity and death in suckling piglets injected with the standard dose of iron to prevent anemia.

Copper (Cu). Copper is required for the activity of enzymes involved in iron metabolism and for connective tissue formation and integrity of the central nervous system. Milk is deficient in copper, but the quantitative requirement is considerably less than that for iron, so copper deficiency in suckling piglets is uncommon. Deficiency signs include anemia (due to reduced iron utilization), bone abnormalities, loss of hair pigmentation, and incoordination. Copper toxicity is unlikely in swine. In fact, levels of up to 250 parts per million (ppm) of dietary copper in growing pig diets (25 times the metabolic requirement) have been fed for several wk to replace antibiotics in growth promotion without deleterious effects. Copper toxicity due to diet-mixing errors results in excessive liver copper accumulation and death.

Iodine (I). Iodine is required for normal function of the thyroid gland as a constituent of the thyroid hormones thyroxin (T4) and triiodothyronine (T3). The thyroid hormones are required to maintain normal basal metabolic rate in all higher animals. Iodine deficiency results in an enlarged thyroid (goiter) and lower metabolic rate (hypothyroidism). Iodine is routinely added to swine diets as iodized salt. Excessive iodine intake results in increased metabolic rate (hyperthyroidism).

Manganese (Mn). Manganese is a constituent of several enzyme systems involved with carbohydrate and lipid metabolism and is required for the formation of chondroitin sulfate, a component of mucopolysaccharide of bone. Deficiency signs include an array of skeletal abnormalities and low reproductive rate. Manganese toxicity is related more to interference with the utilization of other mineral elements than to a specific effect of manganese itself. Calcium-phosphorus-deficiency rickets in laboratory animals and abnormal tooth enamel and reduced growth have been observed in pigs fed excess manganese. The effect on growth appears to be a reflection of appetite.

Selenium (Se). Selenium is a constituent of the group of enzymes known as glutathione peroxidase. It is involved in the breakdown of peroxides arising from tissue lipid oxidation. Thus, it plays a major role, along with vitamin E, in maintaining the integrity of cell membranes. Selenium deficiency in pigs results in nutritional muscular dystrophy, liver necrosis, and mulberry heart disease (cardiomyopathy). All of these diseases involve a failure in the integrity of cell membranes within these organs, preventable by either vitamin E or selenium in the diet. The relatively low quantitative requirement for selenium, and the narrow margin of safety between deficient and toxic levels of dietary selenium, have led to controversy surrounding its use as a supplement in swine diets. Selenium occurs in plant and animal tissues as a constituent of sulfur-containing amino acids (selenium replaces sulfur in some of the molecules). Thus, it can be supplemented in the diet in inorganic form (usually as sodium selenite) or in organic form as a component of plant or animal tissues. Selenium was recognized as a toxic element long before it was known to be required by animals. Toxicity in pigs results in soreness and sloughing of hooves, lameness, abnormal embryonic development, blindness, paralysis, and death.

Zinc (Zn). Zinc is a constituent of many metalloenzymes involved in protein, RNA, and DNA metabolism. Zinc is required for normal protein synthesis and metabolism. It is a component of the hormone insulin, and functions in carbohydrate metabolism. The most obvious signs of zinc deficiency in swine are growth retardation and thickening, roughening, and hyperkeratinization of the skin (parakeratosis). These lesions are reversed within a few d when adequate zinc is returned to the diet. Zinc deficiency also drastically impairs male reproduction and retards bone formation and proliferation of cartilage cells in the epiphyseal growth plate. Zinc deficiency can be induced by high dietary calcium and/or the presence of a high level of phytic acid in the grain or protein supplement.

A wide margin of safety exists between the required intake of zinc (50 ppm) and the toxic level. Some reports have shown an improvement in weight gain among pigs fed 1,000 or 2,000 ppm zinc in the diet. This effect has been suggested to be related to a potential antimicrobial effect of these dietary levels. Zinc supplementation at 0.4% (4,000 ppm) of the diet produces anemia, depressed growth, stiffness, hemorrhaging around bone joints, and excessive bone resorption.

NUTRITION

STRUCTURE OF THE GASTROINTESTINAL TRACT

The pig digestive tract, starting at the mouth and ending at the anus consists of: esophagus, stomach, small intestine (first section—duodenum; second section—jejunum; third section—ileum), cecum (blind sac at junction of ileum and large intestine), large intestine, and rectum (see Figure 10–7).

The pancreas produces the hormones insulin and glucagon (involved in glucose homeostasis) and several digestive enzymes (including amylase, lipase, trypsin, chymotrypsin, and carboxypeptidases). The pancreas lies adjacent to the upper small intestine and secretes digestive enzymes into the lumen of the small intestine via the pancreatic duct, which enters the duodenum just below the stomach. The bile duct,

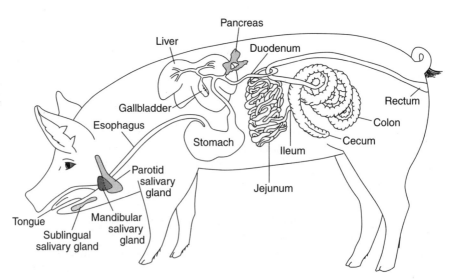

FIGURE 10–7
Diagram of the Digestive System of the Pig.

Source: J. T. Yen (2001) in Ch. 8, Biology of the Domestic Pig, p. 391, Cornell University Press, Ithaca, NY with permission of Cornell University Press.

which also enters the duodenum, carries bile, produced in the liver, to the small intestinal lumen for use in emulsifying fat and enhancing fat digestion and absorption. In addition to the pancreas, digestive enzymes are produced in the stomach and small intestine. The principal enzymes secreted by the swine gastrointestinal tract, the substrates attacked, and the end products of digestion are summarized in Table 10–5. In addition to the importance of digestive enzymes in nutrient utilization, several peptide hormones are important in the functioning of the swine gastrointestinal tract. These hormones include gastrin, secretin, cholecystokinin, somatostatin, gastric-inhibitory polypeptide (GIP), and vasoactive intestinal peptide (VIP).

The proteolytic enzymes, pepsin and trypsin, are secreted in inactive forms, probably as a means of protecting the cells lining the intestine from being destroyed. Pepsinogen is activated to pepsin by hydrochloric acid produced in the stomach. Trypsinogen is activated by an intestinal enzyme, enterokinase, and by trypsin. The inactive forms of chymotrypsin, carboxypeptidases, and other peptidases are activated by trypsin. This cascade of factors controlling protein digestion in the intestinal tract illustrates the complexity of nutrient utilization by the animal from the feeds consumed.

TABLE 10–5
Principal Digestive Enzymes, the Substrates they Attack, and Their End-Products.[a]

ENZYME	ORGIN	SUBSTRATE	END-PRODUCTS
Salivary amylase	Saliva	Starch, dextrins	Dextrins, glucose
Pancreatic amylase	Pancreas	Starch, dextrins	Glucose
Lactase	Small intestine	Lactose	Glucose and galactose
Sucrase	Small intestine	Sucrose	Glucose and fructose
Pancreatic lipase	Pancreas	Triglycerides	Monoglyceride and two fatty acids
Intestinal lipase	Small intestine	Triglycerides	Glycerol and three fatty acids
Pepsin	Stomach	Native proteins	Polypeptides
Trypsin	Pancreas	Native proteins and peptides	Polypeptides
Chymotrypsin	Pancreas	Native proteins and peptides	Polypeptides
Carboxypeptidase A	Pancreas	Peptides	Small peptides and amino acids
Carboxypeptidase B	Pancreas	Peptides with terminal	Basic amino acids arginine or lysine
Amino peptidases	Small intestine	Peptides	Amino acids
Dipeptidases	Small intestine	Dipeptides	Amino acids
Nucleases	Pancreas, small intestine	Nucleic acids	Nucleotides
Nucleotidases	Small intestine	Nucleotides	Purine and pyimidine bases, pentose sugars, phosphoric acid

[a]There are many others, but these are the most common in the pig.
Source: Adapted from Pond et al. (1995), Basic Animal Nutrition and Feeding, John Wiley and Sons, Inc., NY, p. 36.

The end products of digestion (i.e., glucose, fatty acids, and amino acids), are transported from the lumen of the small intestine across the epithelial lining and into the blood and lymph for use by the body. Without this absorptive process, the digestive end products would be useless to the animal.

ROLE OF THE DIGESTIVE TRACT IN NUTRIENT TRANSPORT

Absorption of nutrients in the pig occurs mostly in the duodenum and jejunum and, to a lesser extent, in the ileum and large intestine. The passage of ingesta through the entire length (several meters, even in the growing pig) of the digestive tract takes only 36 to 60 or so hr in swine, so the opportunity for digestion and absorption is limited. Absorption of nutrients is augmented by the enormous surface area of the intestinal lumen due to the presence of folds, villi, and microvilli, which results in a 1,000-fold increase in the absorptive surface compared with the surface area of a simple cylinder of the same length. The transport of nutrients from the lumen into the epithelial cells lining the lumen and then into the blood or lymph may occur by simple diffusion, active transport (requiring energy), or pinocytosis (engulfing large particles in a way similar to the way an *Amoeba* obtains its food). Pinocytosis occurs in newborn pigs to allow the absorption of immunoglobulins and other intact proteins from colostrum. This ability of the newborn piglet to absorb intact proteins is lost within 48 to 72 hr after birth. Most organic nutrients are absorbed by active transport and therefore require energy for assimilation, whereas water and some inorganic minerals may be absorbed by diffusion.

ABSORBABILITY VERSUS DIGESTIBILITY

Digestibility can be defined as the preparation of feed for absorption into the body. The absorption of nutrients from the lumen into the blood and tissues (absorbability) is technically a separate process in the utilization of ingested food. However, in common usage, digestibility is taken to mean disappearance of ingesta from the digestive tract (the amount of a nutrient consumed minus the amount of that nutrient appearing in the feces).

APPARENT VERSUS TRUE DIGESTIBILITY

Apparent digestibility is the established way of expressing the digestibility of nutrients. It is calculated by subtracting the amount of a specific nutrient appearing in the feces from the amount of that nutrient consumed during a specific period of time, typically 4 to 5 d; the difference represents the digested fraction. The absorbability (digestibility) of the nutrient is expressed as a percentage of that ingested. An apparent digestibility of 85% is considered an acceptable value for most nutrients. The feces contain not only the undigested residue from the feed consumed, but also sloughed intestinal mucosal cells, enzymes, and fluids from various portions of the gastrointestinal tract, and waste products excreted into the intestinal lumen. Thus, the true digestibility of a nutrient may be higher than would be assumed from considering its apparent digestibility. Nevertheless, apparent digestibility is a more useful measure because it takes into account not only the direct loss of the ingested nutrient into the feces, but also the losses of that nutrient following its use in metabolism.

Surgical procedures involving the insertion of a plastic tube into the lumen of the ileum (lower small intestine) allow sampling of the ingested nutrients before they reach the large intestine. Scientists have developed sampling techniques that permit calculation of the ileal digestibility of nutrients. Ileal digestibility provides more accurate measurements of nutrient utilization than does the traditional apparent digestibility measurement, which is based on fecal collection. Nutritionists now commonly use these ileal digestibility values in formulating diets for pigs.

CALCULATION OF DIGESTIBLE ENERGY

Carbohydrates, lipids, and proteins all contribute energy to the diet. Therefore, the sum of their individual contributions to the energy content of a specific diet constitute the energy value of that diet. The fundamental unit of energy is the calorie—the amount of heat required to raise the temperature of 1 g of water 1°C (usually from 4° to 5°C). The caloric values of carbohydrates, fats, and proteins, expressed as calories/g, are:

carbohydrates	4.0
proteins	4.0
fats	9.0

Note that fat contains 2.25 times as much energy/g as carbohydrates and fats.

The digestible energy value of a particular diet is calculated by multiplying the digestible calorie concentration of each ingredient by the weight percentage of that ingredient in the diet, then summing the fractional contributions of each ingredient to determine the total digestible calorie content of the diet (Kcal/kg).

DETERMINATION OF APPARENT PROTEIN DIGESTIBILITY

To calculate apparent protein digestibility, nitrogen (N) content of feed and feces is determined and apparent digestibility of nitrogen is calculated as just described. Protein digestibility is then calculated by multiplying N $\times$ 6.25 (average N content of protein is 16%; 16 $\times$ 6.25 = 100%). Digestibility of individual amino acids can be determined by analysis of feed and feces or by surgical cannulation of the lower small intestine. This allows collection of the ingesta at the ileum (the lower portion of the small intestine) before entrance into the large intestine where microbial action may affect the amount of amino acid that is excreted into the feces. Therefore, ileal digestibility of amino acids has become a standard method of expressing digestibility of individual amino acids.

DIGESTIBLE ENERGY (DE) VERSUS METABOLIZABLE ENERGY (ME)

Digestible energy (DE) represents the energy available to the body after absorption from the digestive tract. Some losses of DE occur during the processing of the absorbed nutrients. The greatest loss is the incomplete oxidation of protein and amino acids in normal metabolism. The final breakdown product of protein metabolism in the pig is urea, which is excreted in the urine. Whereas the complete combustion of protein would yield about 5 kcal of energy/g, the energy contained in urea represents about 20% of the total energy content of protein, resulting in only about 4 kcal of energy available for use by the pig. The energy lost in urine, together with that lost in combustible gases produced in the intestinal tract during digestion (minor in pigs compared with ruminants), is unavailable for use by the pig. Metabolizable energy (ME) therefore represents energy consumed minus the sum of fecal, urinary, and gaseous energy losses. The ME of most feedstuffs for the pig is about 90% of DE.

ABSORPTION AND EXCRETION OF MINERAL ELEMENTS

Some mineral elements are absorbed efficiently from the intestinal tract, but the absorption of others, such as iron, is limited by homeostatic factors that inhibit excessive absorption. These homeostatic mechanisms operate so as to optimize the tissue levels to meet metabolic needs. Tissue levels of most mineral elements are controlled not only by differences in absorptive efficiency, but by excretion via the kidneys and by re-entry into the lumen of the intestinal tract for fecal excretion. Excessive intake of some trace elements, particularly copper, may result in toxicity due to liver accumulation. Details of absorption and excretion of mineral elements by the pig are beyond the scope of this discussion. It must be recognized, however, that mineral element imbalances are of concern in swine nutrition, because of interactions between and among essential mineral elements in the diet and those with no known function.

ABSORPTION AND EXCRETION OF VITAMINS

Vitamins are required in such minute amounts that direct methods of absorption are not meaningful. Most fat-soluble vitamins are stored in the liver and other tissues for later use, whereas water-soluble vitamins are excreted in the urine if consumed in excess of current needs. Thus, there is little danger of toxicity of the water-soluble vitamins when they are consumed in excess of requirements. On the other hand, excessive intake of the fat-soluble vitamins can result in the accumulation of toxic tissue levels due to the inability of the pig to excrete the excess.

LIFE-CYCLE FEEDING

Efficient pig production requires that nutrient requirements be provided at the lowest cost compatible with optimum growth, reproduction, and lactation. Quantitative requirements change greatly from birth to maturity in all breeds and environments. In addition, requirements differ between gender and among pigs of different genetic background, and with changes in environmental temperature, health status, housing density, and many other factors. With the recent dramatic increases in size of individual pork production units and the trend toward corporate ownership of these large units, several important management practices have evolved related to life-cycle feeding. These include all-in-all-out (AIAO) production schedules, segregated early weaning (SEW), and split-sex feeding. Each of these practices is discussed in more detail in later chapters of this book that address swine management practices during different phases of the life cycle.

This section discusses nutritional requirements according to finite stages in the life cycle corresponding to discrete phases of housing and husbandry. These stages are suckling, nursery, growing, finishing, pregnancy, post-weaning sows, and lactation. All of the nutrients required by swine can be supplied by a simple mixture of an energy source (e.g., corn or cereal grain) and a protein supplement fortified with minerals and vitamins. An example of such a diet for growing swine is shown in Table 10–6 (NRC, 1998).

SUCKLING

The suckling period includes birth through the nursing stage, typically 2 to 4 wk. The newborn pig has almost no immunity to infectious diseases because the placenta is almost completely impervious to the transfer of immune antibodies to the fetus. Therefore, the ingestion of colostrum immediately after birth is critical, not only to provide nutrients, but to provide passive immunity to disease organisms. Colostrum is high in fat and protein, including immunoglobulins (IGs), of which gammaglobulin is a main component. These large, globular proteins must be ingested within a few hours after birth because their effectiveness in providing immunity depends on their absorption intact. The intestine loses its ability to transport the intact molecules of immunoglobulin within 48 to 72 hr after birth. Therefore, it is important that the neonate nurse as soon as possible to ensure an adequate immune protection and nutrient intake. Colostrum contains higher levels of total solids (dry matter) and protein than milk produced later in lactation. Colostrum collected immediately after farrowing contains 25% to 30% total solids, 15% to 20% protein, 5% to 7% fat, 2% to 3% lactose, and 0.63% ash (minerals) (Klobasa et al., 1987). By d 2 of lactation, total solids decrease to 18% to 20%, fat and lactose increase slightly, and protein decreases greatly as immunoglobulin secretion declines. Average fat, lactose, ash, and protein at 2 d are about 6.5%, 4% to 8%, 0.66%, and 6.4%,

TABLE 10–6
Fortified Swine Diet.[a]

INGREDIENT	AMOUNT IN DIET (%)	DIGESTIBLE ENERGY (KCAL/KG)	DIGESTIBLE ENERGY CONTRIBUTION (KCAL/KG)
Corn	74.44	3,525	2,624
Soybean meal	23.40	3,685	862
Dicalcium phosphate	0.71		
Ground limestone	0.90		
Sodium chloride	0.25		
Vitamin premix	0.10		
Trace mineral premix	0.10		
Antibiotic premix	0.10		
Total	100.00		3,486

[a]Note that in this example corn and soybean meal are the only sources of energy. All other ingredients are important to provide a complete diet, but they contribute no energy.
Source: National Research Council (1998).

respectively. From 2 d to 6 wk of lactation, the percentages of each constituent in sow milk remain relatively constant. Values reported by Klobasa et al. (1987) at 42 d of lactation were 17.0% total solids, and 5.3%, 5.4%, and 6.0% fat, lactose, and protein, respectively. Total daily milk production increases from farrowing to about 3 wk of lactation when it reaches a peak, then declines at 6 wk and beyond. This shape of the lactation curve explains why many pork producers wean pigs at 3 wk of age or younger. The rapidly growing pig has a voracious appetite and within 3 wk, the total milk required by the litter surpasses the milk-producing ability of the sow. If the pigs are not weaned by 3 wk, a dry prestarter (creep) diet must be provided as a supplement to the milk supply to allow normal growth. Weaning at 2 wk has become a common practice in commercial pork production, not only for nutritional efficiency but also for two other important reasons: (1) to shorten the period to rebreeding, thereby improving reproductive rate (see Chapter 5) and (2) to control infectious disease transfer from sow to litter by removing the pigs to a separate location away from the sow. This latter practice has become referred to in the pork industry as segregated early weaning (SEW) and is now standard practice in most large pork production systems. An example of a suitable starter diet of the type used in SEW systems is shown in Table 10–7. Many variations exist.

TABLE 10–7
A Suitable Starter Diet for Young Pigs Immediately After Weaning[a] (less than 12 lb body weight).

Ingredient	% of Diet
Corn or grain sorghum (ground)	34.65
Soybean meal (46.5% CP)	13.00
Spray-dried animal plasma	6.70
Spray-dried blood meal	1.65
Select menhaden fish meal	6.00
Spray-dried whey	25.00
Lactose	5.00
Choice white grease	6.00
Lysine HCl	0.15
DL-methionine	0.15
Monocalcium phosphate (21% P)	0.75
Limestone	0.45
Salt (NaCl)	0.50
Vitamin premix[b]	
Trace mineral premix[b]	
Antibiotic[c]	

[a]The combination of ingredients used in this example may be changed according to price and availability of specific ingredients. For example, barley or grain sorghum might replace some or all of the corn as an energy source or canola meal or another oilseed meal might replace part of the soybean meal.
[b]Vitamin and mineral premixes contain amounts of required nutrients to meet dietary requirements of weaned pigs of this weight. These added nutrients are in addition to the amounts present in the feed ingredients.
[c]Antibiotics are supplemented at subtherapeutic levels to promote health and growth rate in these young pigs.
Source: Adapted from the suggested SEW diet of Tokach, Dritz, Goodband, and Nelssen (1997). Use of this material is credited to S. S. Dritz, R. D. Goodband, J. L. Nelssen, and M. D. Tokach, Swine Nutrition Guide, Kansas State University, Manhattan, KS.

Phase-feeding programs are often used for early and conventionally weaned pigs. Tokach et al. (1997) provided guidelines for phase feeding based on changes in body weight of the piglet. Weaned pigs weighing 5 to 11 lb would be fed a diet such as the starter diet shown in Table 10–7, whereas those weighing 11 to 15 lb, 15 to 25 lb, and 25 to 50 lb would be fed phase 1, phase 2, and phase 3 diets, respectively. Diets in each phase are of similar composition, but have slight decreases in metabolizable energy, protein, calcium, and phosphorus concentrations with each successive phase. These diets and the feeding systems in which they are used are discussed further in Chapter 12.

NURSERY

The nursery period is a critical stage because it represents the transition from dependence on the sow for food and the transfer of this responsibility to people. Not only is the social structure completely changed, but the diet composition is drastically changed from sow milk to foreign, largely plant sources of nutrient. It is important to provide a highly palatable, nutritious diet to encourage early adequate consumption of the diet to minimize the "growth check" associated with the stress of weaning. To achieve this goal, pre-starter diets usually contain a high proportion of milk products and other animal protein sources, such as blood plasma protein and fish meal, all of which are expensive ingredients. The nursery period usually lasts to 10 wk of age, which corresponds to a body weight of 35 to 50 lb/pig.

GROWING

The daily weight gain of the grower pig continues to increase beyond 100 lb (45.4 kg) body weight. Daily accretion of body protein increases progressively (maximum of up to 140 g/d) to about 135 lb (61 kg) body weight, then gradually declines to about 110 g/d at 250 lb (114 kg). Thus, the protein content of the diet is reduced as the pig grows so as to ensure adequate protein deposition in the body but avoid protein intake above the requirement. This reduction in protein content of the diet is done in a step-feeding sequence involving one or more diet changes during the growing period and additional reductions during finishing. Excess dietary protein represents an extra cost because it is used for energy, which is stored as body fat. In contrast to the shape of the protein accretion curve, daily fat accretion continues to increase throughout the growing and finishing periods to slaughter at 114 to 123 kg (250 to 260 or 270 lb). Body fat is synthesized in the body from carbohydrate and from protein not deposited as body protein.

The dietary protein requirement is directly related to the essential amino acid requirements. Lysine is the first limiting amino acid in most feedstuffs for swine. For this reason, it has become customary to express all other essential amino acid requirements on the basis of the lysine requirement set at 100. The National Research Council (1998) has described, based on modeling procedures, the calculation of individual amino acid requirements of grower pigs. (Similar formulas have been developed for finisher pigs, pregnant sows, and lactating sows.) Amino acid requirements are usually expressed as g/megacalorie (Mcal) or kilocalorie (Kcal) of DE. Nutrient requirements of both grower and finisher pigs are based on the assumption that the diet is full-fed (free access).

FINISHING

The nutrient requirements of pigs during the finishing period are similar to those during the growing period, except that protein and amino acid concentrations required are less than those for growing pigs because of the lower daily protein accretion rate in finisher pigs. One or more steps downward in the diet's protein content are made during the finishing period.

Protein requirements of gilts and intact males are higher than those of barrows. For this reason, split-sex feeding during the growing and finishing periods is used to match protein intake with protein requirements, thereby reducing feed costs. Lean genotypes need a higher percentage of dietary protein than fatter genotypes. Therefore, pigs of differing genetic capacity for lean carcasses should be penned separately and fed diets containing protein levels appropriate for their protein requirement. Pigs lacking the genetic capacity to produce highly lean carcasses use excess protein for energy and fattening, an uneconomical process.

BREEDING (BOARS AND SOWS)

Gilts and boars selected for breeding stock are typically full-fed adequate diets during each of the stages of growth through the finishing period [250 to 270 lb (113.5 to 122.6 kg) live weight at 6 mo] to maximize growth rate and provide a means of identifying individuals with the genetic capacity for lean growth.

Boars and gilts selected for breeding are limited-fed from 6 mo of age to breeding at about 8 mo of age to prevent over-fattening prior to breeding. Sows to be rebred post-weaning are similarly limit-fed from weaning to rebreeding (usually 5 to 7 d following a 2- to 3-wk lactation). The degree of restricted feeding is dictated by the body fatness of the sow following weaning.

PREGNANCY

In gestating sows, feed restriction is used to control weight gain and over-fattening. The amount of DE provided daily is determined by the amount of weight gain desired, based on the sow's body weight at breeding and the expected litter size. The concentrations of amino acids and other nutrients needed in the diet to meet requirements are based on the amount of daily DE to be fed.

POST-BREEDING SOWS

The period between weaning and rebreeding represents a short, but critical, time in the life cycle because of the importance of prompt post-weaning resumption of the estrous cycle. The severe drain on body fatness and nutrient reserves must be reversed to promote conception, yet feed intake should be restricted at the time of weaning to reduce milk flow. The post-breeding feeding schedule, therefore, must be tailored to the nutritional status of the sow at the end of a particular lactation period to enhance subsequent reproductive success.

LACTATION

The estimation of the nutrient requirements of lactating sows is difficult because sows do not consume enough feed to meet the large amount of nutrients needed to maintain maximum milk production and adequate body fat. Weight loss during lactation is therefore common. Extremely thin sows result from inadequate energy intake during lactation. This may jeopardize reproductive performance in the subsequent reproductive cycle. A high prevalence of reproductive failure and "downer syndrome" in sows may occur in herds in which lactation energy needs are not met. Daily energy requirement may be met by increasing energy density of the diet with the addition of fat or by otherwise stimulating intake, such as by adjusting environmental temperature. The total daily energy requirement is determined by the sum of the requirements for body tissue maintenance, milk production, and maintenance of body temperature in response to extremes in environment. As in other stages of the life cycle, dietary amino acids and other nutrient needs are met by formulating the diet of lactating sows on the basis of the levels of amino acids and other nutrients in relation to energy density of the diet (g/Mcal DE).

Nutrient requirements of pigs in all stages of the life cycle are listed in Nutrient Requirements of Swine, 10th ed., published by the National Research Council (NRC, 1998).

SUMMARY

Feed costs represent 50% to 85% of the total cost of swine production. For this reason, adequate nutrition of the pig throughout the life cycle is essential for a profitable pork production enterprise. This chapter discussed the nutrients and their physiological functions, the digestive physiology of the pig, and life-cycle feeding. The six broad classes of nutrients (water, carbohydrates, lipids, proteins, mineral elements, and vitamins) include more than 40 specific nutrients, each with one or more specific physiological functions. The digestive system of the pig is well adapted to efficient utilization of nutrients for growth, reproduction, and lactation. The anatomy and physiology of the gastrointestinal tract in relation to preparation for digestion and utilization of each nutrient and for the release of feed energy (expressed as digestible or metabolizable calories) for each productive function was briefly reviewed.

The concept of life-cycle feeding, in which specific nutritional needs at each stage of life is described, was addressed for each stage: suckling, nursery, starter, growing, finishing, pregnancy, lactation, breeding (boars and sows), and post-breeding sows. Recent dramatic increases in size of individual pork production units have fueled the evolution of several important management practices related to life-cycle feeding, including all-in-all-out production schedules, segregated early weaning, phase feeding, and split-sex feeding.

QUESTIONS AND ACTIVITIES

1. A pork producer called and the specialist went to the farm for a visit. Sows were starting labor in the farrowing crates and 30–60 min later labor would stop and the remaining pigs would either have to be pulled by hand or they would die.

Upon examination of the diets, the farmer had not included a mineral supplement for 6 months as a cost savings measure. Which nutrients may have contributed to the acute problem experienced by the sow? If an injection were to be made, what would you inject to overcome the short-term nutrient need?

2. Certain fatty acids are required for synthesis of prostaglandins. Which fatty acid is required for synthesis of PGF2?

3. If you were to take all of the amino acids required by the NRC and put purified amino acids in the diet, how much total amino acids (equivalent to the % crude protein) would be in a diet for a 100 kg pig intended to be fed 13.8% crude protein? Does this mean we overfeed nitrogen in conventional diets? What happens to that excess nitrogen?

4. Piglets nursing sows indoors on slotted floors are often given an injection of iron. Why? What would be symptoms that a pig would show if the iron injection were accidentally missed in that pig? How could you confirm the accident with a blood sample? That is, what would you measure in the blood?

5. The bioavailability of P is low in grain and soybean meal. Phytase is an enzyme that can be added to the pig's diet to make some of the plant P more available to the pig. How much can this lower the need for P in the diet and how much can phytase lower P in the manure?

6. The lysine requirement is much higher for a high-lean-gain genetic line of pig than for older lines of pigs. Can you find information in the literature to suggest how high-lean and conventional lines might differ in their response to dietary lysine?

LITERATURE CITED

Burrin, D. G. 2001. Nutrient requirements and metabolism. In: W. G. Pond and H. J. Mersmann (eds.) Biology of the Domestic Pig. Chapter 7, pp. 309–389. Cornell University Press, Ithaca, NY.

Close, W. H. and D. J. A. Cole. 2000. Nutrition of Sows and Boars. Nottingham University Press, Nottingham, UK.

Cole, D. J. A., W. Haresign, and P. C. Garnsworthy. 2000. Recent Developments in Pig Nutrition 2. Nottingham University Press, Nottingham, UK.

Garnsworthy, P. C. and J. Wiseman, 2001. Recent Developments in Pig Nutrition 3. Nottingham University Press, Nottingham, UK.

Kellems, R. O. and D. C. Church. 2001. Livestock Feeds and Feeding, 5th ed. Prentice-Hall, New York.

Klobasa, F., E. Werhahn, and J. E. Butler. 1987. Composition of sow milk during lactation. J. Anim. Sci. 64:1458–1466.

Lewis, A. J. and L. L. Southern (eds.). 2001. Swine Nutrition. 2nd ed. CRC Press, Boca Raton, FL.

Miller, E. R., D. E. Ullrey, and A. J. Lewis. 1991. Swine Nutrition. Butterworth-Heinemann, New York.

National Research Council. 1998. Nutrient Requirements of Swine. 10th rev. ed. National Academy Press, Washington, DC.

National Research Council. 1980. Mineral Tolerance of Domestic Animals. National Academy Press, Washington, DC.

Pond, W. G., J. H. Maner, and D. L. Harris. 1991. Pork Production Systems. Van Nostrand Reinhold, New York.

Pond, W. G., D. C. Church, and K. R. Pond. 1995. Basic Animal Nutrition and Feeding. John Wiley & Sons, Inc., New York.

Tokach, M. D., S. S. Dritz, R. D. Goodband, and J. L. Nelssen. 1997, October. Starter pig recommendations. MF2300, Swine Nutrition Guide, Kansas State University, Manhattan, KS.

Yen, J. T. 2001. Digestive system. In: W. G. Pond and H. J. Mersmann eds. Biology of the Domestic Pig, Chapter 8, pp. 390–453. Cornell University Press, Ithaca, NY.

INTERNET RESOURCES

Nutrient Requirements of Swine, Tenth Revised Edition, 1998:
http://www.nap.edu/catalog/6016.html

Kansas State University ASI Swine Information:
http://www.oznetksu.edu/dp_ansci/swine/swine.htm

Nebraska Extension Swine Nutrition Guide, EC95-273:
http://ianr.unl.edu/PUBSswine/ec273.htm

Tri-State Swine Nutrition Guide, Bulletin 869-98:
http://ohioline.ag.ohio-state.edu/b689/b689_83.html

Swine Nutrition at Oklahoma State University:
http://www.ansi.okstate.edu/course/4643/nutr.htm

Swine Nutrition at Iowa State University:
http://ans.iastate.edu/current/holden_ASAS.html

Swine-Nutrition, OMAFRA Livestock:
http://www.gov.on.ca/OMAFRA/english/livestock/swine/nutritio.html

Babcock's Nutrition Guidelines:
http://www.babcockswine.com/nutritio.htm

Moorman's Nutrition Guidelines:
http://www.moormans.com

Purina Mills Nutrition Guidelines:
http://www.purinamills.com

11
FEED RESOURCES

INTRODUCTION

Hundreds of feedstuffs are available worldwide from which to formulate balanced diets for all stages of the pig's life cycle. Detailed information on feedstuff composition and characterization is available from several sources, including Morrison's Feeds and Feeding (Morrison, 1956), which was widely used throughout the first half of the 20th century and contains an enormous amount of information still useful today. Other information on feedstuffs for swine is available in FAO United Nation, Animal Feed Resources Information System (2001); Lewis and Southern (2001); Miller et al. (1991); the National Research Council (1998); Patience and Thacker (1989); and Pond and Maner (1974, 1984). As globalization of pork production and marketing continues, an increase can be expected in the use of a variety of nontraditional feedstuffs from around the world. Myer and Brendemuhl (2001) reviewed the nature and composition of an array of some of these miscellaneous feedstuffs. This chapter briefly describes the broad classes of feedstuffs for swine: energy sources, protein sources, and vitamin and mineral sources. The composition and special characteristics of some of the most commonly used feedstuffs within each class are also discussed.

ENERGY SOURCES

Carbohydrates are the most abundant source of energy in plant materials. Plant sources of carbohydrates are the most widely used energy sources for swine. In addition to the energy feedstuffs discussed here, many by-products of agriculture and industry have been underutilized as feed resources despite their potential value. These by-products include food processing wastes, nonfood industrial wastes, forest residues, animal wastes, crop residues, and aquatic plants. Future use of these resources in swine feeding can be expected to increase in a world with increased population and finite resources.

GRAINS

Grains and their by-products are the most important sources of energy in swine feeding. Like almost all feedstuffs, they also contribute significant amounts of proteins, minerals, and vitamins. Corn (maize) and grain sorghum (milo) are the most common grains used in pork production in the United States. Wheat, barley, oats, rye, and rice are in demand for direct human consumption in many parts of the world, but significant amounts are also used in pig diets in some areas of the United States and in other countries. As a group, corn and the cereal grains are high in starch, low in fiber, and low in protein (8% to 15%). Lysine is the most limiting amino acid, followed by tryptophan and threonine. Table 11–1 shows the partial chemical composition of corn and some of the commonly used grains and other energy sources in pig diet formulation. Other grains used less commonly for swine include triticale (cross between wheat and rye), buckwheat (not a true grain), and grain amaranth. Information on the chemical composition of energy sources and protein supplements used in swine feeding in the USA is available in the National Research Council publication, Nutrient Requirements of Swine, 10th ed. (NRC, 1998). Tables 11–2a and b contain chemical composition and amino acid composition of many feedstuffs (National Research Council, 1998). A vast amount of worldwide animal feed information can be obtained on the Internet (FAO United Nations, Animal Feed Resources Information System, 2001) (www.fao.org).

TABLE 11–1
Composition of Barley, Corn, Grain Sorghum, Oats, Rice, Rye, and Wheat.

Grain	Dry Matter %	DE kcal/kg	Protein %	Ca %	P total/ %	P Bioav %	Vit. E mg/kg	Beta Carotene mg/kg	Lys %	Thr %	Try %
Barley	89	3050	11.3	0.06	0.35	30	7.4	4.1	0.41	0.35	0.11
Corn	89	3525	8.3	0.03	0.28	14	8.3	8.0	0.17	0.29	0.06
Grain Sorghum	89	3380	9.2	0.03	0.29	20	5.0	NA*	0.22	0.31	0.10
Oats	89	2770	11.5	0.07	0.31	22	7.0	3.7	0.40	0.44	0.14
Rice	89	3565	7.9	0.04	0.18	NA	2.0	NA	0.30	0.26	0.10
Rye	88	3270	11.8	0.06	0.33	NA	9.0	NA	0.38	0.32	0.12

*NA = no data available.
Source: National Research Council (1998). Nutrient Requirements of Swine, 10th Revised Edition.

TABLE 11–2a
Chemical Composition of Some Feed Ingredients Commonly Used for Swine (data on as-fed basis).[a]

Entry Number	Description	International Feed Number[b]	Dry Matter (%)	DE (kcal/kg)	ME (kcal/kg)	NE (kcal/kg)	Crude Protein (%)	Crude Fat (%)	Linoleic Acid (%)	NDF (%)	ADF (%)	Calcium (%)	Phosphorus (%)	Bioavailability of Phosphorus[c] (%)
	Alfalfa													
01	meal dehydrated, 17% CP	1-00-023	92	1,830	1,650	910	17.0	2.6	0.35	41.2	30.2	1.53	0.26	100
02	meal dehydrated, 20% CP	1-00-024	92	2,095	1,885	1,290	19.6	3.3	0.44	38.8	26.4	1.61	0.28	—
	Bakery Waste													
03	dried bakery product	4-00-466	91	3,940	3,700	2,415[d]	10.8	11.3	5.70	2.0	1.3	0.13	0.25	—
	Barley													
04	grain, two row	4-00-572	89	3,050	2,910	2,340	11.3	1.9	0.88	18.0	6.2	0.06	0.35	—
05	grain, six row	4-00-574	89	3,050	2,910	2,310	10.5	1.9	0.91	18.6	7.0	0.06	0.36	30
06	grain, hulless	4-00-552	88	3,360	3,320	2,650	14.9	2.1	1.14	10.1	2.2	0.04	0.45	—
	Beet, Sugar													
07	pulp, dried	4-00-669	91	2,865	2,495	1,860	8.6	0.8	—	42.4	24.3	0.70	0.10	—
	Blood													
08	meal, conventional	5-00-380	92	2,850	2,350	1,950	77.1	1.6	0.09	13.6	1.8	0.37	0.27	—
09	meal, flash dried	5-26-006	92	2,300	1,950	1,385[d]	87.6	1.6	—	—	—	0.21	0.21	—
10	meal, spray or ring dried	5-00-381	93	3,370	2,945	2,070	88.8	1.3	0.17	—	—	0.41	0.30	92
11	plasma, spray dried[e]	—	91	—	—	—	78.0	2.0	—	—	—	0.15	1.71	—
12	cells, spray dried[e]	—	92	—	—	—	92.0	1.5	—	—	—	0.02	0.37	—
	Brewers' Grain													
13	dried	5-02-141	92	2,100	1,960	1,630	26.5	7.3	3.14	48.7	21.9	0.32	0.56	34
	Buckwheat, Common													
14	grain	4-00-994	88	2,825	2,640	1,620	11.1	2.4	0.53	17.8	14.3	0.09	0.31	—
	Canola (Rapeseed)													
15	meal, sol. extr.	5-06-145	90	2,885	2,640	1,610	35.6	3.5	0.42	21.2	17.2	0.63	1.01	21
	Casein													
16	dried	5-01-162	91	4,135	3,535	2,555	88.7	0.8	0.03	—	—	0.61	0.82	—
	Cassava (Tapioca or Manioc)													
17	meal, dehydrated	4-01-152	88	3,385	3,330	2,330	3.3	0.5	—	7.7	4.6	0.22	0.13	—
	Coconut (Copra)													
18	meal, sol. extr.	5-01-573	92	3,010	2,565	1,695	21.9	3.0	0.03	51.3	25.5	0.16	0.58	—

#	Feed	IFN	%											
	Corn, Yellow													
19	distillers' grain	5-02-842	94	3,100	2,715	1,170[d]	24.8	7.9	4.46	40.4	17.5	0.10	0.40	—
20	distillers' grain with solubles	5-02-843	93	3,200	2,820	2,065	27.7	8.4	2.15	34.6	16.3	0.20	0.77	7721
21	distillers' solubles	5-02-844	92	3,325	2,945	2,250	26.7	9.1	5.36	24.8	7.5	0.29	1.03	—
22	gluten feed	5-02-903	90	2,990	2,605	1,740	21.5	3.0	1.43	33.3	10.7	0.22	0.83	59
23	gluten meal, 60% CP	5-28-242	90	4,225	3,830	2,550	60.2	2.9	1.17	8.7	4.6	0.05	0.44	15
24	grain	4-02-935	89	3,525	3,420	2,395	8.3	3.9	1.92	9.6	2.8	0.03	0.28	14
25	grits by-product (Hominy Feed)	4-03-011	90	3,355	3,210	2,260	10.3	6.7	2.97	28.5	8.1	0.05	0.43	14
	Cottonseed													
26	meal, mech. extr. 41% CP	5-01-617	92	2,945	2,690	1,870	42.4	6.1	3.15	25.7	18.0	0.23	1.03	—
27	meal, sol. extr. 41% CP	5-07-872	90	2,575	2,315	1,325	41.4	1.5	0.51	28.4	19.4	0.19	1.06	1
	Fababean (Broadbean)													
28	seeds	5-09-262	87	3,245	3,045	2,000	25.4	1.4	0.62	13.7	9.7	0.11	0.48	—
	Feather													
29	meal, hydrolyzed	5-03-795	93	2,990	2,485	2,250	84.5	4.6	0.83	—	—	0.33	0.50	31
	Fish													
30	Anchovy meal, mech. extr.	5-01-985	92	3,230	2,695	1,695[d]	64.6	7.9	0.27	—	—	3.93	2.55	—
31	Herring meal, mech. extr.	5-02-000	93	3,960	3,260	2,020	68.1	9.2	0.15	—	—	2.40	1.76	—
32	Menhaden meal, mech. extr.	5-02-009	92	3,770	3,360	2,335	62.3	9.4	0.12	—	—	5.21	3.04	94
33	White meal, mech. extr.	5-02-025	91	3,395	2,810	2,020	63.3	4.8	0.08	—	—	6.65	3.59	—
34	solubles, condensed	5-01-969	51	1,910	1,625	995[d]	32.7	5.6	—	—	—	0.22	0.59	—
35	solubles, dried	5-01-971	92	3,310	3,045	1,770	64.2	7.4	0.12	—	—	0.55	1.25	—
	Flax (Linseed)													
36	meal, sol. extr.	5-02-048	90	3,060	2,710	1,840	33.6	1.8	0.36	23.9	15.0	0.39	0.83	—
	Lentil													
37	seeds	5-02-506	89	3,540	3,450	2,205	24.4	1.3	0.41	10.1	5.4	0.10	0.38	—
	Lupin (Sweet White)													
38	seeds	5-27-717	89	3,450	3,305	2,130	34.9	9.2	1.62	20.3	16.7	0.22	0.51	—
	Meat													
39	meal rendered	5-00-385	94	2,695	2,595	2,175	54.0	12.0	0.80	31.6	8.3	7.69	3.88	—
40	meal rendered with bone	5-00-388	93	2,440	2,225	1,355	51.5	10.9	0.72	32.5	5.6	9.99	4.98	90[f]
	Milk (Cattle)													
41	skim, dried	5-01-175	96	3,980	3,715	2,360	34.6	0.9	0.01	—	—	1.31	1.00	91
	Millet (Proso)													
42	grain	4-03-120	90	3,020	2,950	2,095	11.1	3.5	1.92	15.8	13.8	0.03	0.31	—

(continued)

TABLE 11–2a (continued)

Entry Number	Description	International Feed Number[b]	Dry Matter (%)	Calcium (%)	Phosphorus (%)	Sodium (%)	Chlorine (%)	Potassium (%)	Magnesium (%)	Sulfer (%)	Copper (mg/kg)	Iron (mg/kg)	Manganese (mg/kg)	Selenium[c] (mg/kg)	Zinc (mg/kg)
	Oat														
43	grain	4-03-309	89	0.07	0.31	0.08	0.10	0.42	0.16	0.21	6	85	43	0.30	38
44	grain, naked	4-25-101	86	0.08	0.38	0.02	0.11	0.36	0.12	0.14	4	58	37	0.09	34
45	groat	4-03-331	90	0.08	0.41	0.05	0.09	0.38	0.11	0.20	6	49	32	—	—
	Pea														
46	seeds	5-03-600	89	0.11	0.39	0.04	0.05	1.02	0.12	0.20	9	65	23	0.38	23
	Peanut (Groundnut)														
47	meal, mech. extr.	5-03-649	92	0.17	0.59	0.06	0.03	1.20	0.33	0.29	15	285	39	0.28	47
48	meal, sol. extr.	5-03-650	92	0.22	0.65	0.07	0.04	1.25	0.31	0.30	15	260	40	0.21	41
	Potato														
49	protein concentrate	5-25-392	91	0.17	0.19	0.03	0.20	0.80	0.05	0.23	13	40	5	1.00	25
	Poultry														
50	by-product, meal rendered	5-03-798	93	4.46	2.41	0.49	0.49	0.53	0.18	0.52	10	442	9	0.88	94
	Rice														
51	bran														
52	grain, polished and broken (Brewers' Rice)	4-03-928	90	0.07	1.61	0.03	0.07	1.56	0.90	0.18	9	190	228	0.40	30
53	polishings	4-03-932	89	0.04	0.18	0.04	0.07	0.13	0.11	0.06	21	18	12	0.27	17
		4-03-943	90	0.09	1.18	0.06	0.11	1.11	0.65	0.17	6	160	12	—	26
	Rye														
54	grain	4-04-047	88	0.06	0.33	0.02	0.03	0.48	0.12	0.15	7	60	58	0.38	31
	Safflower														
55	meal, sol. extr.	5-04-110	92	0.34	0.75	0.05	0.08	0.76	0.35	0.13	10	495	18	—	41
56	meal without hulls, sol. extr.	5-07-959	92	0.37	1.31	0.04	0.16	1.00	1.02	0.20	9	484	39	—	33
	Sesame														
57	meal, mech. extr.	5-04-220	93	1.90	1.22	0.04	0.07	1.10	0.54	0.56	34	93	53	0.21	100
	Sorghum														
58	grain	4-20-893	89	0.03	0.29	0.01	0.09	0.35	0.15	0.08	5	45	15	0.20	15

#	Feed	Code												
	Soybean													
59	meal, sol. extr.	5-04-604	89	0.32	0.65	0.01	0.05	1.96	0.27	0.43	20	202	29	0.32	50
60	meal without hulls, sol. extr.	5-04-612	90	0.34	0.69	0.02	0.05	2.14	0.30	0.44	20	176	36	0.27	55
61	protein concentrate	—	90	0.35	0.81	0.05	—	2.20	0.32	—	13	110	—	—	30
62	protein isolate	5-08-038	92	0.15	0.65	0.07	0.02	0.27	0.08	0.71	14	137	5	0.14	34
63	seeds, heat processed	5-04-597	90	0.25	0.59	0.03	0.03	1.70	0.28	0.30	16	80	30	0.11	39
	Sunflower														
64	meal, sol. extr.	5-09-340	90	0.36	0.86	0.02	0.10	1.07	0.68	0.30	26	254	41	0.50	66
65	meal without hulls, sol. extr.	5-04-739	93	0.37	1.01	0.04	0.13	1.27	0.75	0.38	25	200	35	0.32	98
	Triticale														
66	grain	4-20-362	90	0.05	0.33	0.03	0.03	0.46	0.10	0.15	8	31	43	—	32
	Wheat														
67	bran	4-05-190	89	0.16	1.20	0.04	0.07	1.26	0.52	0.22	14	170	113	0.51	100
68	grain, hard red spring	4-05-258	88	0.05	0.36	0.02	0.09	0.41	0.16	0.17	7	64	42	0.30	43
69	grain, hard red winter	4-05-268	88	0.06	0.37	0.01	0.06	0.49	0.13	0.15	6	39	34	0.33	40
70	grain, soft red winter	4-05-294	88	0.04	0.39	0.01	0.08	0.46	0.11	0.16	8	32	38	0.28	47
71	grain, soft white winter	4-05-337	89	0.05	0.35	0.01	0.07	0.44	0.15	0.18	7	60	37	0.26	28
72	middlings, < 9.5% fiber	4-05-205	89	0.12	0.93	0.05	0.04	1.06	0.41	0.17	10	84	100	0.72	92
73	red dog, < 4% fiber	4-05-203	88	0.07	0.57	0.04	0.10	0.63	0.16	0.24	6	46	55	0.30	65
74	shorts, < 7% fiber	4-05-201	88	0.09	0.84	0.02	0.04	1.06	0.25	0.20	12	100	89	0.75	100
	Whey														
75	dried	4-01-182	96	0.75	0.72	0.94	1.40	1.96	0.13	0.72	13	130	3	0.12	10
76	low lactose, dried	4-01-186	96	2.00	1.37	1.85	3.43	4.68	0.25	1.59	3	85	8	0.06	11
77	permeate, dried	—	96	0.86	0.66	1.00	2.23	2.10	—	—	—	—	—	—	—
	Yeast, Brewers'														
78	dried	7-05-527	93	0.16	1.44	0.10	0.12	1.80	0.23	0.40	33	215	8	1.00	49
	Yeast, Torula														
79	dried	7-05-534	93	0.58	1.52	0.07	0.12	1.94	0.20	0.55	17	222	13	0.02	99

[a] Dash indicates that no data were available.
[b] First digit is class of feed: 1, dry forages and roughages; 2, pasture, range plants, and forages fed green; 3, silages; 4, energy feeds; 5, protein supplements; 6, minerals; 7, vitamins; 8, additives; the other five digits are the International Feed Number.
[c] Selenium values are extremely dependent on soil conditions and some values may differ substantially from those presented here.

Source: National Research Council (1998) with permission (Table 11–1).

TABLE 11-2b
Amino Acid Composition of Some Feed Ingredients Commonly Used for Swine (data on as-fed basis).[a]

Entry Number	Description	International Feed Number[b]	Dry Matter (%)	Crude Protein (%)	Argi- nine (%)	His- ti- dine (%)	Iso- leu- cine (%)	Leu- cine (%)	Lysine (%)	Me- thi- onine (%)	Cys- tine (%)	Phenyl ala- nine (%)	Tyro- sine (%)	Thre- onine (%)	Tryp- to- phan (%)	Valine (%)
	Alfalfa															
01	meal dehydrated, 17% CP	1-00-023	92	17.0	0.71	0.37	0.68	1.21	0.74	0.25	0.18	0.84	0.55	0.70	0.24	0.86
02	meal dehydrated, 20% CP	1-00-024	92	19.6	0.91	0.38	0.89	1.40	0.90	0.34	0.26	0.93	0.60	0.82	0.35	1.05
	Bakery Waste															
03	dried bakery product	4-00-466	91	10.8	0.46	0.24	0.38	0.80	0.27	0.18	0.23	0.50	0.36	0.33	0.10	0.46
	Barley															
04	grain, two row	4-00-572	89	11.3	0.54	0.25	0.39	0.77	0.41	0.20	0.28	0.55	0.29	0.35	0.11	0.52
05	grain, six row	4-00-574	89	10.5	0.48	0.22	0.37	0.68	0.36	0.17	0.20	0.49	0.32	0.34	0.13	0.49
06	grain, hulless	4-00-552	88	14.9	0.56	0.23	0.41	0.77	0.44	0.16	0.24	0.61	0.40	0.40	0.13	0.55
	Beet, Sugar															
07	pulp, dried	4-00-669	91	8.6	0.32	0.23	0.31	0.53	0.52	0.07	0.06	0.30	0.40	0.38	0.10	0.45
	Blood															
08	meal, conventional	5-00-380	92	77.1	3.34	5.06	0.91	10.99	7.04	0.99	1.09	5.34	2.29	4.05	1.08	7.05
09	meal, flash dried	5-26-006	92	87.6	3.37	4.57	0.88	11.48	7.56	0.95	1.20	6.41	2.32	4.07	1.06	8.03
10	meal, spray or ring dried	5-00-381	93	88.8	3.69	5.30	1.03	10.81	7.45	0.99	1.04	5.81	2.71	3.78	1.48	7.03
11	plasma, spray dried	—	92	78.0	4.55	2.55	2.71	7.61	6.84	0.75	2.63	4.42	3.53	4.72	1.36	4.94
12	cells, spray dried	—	92	92.0	3.77	6.99	0.49	12.70	8.51	0.81	0.61	6.69	2.14	3.38	1.37	8.50
	Brewers' Grains															
13	dried	5-02-141	92	26.5	1.53	0.53	1.02	2.08	1.08	0.45	0.49	1.22	0.88	0.95	0.26	1.26
	Buckwheat, Common															
14	grain	4-00-994	88	11.1	0.92	0.25	0.40	0.64	0.57	0.19	0.23	0.45	0.31	0.41	0.17	0.56
	Canola (Rapeseed)															
15	meal, sol. extr.	5-06-145	90	35.6	2.21	0.96	1.43	2.58	2.08	0.74	0.91	1.43	1.13	1.59	0.45	1.82
	Casein															
16	dried	5-01-162	91	88.7	3.26	2.82	4.66	8.79	7.35	2.70	0.41	4.79	4.77	3.98	1.14	6.10
	Cassava (Tapioca or Manioc)															
17	meal	4-01-152	88	3.3	0.18	0.08	0.11	0.19	0.12	0.04	0.05	0.15	0.04	0.11	0.04	0.14
	Coconut (Copra)															
18	meal, sol. extr.	5-01-573	92	21.9	2.38	0.39	0.75	1.36	0.58	0.35	0.29	0.84	0.58	0.67	0.19	1.07
	Corn, Yellow															
19	distillers' grain	5-02-842	94	24.8	0.90	0.63	0.95	2.63	0.74	0.43	0.28	0.99	0.82	0.62	0.20	1.24
20	distillers' grain with solubles	5-02-843	93	27.7	1.13	0.69	1.03	2.57	0.62	0.50	0.52	1.34	0.83	0.94	0.25	1.30
21	distillers' solubles	5-02-844	92	26.7	0.90	0.66	1.21	2.25	0.82	0.51	0.46	1.38	0.80	1.03	0.23	1.50
22	gluten feed	5-02-903	90	21.5	1.04	0.67	0.66	1.96	0.63	0.35	0.46	0.76	0.58	0.74	0.07	1.01
23	gluten meal, 60% CP	5-28-242	90	60.2	1.93	1.28	2.48	10.19	1.02	1.43	1.09	3.84	3.25	2.08	0.31	2.79
24	grain	4-02-935	89	8.3	0.37	0.23	0.28	0.99	0.26	0.17	0.19	0.39	0.25	0.29	0.06	0.39
25	grits by-product (Hominy Feed)	4-03-011	90	10.3	0.56	0.28	0.36	0.98	0.38	0.18	0.18	0.43	0.40	0.40	0.10	0.52

#	Feed	Code														
	Cottonseed															
26	meal, mech. extr. 41% CP	5-01-617	92	42.4	4.26	1.11	1.29	2.45	1.65	0.67	0.69	1.97	1.23	1.34	0.54	1.76
27	meal, sol. extr. 41% CP	5-07-872	90	41.4	4.55	1.17	1.30	2.47	1.72	0.67	0.70	2.20	1.22	1.36	0.48	1.78
	Fababean (Broadbean)															
28	seeds	5-09-262	87	25.4	2.28	0.67	1.03	1.89	1.62	0.20	0.32	1.03	0.87	0.89	0.22	1.14
	Feather															
29	meal, hydrolyzed	5-03-795	93	84.5	5.62	0.93	3.86	6.79	2.08	0.61	4.13	4.01	2.41	3.82	0.54	5.88
	Fish															
30	Anchovy meal, mech. extr.	5-01-985	92	64.6	3.68	1.56	3.06	5.00	5.11	1.95	0.61	2.66	2.15	2.82	0.76	3.51
31	Herring meal, mech. extr.	5-02-000	93	68.1	4.01	1.52	2.91	5.20	5.46	2.04	0.66	2.75	2.18	3.02	0.74	3.46
32	Menhaden meal, mech. extr.	5-02-009	92	62.9	3.66	1.78	2.57	4.54	4.81	1.77	0.57	2.51	2.04	2.64	0.66	3.03
33	White meal, mech. extr.	5-02-025	91	63.3	4.04	1.34	2.61	4.39	4.51	1.76	0.68	2.32	2.03	2.60	0.66	3.06
34	solubles, condensed	5-01-969	51	32.7	1.61	1.56	1.06	1.86	1.73	0.50	0.30	0.93	0.40	0.86	0.31	1.16
35	solubles, dried	5-01-971	92	64.2	2.67	1.23	1.56	2.68	2.84	0.98	0.49	1.22	0.62	1.40	0.34	1.94
	Flax (Linseed)															
36	meal sol. extr.	5-02-048	90	33.6	2.97	0.68	1.56	2.06	1.24	0.59	0.59	1.57	1.03	1.26	0.52	1.74
	Lentil															
37	seeds	5-02-506	89	24.4	2.05	0.78	1.00	1.84	1.71	0.18	0.27	1.29	0.70	0.84	0.21	1.27
	Lupin (Sweet White)															
38	seeds	5-27-717	89	34.9	3.38	0.77	1.40	2.43	1.54	0.27	0.51	1.22	1.35	1.20	0.26	1.29
	Meat															
39	meal rendered	5-00-385	94	54.0	3.60	1.14	1.60	3.84	3.07	0.80	0.60	2.17	1.40	1.97	0.35	2.66
40	meal rendered with bone	5-00-388	93	51.5	3.45	0.91	1.34	2.98	2.51	0.68	0.50	1.62	1.07	1.59	0.28	2.04
	Milk (Cattle)															
41	skim, dried	5-01-175	96	34.6	1.24	1.05	1.87	3.67	2.86	0.92	0.30	1.78	1.87	1.62	0.51	2.33
	Millet (Proso)															
42	grain	4-03-120	90	11.1	0.41	0.20	0.46	1.24	0.23	0.31	0.18	0.56	0.31	0.40	0.16	0.57
	Oat															
43	grain	4-03-309	89	11.5	0.87	0.31	0.48	0.92	0.40	0.22	0.36	0.65	0.41	0.44	0.14	0.66
44	grain, naked	4-25-101	86	17.1	0.77	0.26	0.48	0.86	0.47	0.19	0.32	0.60	0.42	0.40	0.16	0.63
45	groat	4-03-331	90	13.9	0.85	0.24	0.55	0.98	0.48	0.20	0.22	0.66	0.51	0.44	0.18	0.72
	Pea															
46	seeds	5-03-600	89	22.8	1.87	0.54	0.86	1.51	1.50	0.21	0.31	0.98	0.71	0.78	0.19	0.98

(continued)

TABLE 11-2b (continued)

Entry Matter Number	Description	International Feed Number[b]	Dry (%)	Crude Protein (%)	Agri- nine (%)	His- ti- dine (%)	Iso- leu- cine (%)	Leu- cine (%)	Lysine (%)	Me- thi- onine (%)	Cys- tine (%)	Phenyl ala- nine (%)	Tyro- sine (%)	Thre- onine (%)	Tryp- to- phan (%)	Valine (%)
	Peanut (Groundnut)															
47	meal, mech. extr.	5-03-649	92	43.2	4.79	1.01	1.41	2.77	1.48	0.50	0.60	2.02	1.74	1.16	0.41	1.70
48	meal, sol. extr.	5-03-650	92	49.1	5.09	1.06	1.78	2.83	1.66	0.52	0.69	2.35	1.80	1.27	0.48	1.98
	Potato															
49	protein concentrate	5-25-392	91	73.8	3.80	1.71	4.09	7.61	5.83	1.68	1.20	4.89	4.27	4.30	1.02	4.89
	Poultry															
50	by-product, meal rendered	5-03-798	93	64.1	3.94	1.25	2.01	3.89	3.32	1.11	0.65	2.26	1.56	2.18	0.48	2.51
	Rice															
51	bran grain, polished + broken	4-03-928	90	13.3	1.00	0.34	0.44	0.92	0.57	0.26	0.27	0.56	0.40	0.48	0.14	0.68
52	(Brewers' Rice)	4-03-932	89	7.9	0.52	0.18	0.34	0.67	0.30	0.18	0.11	0.39	0.38	0.26	0.10	0.49
53	polishings	4-03-943	90	13.0	0.82	0.28	0.43	0.82	0.58	0.23	0.22	0.49	0.44	0.44	0.13	0.75
	Rye															
54	grain	4-04-047	88	11.8	0.50	0.24	0.37	0.64	0.38	0.17	0.19	0.50	0.26	0.32	0.12	0.51
	Safflower															
55	meal, sol. extr.	5-04-110	92	23.4	2.04	0.59	0.67	1.52	0.74	0.34	0.38	1.07	0.77	0.64	0.33	1.18
56	meal without hulls, sol. extr.	5-07-959	92	42.5	3.59	1.07	1.69	2.57	1.17	0.66	0.69	2.00	1.08	1.28	0.54	2.33
	Sesame															
57	meal, mech. extr.	5-04-220	93	42.6	4.86	0.98	1.47	2.74	1.01	1.15	0.82	1.77	1.52	1.44	0.54	1.85
	Sorghum															
58	grain	4-20-893	88	9.2	0.38	0.23	0.37	1.21	0.22	0.17	0.17	0.49	0.35	0.31	0.10	0.46
	Soybean															
59	meal, sol. extr.	5-04-604	89	43.8	3.23	1.17	1.99	3.42	2.83	0.61	0.70	2.18	1.69	1.73	0.61	2.06
60	meal without hulls	5-04-612	90	47.5	3.48	1.28	2.16	3.66	3.02	0.67	0.74	2.39	1.82	1.85	0.65	2.27
61	protein concentrate	—	90	64.0	5.79	1.80	3.30	5.30	4.20	0.90	1.00	3.40	2.50	2.80	0.90	3.40
62	protein isolate	5-08-038	92	85.8	6.87	2.25	4.25	6.64	5.26	1.01	1.19	4.34	3.10	3.17	1.08	4.21
63	seeds, heat processed	5-04-597	90	35.2	2.60	0.96	1.61	2.75	2.22	0.53	0.55	1.83	1.32	1.41	0.48	1.68

	Sunflower														
64	meal, sol. extr.	5-09-340	90	26.8	2.38	0.66	1.29	1.86	1.01	0.59	0.48	1.23	0.76	1.04	0.38	1.49
65	meal without hulls, sol. extr.	5-04-739	93	42.2	2.93	0.92	1.44	2.31	1.20	0.82	0.66	1.66	1.03	1.33	0.44	1.74
	Triticale															
66	grain	4-20-362	90	12.5	0.57	0.26	0.39	0.76	0.39	0.20	0.26	0.49	0.32	0.36	0.14	0.51
	Wheat															
67	bran	4-05-190	89	15.7	1.07	0.44	0.49	0.98	0.64	0.25	0.33	0.62	0.43	0.52	0.22	0.72
68	grain, hard red spring	4-05-258	88	14.1	0.67	0.34	0.47	0.93	0.38	0.23	0.30	0.67	0.40	0.41	0.16	0.61
69	grain, hard red winter	4-05-268	88	13.5	0.60	0.32	0.41	0.86	0.34	0.20	0.29	0.60	0.38	0.37	0.15	0.54
70	grain, soft red winter	4-05-294	88	11.5	0.50	0.20	0.45	0.90	0.38	0.22	0.27	0.63	0.37	0.39	0.26	0.57
71	grain, soft white winter	4-05-337	89	11.8	0.55	0.27	0.44	0.79	0.33	0.20	0.28	0.55	0.36	0.35	0.15	0.53
72	middlings, < 9.5% fiber	4-05-205	89	15.9	0.97	0.44	0.53	1.06	0.57	0.26	0.32	0.70	0.29	0.51	0.20	0.75
73	red dog, < 4% fiber	4-05-203	88	15.3	0.96	0.41	0.55	1.06	0.59	0.23	0.37	0.66	0.46	0.50	0.10	0.72
74	shorts, < 7% fiber	4-05-201	88	16.0	1.07	0.43	0.58	1.02	0.70	0.25	0.28	0.70	0.51	0.57	0.22	0.87
	Whey															
75	dried	4-01-182	96	12.1	0.26	0.23	0.62	1.08	0.90	0.17	0.25	0.36	0.25	0.72	0.18	0.60
76	low lactose, dried	4-01-186	96	17.6	0.53	0.33	1.16	1.61	1.51	0.39	0.46	0.63	0.52	1.17	0.31	1.15
77	permeate, dried	—	96	3.8	0.06	0.05	0.17	0.22	0.18	0.03	0.04	0.06	—	0.14	0.03	0.13
	Yeast, Brewers															
79	dehydrated	7-05-527	93	45.9	2.20	1.09	2.15	3.13	3.99	0.74	0.50	1.83	1.55	2.20	0.56	2.39
	Yeast, Torula															
79	dehydrated	7-05-534	93	46.4	2.48	1.09	2.50	3.32	3.47	0.69	0.50	2.33	1.65	2.30	0.51	2.60

[a] Dash indicates that no data were available.
[b] First digit is class of feed: 1, dry forages and roughages; 2, pasture, range plants, and forages fed green; 3, silages; 4, energy feeds; 5, protein supplements; 6, minerals; 7, vitamins; 8, additives; the other five digits are the International Feed Number.

Source: National Research Council (1998) with permission (Table 11–4).

Plant geneticists have continued to develop cultivars (varieties) of corn and cereal grains that have improved nutritional value. For example, high-lysine corn (e.g., opaque-2 corn) contains a higher level of lysine than conventional corn, making it possible to use less high-protein feedstuffs, which are more expensive than grains, to provide the needed intake of lysine. Similarly, high-lysine mutants of barley are available. The lower per-acre yields of high-lysine corn and barley compared with yields of conventional cultivars has discouraged the widespread use of these genetically improved grains. More recently, genetically produced high-oil corn containing more energy and genetically produced low-phytate corn containing more bioavailable phosphorus have been developed. Such changes in composition of grains brought about by modern biotechnology (genetically modified [GM] plants) in plant genetics can be expected to continue; these improvements should reduce feed costs in pig production.

As a general rule, the energy value of corn for swine is considered the standard of comparison for alternative grain sources. If corn is given a value of 100, the relative energy value of grain sorghum is considered to be about 96, wheat, 99; barley, 88; and oats, 79. Such comparisons are intended only as guidelines because of the variability within each grain associated with cultivar and growth environment (temperature, moisture, soil fertility, and other factors). The special characteristics of corn and the individual cereal grains also must be taken into account. For example, the higher vitamin A activity of yellow corn than of the cereal grains and the differences in palatability associated with the presence of tannin and other unpalatable constituents in various cultivars of grain sorghum and other grains are factors to consider.

GRAIN BY-PRODUCTS

By-products of corn and cereal grain processing are available in large quantities for use in livestock feeding. Although their usefulness for swine is limited due to their higher fiber content and low protein quality, these products contribute to lowering swine feed costs by their judicious inclusion in grower and gestation diets. The utilization of cereal by-products in swine feeding was reviewed by Sauber and Owens (2001).

Milling By-products

The manufacture of flour from cereal grains and corn for human consumption results in several by-products available for pig feeding. The seed is covered with high-fiber layers that are higher in protein and minerals than the inner substance, consisting of mainly starch. The bran fraction is contained in the outermost layer of the seed. The aleurone layer separates the bran layer from the inner endosperm, which consists of thin-walled cells packed with starch intermixed with gluten. At the base of the seed is the germ, which is high in fat, fat-soluble vitamins, and minerals.

The objective of the milling process is to remove as much of the starch and gluten as possible while excluding most of the germ, aleurone, and bran fractions. Because of variability in the degree to which this separation is accomplished, flour by-products contain a broad range of fiber (bran and aleurone fractions). The currently accepted labeling of wheat by-products is based on the fiber content. Thus, the wheat flour by-product containing more than 9.5% fiber is wheat bran; that with 7.1% to 9.5% fiber is middlings or mill run; that with 1.5% to 7.0% fiber is red dog, white shorts, or white middlings; and that with 1.5% or less fiber is feed flour. There is wide variation in the nomenclature describing these wheat by-products, depending on location and sampling variations.

Although wheat by-products are used throughout the world and can supply a significant proportion of the energy and protein in swine diets, the variation in nutritive value is of concern in diet formulation. Most of the variation is in fiber content, which affects digestible energy (DE) content of the mixed diet. Corn and rice milling by-products are similar to those described for wheat, and the same concerns related to variation in fiber content exist.

Distillery By-products

Alcoholic beverages and ethanol for motor vehicle fuel are the main products of grain distillation. The production of distilled liquors and ethanol from cereal grains or corn involves grinding, cooking, adding enzymes to hydrolyze the starch to simple sugars, and adding yeast to cause fermentation. After fermentation, the alcohol is removed and the residue is available for feeding to swine and other livestock. The residue can be fractionated into several components or used without further processing (the unprocessed residue is termed *stillage* or *distillery slop*). The coarse particles are usually removed from the stillage and are termed *wet distillers'* grain. After dehydration, this product is sold as dried distillers' grains. The fine particles remaining after removal of coarse particles are usually dehydrated to form dried distillers' solubles. The addition of dried distillers' solubles to wet distillers' grain produces distillers' grain with solubles that, after drying, are sold as distillers' dried grains with solubles. The Association of American Feed Control Officials (AAFCO, 1987) has standardized the nomenclature so that each of these products is further identified by a prefix naming the cereal or corn used in its manufacture.

The nutritional value of distillery by-products mimics that of the grain from which they were derived, except that the protein, mineral, and fiber levels are increased and the digestible energy value is reduced compared to the unprocessed grain. Distillers' dried grains with solubles (DDGS) has been shown to be a useful feed for swine. Its favorable levels of protein, amino acids, water-soluble vitamins, and trace minerals make it an economical supplement for growing and pregnant swine. The addition of up to 15% of DDGS appears to be compatible with normal growth and feed efficiency in growing pigs. The higher fiber content of DDGS, than of the unprocessed grain from which it came, results in a DE of about 2,940 kcal/kg compared with 3,525 kcal/kg for corn.

Brewery By-products

The American Association of Feed Control Officials (AAFCO, 1987) described the origin, manufacture, and composition of brewers' dried grains as "the dried extracted residue of barley malt alone or in mixture with other cereal grains or grain products resulting from the manufacture of wort or beer and may contain pulverized dried spent hops in an amount not to exceed 3%, evenly distributed." Its high fiber content (13% to 26%) and low quality protein (deficient in lysine, threonine, and tryptophan), despite its high protein content (25%), limits its use in swine feeding to gestation diets.

Corn Gluten Feed

The residue of the corn kernel remaining after the extraction of most of the starch, gluten, and germ by wet milling employed in the manufacture of corn starch or corn syrup is termed *corn gluten feed* (AAFCO, 1987). Although it contains about 20% protein, its severely deficient lysine and tryptophan content limits its use to a source of energy rather than protein in swine diets. It has been used successfully as the primary source of energy in pregnant sow diets. Corn gluten meal is produced by the same process as corn gluten feed, but contains 40% or 60% protein. Its severe amino acid deficiencies preclude its use as a swine feed. Its high carotenoid pigment content puts it in demand as a pigment source for use in poultry.

Hominy Feed

This relatively high-energy corn by-product is produced in the manufacture of hominy grits and pearl hominy for human use. It is a mixture of corn bran, corn germ, and part of the starch of white or yellow corn or a mixture of the two. Fat content must be at least 4%; protein content is 10%. Hominy feed can replace all of the corn in grower or finisher diets without an adverse effect on pig performance.

ROOTS AND TUBERS

A major constraint on the use of roots and tubers for swine feeding is the difficulty of processing and storage. The subject was reviewed by Plunkett (1979).

Cassava

The root crop, cassava (*Manihot esculenta* Crantz; also known as yuca, manioc, tapioca, mandioca), is grown in large quantities in the lowland tropics throughout the world. Dried and processed cassava is used as a low-cost energy source in commercial pig diets in many countries and, to a very limited degree, in the United States. Fresh cassava root contains only about 35% dry matter and toxic levels of hydrocyanic acid (HCN). Drying destroys the hydrocyanic acid. The dry meal contains less than 3% protein, but is equal to corn in DE and contains seven times as much calcium as corn.

Potatoes

Potatoes (*Solanum tuberosum*) are widely used in livestock feeding in Europe. Surplus or small potatoes unfit for human consumption and potato wastes from potato processing industries are available in many regions for swine feeding. Composition varies widely according to variety and environmental conditions during growth, storage, and processing. Water content ranges from 60% to more than 85%. On a dry matter basis, potatoes contain 4% to 6% ash, 10% to 11% protein, 4% fiber, 0.2% fat, and 82% to 83% soluble carbohydrates (nitrogen-free extract). Raw potatoes contain alkaloids, protease inhibitors that depress growth and feed consumption. Cooking is required to destroy these heat-labile inhibitors and to alter the starch fraction so as to improve starch digestion. Steaming for 20 min at 100°C (212°F) is adequate to maximize starch digestibility and destroy the protease inhibitor. The DE of cooked potato is comparable to that of corn. Amino acid balance of the protein is superior to that of corn. Various processed dry potato products are available, including potato meal, potato flakes, potato slices, and potato pulp. Each of these products is prepared with differing and often variable heat and physical treatments; therefore, processing conditions and nutritive value of a particular batch of one or more of these products should be ascertained before using the product in formulating a pig diet.

Sweet Potatoes

Sweet potatoes (*Ipomaea batata* L.) have been used in pig feeding in many forms (raw, cooked, dehydrated) as silage. The fresh root contains about 1.35% protein (at 32% dry matter) or 4.2% of the dry matter. When fed raw,cooked, or as silage, and wet, sweet potatoes are bulky and have a feeding value 25% to 33% that of corn.

OTHER PLANT SOURCES OF ENERGY

Bananas and Plantains

Bananas (*Musa* species, more than 32 in number) and plantains (*Musa paradisica*) are used mainly for direct human consumption, but surpluses, damaged, and rejected supplies find their way into pig diets in fresh or dried form. The fresh banana, including the peel, contains about 80% water, 1.0% protein, 1.0% fiber, 0.2% fat, 1.0% ash, and 16.8% soluble carbohydrate (nitrogen-free extract). Composition of bananas and plantains varies considerably from these values. Dried, green banana meal can replace approximately 50% of the corn in the diet of lactating sows fed a 16% protein diet based on corn, wheat, and fish meal.

Cane Sugar and Beet Sugar

Crude and refined sugars, although not commonly used, are an excellent source of energy for swine feeding. Because these products are devoid of protein and contain low levels of vitamins and minerals, they must be supplemented with all noncarbohydrate nutrients at high levels. Substitution of up to 60% of the corn or cereal grains in the diet with cane sugar does not appear to be associated with reduced performance of growing pigs, but suckling age pigs (less than 3 wk of age) cannot tolerate cane sugar because their secretion of the digestive enzyme, sucrase, is not yet available in sufficient quantity to hydrolyze the sucrose contained in cane sugar to its constituent simple sugars, glucose and fructose.

Cane Molasses

Molasses, a major by-product of sugar production, is a satisfactory source of energy for swine. It is high in water (15% to 25%) and water-soluble carbohydrates (65%). The sugars present in molasses are highly digested by pigs of all ages, except in pigs younger than 3 wk old, whose digestive system does not yet produce enough sucrase to utilize the 30% sucrose content of cane molasses. Molasses is a very poor source of protein (less than 3% amino acid nitrogen). Molasses is an effective laxative agent to prevent constipation in sows. Dietary levels of 15% to 25% molasses increase fecal moisture and increase the rate of passage of ingesta through the digestive tract.

Dried Bakery Products

Reclaimed bakery products are often blended and processed to provide a product that contains 9% to 10% crude protein, 11% to 13% fat, 1% to 1.5% crude fiber, 3% to 4% ash, and a DE of 4.5 kcal/g. These products are highly palatable and are therefore an excellent ingredient at up to 30% by weight in starter diets for young pigs.

FATS AND OILS

The main nutrient supplied by fats and oils is energy. The calorie content of fats and oils is about 2.25 times that of carbohydrates and proteins. Fats are often used to increase the dietary caloric density, particularly for lactating sows. The increased caloric density results in less total feed consumption by the animal. Therefore, in diets containing added fats or oil, the percentage of other nutrients should be increased to obtain an adequate ratio of calories to other nutrients. The pig tends to eat to satisfy energy requirements, and a deficiency of protein or other nutrients (vitamins, minerals) can result if the fat content of the diet is increased without increasing the concentration of other nutrients in the diet.

Animal fats, including tallow, lard, and fish oils, as well as vegetable fats such as coconut, soybean, cottonseed, corn, canola, safflower, sesame, and other by-products of other food and industrial uses, are available for use in swine diets. AAFCO has established the following categories and definitions of fats and oils used in animal feeds.

Animal Fats

These fats are obtained from mammals and/or poultry in the commercial processes of rendering or extracting. Fish oils are also animal fats, but their origin and processing methods are different than defined by AAFCO for other animal fats. If the product bears a name descriptive of its kind or origin, that name must correspond exactly to the content.

Hydrolyzed Fat or Oil (Feed Grade)

This product is obtained in the fat-processing methods commonly used in soap making. It must contain not less than 85% total fatty acids, but not more than 6% unsaponifiable (not hydrolyzed by alkali) substances and not more than 1% insoluble materials. It must be described according to source (hydrolyzed animal fat or hydrolyzed vegetable fat or a mixture).

Fat Product (Feed Grade)

Any fat product not meeting definitions for animal or vegetable fat, hydrolyzed fat, or fat ester belongs to this category. Minimum percentage of total fatty acids and maximum percentages of unsaponifiable and insoluble fractions must be specified. The fat product must be adequately tested for safety in animal feeding before being marketed and its components must be listed.

Vegetable Fat or Oil

These products are obtained by extracting the oil from seeds or fruits commonly processed for human food. They must contain not less than 90% total fatty acids, not more than 2% unsaponifiable matter, and not more than 1% insoluble matter. The label must specify the source (e.g., soybean oil, coconut fat).

Vegetable Oil Refinery Lipid (Soapstock) (Feed Grade)

This product is obtained in alkaline refining of vegetable oil for human use. It may not contain more than 17% ash on a dry basis. It is a product of alkaline treatment, so it must be neutralized with acid before use in animal feeds. The product consists primarily of salts of fatty acids, glycerides, and phosphatides. Its feeding value for pigs has not been thoroughly studied, but available evidence suggests a significant reduction in growth rate of pigs fed diets in which soapstock supplies 15% of the calories (about 7% of the dry weight of the diet).

PROTEIN SOURCES

Among nutrient classes needed by swine, protein sources are second only to energy sources in the amounts needed, expressed as percentages of the diet. The very young pig requires more than 25% protein in its diet. The protein requirement, expressed as a percentage of the diet, declines steadily to maturity, but even the adult requires more than 10% protein in the diet to maintain long-term normal body functions. Protein sources are generally more expensive than energy sources. Because most feedstuffs contain some protein, an economically based choice of feedstuffs should result in a mixture that provides the smallest amount of protein compatible with normal productive function (i.e., growth, reproduction, and lactation).

A broad array of protein supplements of animal and plant origin is available (Chiba, 2001; Pond and Maner, 1974, 1984). The chemical composition of the most commonly used high-protein sources is shown in Table 11–2a and their amino acid composition is listed in Table 11–2b. Note that Tables 11–2a and 11–2b also include the composition of common energy sources. This allows one to contrast the general differences between energy and protein feedstuffs in overall composition.

The characteristics, deficiencies, and special supplementary qualities of protein sources need to be considered for use in correcting specific amino acid deficiencies of available energy sources. As a group, animal protein sources tend to be higher than plant sources in total protein, amino acid adequacy, vitamin content, mineral content, and mineral bioavailability. Processing conditions, including heat and other treatments, have important effects on the nutritional value of animal and plant protein sources. Aspects of food and feed processing of both animal and plant products as related to bioavailability of nutrients need to be addressed in selecting feed ingredients in pig diets.

ANIMAL PROTEIN SOURCES

Some of the animal proteins most common in pig diets are blood meal, blood plasma, and blood cells; casein; fish meal (including anchovy, herring, menhaden, whitefish); fish solubles; meat meal; meat and bone meal; poultry by-product meal; dried skim milk; dried whey; dried brewers' yeast; and dried torula yeast. Recent concerns about the use of tissues from cattle and sheep in animal feeds because of the threat of spread of mad cow disease (bovine spongioform encephalopathy, BSE) have resulted in legal restrictions on their use in feed in the United States and many other countries. Importation of animal products for use in livestock feeding in the United States has been banned until the BSE problem is brought under control. The continued use of these slaughter by-products in swine diets is therefore in jeopardy. Nevertheless, some animal by-products for swine remain in use.

Blood Meal

Dried blood meal contains more than 80% protein and 9% total lysine, but is seriously deficient in isoleucine. Conventional drying methods have used a high-temperature (up to 165°C), batch-drying process that has resulted in low palatability and poor bioavailability of some amino acids, particularly lysine. Newer methods of drying blood, such as flash drying, spray drying, and ring drying, have improved both palatability and amino acid bioavailability. The high red blood cell content of blood meal makes it an excellent source of iron. Levels of 5% to 10% blood meal dried by modern processes can be used effectively in pig diets. At levels above 5% dried blood meal in the diet, care must be taken to ensure an adequate level of isoleucine.

Blood Plasma

This spray-dried fraction of blood contains the plasma proteins, but not the blood cells. Blood plasma is an excellent source of dietary protein in terms of amino acid adequacy and bioavailability. It is commonly used as an important constituent of pig starter diets. Plasma may improve weaning pig health.

Blood Cells

This fraction of blood contains red and white cells removed in the process of separating plasma (or serum, if no anticoagulant is used) from whole blood. It contains more than 90% protein, but is deficient in isoleucine as in the case of blood meal. Blood cells are high in hemoglobin, so they are an excellent source of iron of high bioavailability.

Meat Meal

Meat meal is, by definition, the finely ground, dried rendered residue from animal tissues, exclusive of hair, hoof, hide, trimmings, blood, and digestive tract contents. It contains an average of 54% to 55% protein, 8% calcium, and 3.9% phosphorus. If meat meal contains more than 4.0% phosphorus, it is labeled meat and bone meal (meat rendered with bone). The quality of meat meal is greatly affected by the degree of dilution of muscle protein with tendon and bone—this protein is high in collagen. Collagen, also present in skin, connective tissue, and cartilage, is devoid of the amino acid tryptophan. Meat meal is more abundant than plant sources of protein in biologically available iron and zinc. The recent concern in Europe for transmission of prion diseases (BSE, or mad cow disease) through meat meal limits its use in animal feed in some countries, including the United States.

Meat and Bone Meal

The production of meat and bone meal is similar to that of meat meal, except for the inclusion of a higher level of bone and connective tissue. Protein content of meat and bone meal is lower (50% versus 54% to 55%) and calcium and phosphorus contents are higher (8% to 15% versus 7% to 8% calcium and 4.5% to 6% versus 3.5% to 4.5% phosphorus, respectively). Like meat meal, meat and bone meal is higher in bioavailable iron and zinc than plant sources of protein.

Casein

This highly nutritious dried product is the major protein of milk. Purified casein contains about 88% protein of high biological value. Casein is somewhat limited in sulfur-contained amino acids, but when supplemented with methionine, it may be used as the sole protein source in semi-purified diets in baby pig research. It is high in calcium and phosphorus, but low in most other minerals. Because of its high degree of purification, dried casein is a relatively poor source of vitamins.

Hydrolyzed Feather Meal

This by-product of the poultry industry is high in protein (84.5%), but low in protein digestibility (63%) and amino acids, particularly lysine digestibility (40%). Therefore, it is of limited value as a constituent of swine diets.

Fish Meal

Fish meals are of two types: those made from wastes of the fishery industry and those made from whole fish caught for processing to obtain specific products, often fish oil. Several species of fish (anchovy, herring, menhaden, and white fish) are used to produce fish meal. The protein content ranges from about 63% to 68% and amino acid adequacy and bioavailability are high in each. Freshly caught fish contain 68% to 85% water, 15% to 25% protein, and 1% to 25% oil. For storage stability, water content must be reduced by drying to about 10%. Stability also requires that the fat content be reduced to about 10%, and even a meal with this level of fat may become rancid in storage. In major industrial fisheries, high-oil fish is reduced to meal as a by-product of fish oil production. Fish meal contains significant oil residues (5% to 10% fat). The soft fish oil in the pig diet increases the degree of unsaturation of the pork fatty acids and may impart a fishy flavor to the meat. Fish meal is an excellent source of calcium, phosphorus, selenium, zinc, and vitamin E.

Fish Solubles

Condensed fish solubles contain about 51% dry matter. Except for the difference in the amount of water removed, condensed fish solubles and dried fish solubles are similar in composition. Dried fish solubles contain about 64% protein compared with 33% for condensed fish solubles.

Milk, Dried

Dried whole and skim cows' milk are excellent sources of protein for pigs. Normally, the demand for whole and skim milk for human consumption limits their use in diets for pigs. The only major difference between whole and skim milk is that most of the fat and fat-soluble vitamins are removed from skim milk and all other constituents are increased proportionately. The only nutrients likely to be deficient in a diet containing a large proportion of dried whole milk are iron and copper. Skim milk is also deficient in vitamin D if not fortified in processing. Dried skim milk contains about 34.5% protein of excellent nutritional value. The protein is a mixture of casein, lactalbumin, and other proteins; this mixture provides an amino acid balance superior to that of purified casein, the major protein of milk.

Poultry By-product Meal

The ground, rendered, and dried parts of the poultry carcass not used for human consumption (including neck, feet, and emptied digestive tract, but excluding feathers) are used to produce poultry by-product meal. It is an excellent protein source for pig diets. The protein content is about 64%, the fat content 13%, and the calcium and phosphorus contents are about 4.5% and 2.4%, respectively. Poultry by-product meal is also a good source of iron, zinc, and most of the water-soluble vitamins.

Whey

Liquid whey, a by-product of the manufacture of cheese and lactic acid-precipitated casein from milk, is available for feeding to pigs wherever cheese is produced. Fresh whey contains about 93% to 96% water, 0.7% to 0.9% protein, 5% lactose, and 0.7% to 1.0% ash. To facilitate storage and transportation, fresh whey is often dried. Dried whey contains 12% protein, 0.75% calcium, and 0.72% phosphorus, and is an excellent source of the water-soluble vitamins. The high content of lactose in dried whey limits its use mostly to diets for suckling-age and growing pigs during the early post-weaning period, due to the decrease in lactase production in some older pigs and consequent inability to utilize high lactose intake. Dried whey, from which much of the lactose has been removed (low-lactose dried whey), contains more protein (17% to 18%), calcium (2%), and phosphorus (1.4%) than regular dried whey. Another variant of dried whey is permeate dried whey, whose protein content is less than that of regular dried whey (3.8% versus 12%). It contains levels of calcium and phosphorus comparable to those of regular dried whey.

Other Animal Proteins

Many animal protein sources are produced locally or seasonally and therefore may not be universally available. Examples are fish protein concentrates, fish silage, liquified fish, shrimp meal, krill meal, king crab meal, hatchery waste, hog hair meal, and animal and poultry wastes. In general, animal protein sources are superior to plant protein sources in amino acid balance and bioavailability and in the level and bioavailability of minerals and vitamins.

PLANT PROTEIN SOURCES

Some of the plant proteins most common in pig diets are the oilseed meals, seed legumes, alfalfa, and other legume and grass forages (see Table 11–2 for the composition of these supplements).

Oilseed Meals

This group of plant protein supplements is the result of oil extraction of the seed for use in human food and industry. The most important worldwide is soybean meal. Others of major importance are canola, coconut, cottonseed, linseed, peanut, safflower, sesame, and sunflower meals.

Soybean Meal. Properly processed soybean meal contains a protein of excellent quality. Depending on completeness of oil extraction and degree of removal of hulls, soybean meal contains between 43% and 50% protein and has an excellent balance of amino acids that are well digested except by the very young pig. It is not severely deficient in any amino acid except methionine. Relative to the requirement of the pig, most amino acids are present in excess of the requirement when soybean meal is used to supply all of the dietary protein. However, in practice, it is used as a protein supplement in combination with high-energy grains such as corn or grain sorghum and by-product, high-energy feedstuffs. A combination of corn and soybean meal provides an excellent balance of amino acids, with lysine as the first limiting amino acid. Supplementation of the diet with synthetic lysine allows the mixed diet to be fed at a lower level of protein while still meeting all essential amino acid requirements at a reduction in total diet cost, compared with the diet containing more protein but no synthetic lysine.

Soybean oil is removed from the seed by either solvent extraction (the usual method) or by a mechanical expeller process. Both processes involve heat, which serves to destroy several antinutritional compounds present in the raw seed. These factors include trypsin inhibitors, hemagglutinins, saponins, isoflavones, and goitrogens. The growth inhibition in pigs fed raw soybeans appears to result from reduced feed intake and inhibition of protein digestion.

Removal of the soybean hull increases the protein content of the processed meal to about 50% by weight. If the hulls are added back after fat extraction, the protein content of the meal is reduced to about 44% and the crude fiber content is increased (consequently, the DE is decreased). Soybean meal produced by the expeller process contains the hull and is therefore lower in protein and DE than dehulled solvent-extracted meal.

Soybean meal is low in calcium, phosphorus, and fat-soluble vitamins. As is the case with most plants, the phosphorus in soybean meal is present largely as phytic acid (phytin) phosphorus, which is biologically unavailable to the pig. The nutrient content of soybean meal is relatively constant within and across processing plants, but calcium carbonate is used in some plants to improve flowability of the meal, resulting in a high-calcium product.

Canola (Rapeseed) Meal. Earlier cultivars of rapeseed contained high levels of erucic acid and glucosinolate; both have antinutritional properties for swine. The use of rapeseed meal was limited by the presence of these compounds. Plant breeders have developed cultivars of rapeseed low in erucic acid and glucosinolates; these cultivars are termed *canola*. Thus, canola meal, a by-product of vegetable oil from the canola plant, has supplanted rapeseed meal as a protein supplement for swine. Canola meal contains about 35% to 38% protein and 2% to 4% fat. Its 3% to 4% fiber content gives it a lower DE value than that of soybean meal. Lysine content is somewhat lower than that of soybean meal, but methionine-cystine content is higher, making canola meal a satisfactory protein supplement for pigs.

Coconut Meal. The residual product after extraction of the oil from the coconut, coconut meal (or copra meal), is available as an inexpensive protein supplement in many areas of the world. It contains 20% to 26% protein and about 10% crude fiber, resulting in a DE content less than that of soybean meal. The lysine content of coconut meal is less than that of soybean meal as a percent of protein, but the level of methionine-cystine (sulfur-containing amino acids) is appreciably higher. The digestibility of protein and amino acids in coconut meal is relatively low. Therefore, its use as a protein supplement in pig diets should be restricted to avoid reduced growth efficiency.

Cottonseed Meal. The seeds of the cotton plant, from which cottonseed meal is produced, contain a natural pigment, gossypol, that is toxic to pigs. Improved cultivars of cotton contain lower levels of gossypol, resulting in greater usefulness of the meal for pig feeding. Degossypolized cottonseed meal, or that produced from low-gossypol cultivars and containing less than 0.04% free gossypol, is an acceptable protein supplement for swine. A standard 15% to 16% protein diet for swine should not contain more than 0.01% free gossypol.

Cottonseed meal contains 36% to 41% protein and 1.65% lysine compared with 43% to 50% protein and 2.8% lysine contained in soybean meal. Therefore, a high-lysine protein supplement or synthetic lysine must be added to diets containing cottonseed meal as the primary protein source. Mechanically extracted cottonseed meal contains more residual fat than solvent-extracted meal, so DE is higher. Although cottonseed meal contains a high level of phosphorus (about 1%), most of the phosphorus is present as phytic acid phosphorus, which the pig digestive system cannot break down to release for absorption. The enzyme phytase, produced by microorganisms and some plants, must be added to pig diets to utilize phytin phosphorus. Reduction in environmental pollution with excess phosphorus movement from swine manure into soil and water is an active research area.

Linseed Meal. A by-product of drying oil production, linseed meal comes from flax or linseed. It contains about 34% protein and 9% fiber and is similar to soybean meal in calcium and phosphorus content, but contains less than half as much lysine.

Peanut Meal (Groundnut Meal). The by-product of oil extraction from peanuts, peanut meal contains 43% to 50% protein and 1% to 6.5% fat. Mechanically extracted meal contains more residual fat and less protein than solvent-extracted meal. Peanut meal is severely deficient in lysine and low in tryptophan and methionine. Peanuts are prone to infection with the mold *Aspergillus flavus*, which produces the toxic substance aflatoxin B. The absence of the aflatoxin should be established before peanut meal is fed to pigs.

Safflower Meal. Safflower has long been cultivated as an oilseed crop. The seed contains about 36% to 40% oil; the high-fiber hull makes up about 40% of the weight of the seed.

After the oil is extracted, the meal contains only 18% to 22% protein and 60% hulls. Safflower meal protein is low in lysine and methionine. Therefore, safflower meal is of limited value in pig diets because of its low protein, low lysine, and high fiber content.

Sesame Meal. Sesame is probably one of the world's oldest cultivated oilseed crops. The seed contains 44% to 54% oil and the meal remaining after oil extraction contains 38% to 48% protein, 1% to 11% fat, and 5% to 7% fiber. Sesame meal is low in lysine, but is a good source of methionine and tryptophan.

Sunflower Meal. The composition of sunflower meal varies according to the composition of the seed and, as with other oilseed proteins, the nutritive value of the meal varies with the processing method. Solvent-extracted meals contain more protein and less fiber than those produced by mechanical extraction. Solvent-extracted meals are available with or without the hulls. Solvent-extracted sunflower meal with hulls contains only about 26% to 28% protein and 2,000 kcal DE/kg, whereas meal without the hulls contains about 41% to 43% protein and 2,840 kcal DE/kg. Sunflower meal is very low in lysine, but is a good source of methionine.

OTHER PLANT PROTEIN SOURCES

Seed Legumes

The seed legumes have several advantages and offer a broad range of genetic variability and wide environmental adaptability. With the exception of soybeans and peanuts, which contribute significantly to the total world protein supply, only about 20 of the seed-producing species of the 13,000 that make up the legumes (Family Leguminosae) are used in appreciable quantities for food and feed protein. Because of low yields, most of this limited number of crops cannot compete economically with higher-yielding crops.

The soybean, the prime example of a widely used seed legume, is produced in huge quantities in both temperate and tropical climates. It is used largely for human food in many parts of the world, but in the United States, soybean seeds are used to extract oil for food and industrial uses and the meal is a by-product. However, cooked whole soybeans are an excellent source of both energy and protein for swine. Seed legumes, in general, (soybeans and peanuts are no exception) contain a variety of heat-labile, antinutritional factors that must be destroyed by moist heat to be used successfully in pig diets. Seed legumes that are usually used for direct human consumption often find their way into swine diets. Damaged or otherwise inferior seeds unsuited for human consumption can be used as protein supplements if properly processed for mixing into commercial diets. Examples are dry beans (snap bean, white kidney bean, navy bean, common bean), mung bean, lima bean, chick pea, pigeon pea, cow pea, field bean, field pea (garden pea, arveja), and sweet lupin. Most contain 20% to 30% protein and most are of high protein quality, except for a deficiency of methionine-cystine. These seed legumes are described in Pond and Maner (1984).

Alfalfa

This leafy legume contains 16% to 22% protein and is a good source of carotene, vitamin E, calcium, and most of the water-soluble vitamins. Although it is high in fiber and lignin, it has been used successfully as the chief feedstuff in diets of sows throughout gestation. Dehydrated alfalfa leaf meal is higher in protein and lower in fiber than unprocessed alfalfa forage and can be used at a level of 5% to 10% of the diet of growing pigs without appreciably reducing weight gain.

OTHER SOURCES OF PROTEIN AND AMINO ACIDS

Single-cell Protein

Single-cell protein (SCP) has undergone increasingly active investigation during the past decade with regard to nutritional value and safety in human and animal nutrition. Much of the effort has been focused on SCP grown on methanol and other hydrocarbons and on animal wastes. In the case of yeast, grown mainly on sulfite waste liquor, molasses, whey, or other carbohydrates. Although there are differences in chemical and physical characteristics and nutritional value among fungi, yeast, algae, and bacterial protein, the SCPs as a group are high in protein (>40%), relatively well balanced in essential amino acids, generally good sources of water-soluble vitamins and trace elements (a notable exception is Torula yeast, whose selenium content is extremely low), low in fiber, and high in DE.

Seaweed

There are many species of seaweed, each undergoing seasonal variation in chemical composition. In general, seaweed contains low levels of protein (3% to 15%) and energy (500 to 600 kcal/kg) and high levels of minerals (15% to 50% ash). Therefore, this plentiful and inexpensive plant resource cannot be a major contender as a dietary ingredient for pigs. However, recent research with a species of brown seaweed (*Ascophyllum nodosum*), which grows off the coasts of Canada and Northern Europe, suggests that the seaweed or an extract from it improves immune function in young pigs exposed to infectious disease. The action involved in this immune response has not been identified. The active constituent may be present in the protein, carbohydrate, lipid, or mineral fraction of the seaweed.

Synthetic Amino Acids

During the past 10 years, recombinant DNA (gene splicing) technology has been developed to produce microorganisms capable of producing mass quantities of lysine, tryptophan, threonine, and other amino acids for use in pig diets. The availability of these economically priced, synthetic amino acids has allowed the formulation of diets that are lower in total protein than previously feasible without any reduction in animal performance. This saving in protein has far-reaching implications for pig production efficiency and environmental stability because nitrogen pollution of soil and water will be reduced as a result of improved biological efficiency and enhanced nitrogen utilization.

MINERAL AND VITAMIN SOURCES

Although the largest proportion of the pig diet consists of energy and protein, the mineral elements and vitamins are vital to normal growth and reproduction. Most energy and protein sources provide some vitamins and minerals, but it is customary to supplement the diet with specific mineral and vitamin premixes to ensure adequate intake of all minerals and vitamins. Supplemental calcium and phosphorus are usually provided as inorganic mined products. Common calcium and phosphorus supplements are shown in Table 11–3. Salt (NaCl) is provided at 0.3% to 0.5% of the diet. A typical trace mineral element premix contains copper, iodine, iron (ferrous), manganese, selenium, and zinc.

Complete vitamin supplemental premixes added to the diet eliminates dependence on feedstuff ingredients to supply the pig's vitamin requirements. Vitamin A, vitamin D, vitamin E, vitamin K, vitamin B_{12}, niacin, pantothenic acid, riboflavin, and choline are commonly included in the complete vitamin premix.

VARIABILITY IN COMPOSITION OF INDIVIDUAL FEED INGREDIENTS

Many factors influence nutrient composition of a particular energy or protein source, including variables associated with soil fertility, weather conditions during crop growth, stage of maturity at harvest, and processing methods. As a result of this variability, producers are advised to obtain a proximate analysis of each major

TABLE 11–3
Calcium and Phosphorus Content of Common Mineral Supplements (in percent).

SUPPLEMENT	CALCIUM	PHOSPHORUS
Bone meal, raw	21	9–10
Bone meal, steamed	24–39	12–14
Bone black, spent	27	12
Dicalcium phosphate	23–26	18–21
Defluorinated rock phosphate[a]	31–34	13–17
Raw-rock phosphate[b]	24–29	13–15
Soft phosphate	18	9
Curacao phosphate	35	15
Oyster shell	38	—
Limestone	38	—
Calcite, high grade	34	—
Gypsum	22	—
Dolomite limestone	22	—
Wood ashes	21	—

[a]Usually less than 0.2% F.
[b]2.0–4.0% F.
Source: Adapted from Pond et al. (1991).

ingredient and, in many cases, an analysis of the concentration of specific selected nutrients (e.g., lysine, calcium, and phosphorus) before inclusion of the ingredient in the mixed diet. Also, many energy and protein feedstuffs contain heat-labile toxic compounds and may contain other constituents or have other properties that limit their use in swine diets. Myer and Brendemuhl (2001) presented tables containing suggested maximum percentages of selected energy feedstuffs and protein feedstuffs in the diets of pigs of various ages and of gestating and lactating sows. Some of this information is shown in Table 11–4.

TABLE 11–4
Suggested Maximum Dietary Inclusion Levels of Selected Energy and Protein Feedstuffs for Swine.

FEEDSTUFF	SUGGESTED MAXIMUM AMOUNTS IN DIETS(%)				
	STARTING (<44 LB)	GROWING (44–110 LB)	FINISHING (110–242 LB)	GESTATION	LACTATION
Energy Sources					
Corn (maize)	75	85	90	90	85
Cassava meal	40	40	40	40	40
Molasses, final	10	10	20	10	10
Potato chips/fries	10	30	20[c]	30	30
Rice, broken	30	30	30	30	30
Rice bran	0	20	20	50	0
Rice polishings	0	20	20	20	20
Sugar beet pulp, dried	0	10	10	40	10
Triticale	75	85	90	90	85
Protein Sources					
Soybean meal (48%)	30	30	25	20	30
Beans (Phaseolus), heated, ground	0	10	20	20	10
Canola seeds, raw	15	15	10[c]	10	10
Crab meal	0	5	5	5	5
Faba beans, raw, low tannin	15	20	20	10	10
Lupins (L. albus), raw	10	15	15	15	15
Lupins (L. angustifolius), raw	20	30	30	20	20
Mung beans, raw	0	10	15	10	0
Peanuts, raw	0[a]	20	10[c]	20	20
Peas, raw, dark-seeded	5[b]	10	20	10	10
Peas, raw, white-flowered	15[b]	30	30	20	20
Poultry hatchery by-product meal	0	3	3	3	3
Shrimp meal	0	5	5	5	5
Sunflower seeds, raw	10	10	10	30	30

[a]10% for roasted peanuts.
[b]Can double levels if heat processed.
[c]High levels may result in soft carcass fat.
Source: Adapted from Myer and Brendemuhl (2001), pp. 856 and 857.

SUMMARY

A broad array of feedstuffs is available worldwide for use in swine production. Many resources are available that describe the nutrient composition and feeding characteristics of these feedstuffs. As globalization of pork production and marketing continues, the number of feedstuffs used in pork production can be expected to increase. This chapter describes the common energy and protein sources used in formulating swine diets. Energy sources include grains, grain by-products, roots and tubers, and other plant sources of carbohydrate energy as well as plant and animal fats and oils. Protein sources include both plant and animal products, many of which are by-products of the human food industry.

An important aspect of swine feeding is the utilization of many human food by-products and inedible foods that would otherwise be wasted or added to environmental pollution if not fed to swine and other livestock. Tabular information on the composition of many of these feed resources and suggested maximum dietary inclusion of selected feedstuffs are given in the chapter, as well as extensive references to recent published literature and Internet resources related to ingredient composition for swine feeding.

QUESTIONS AND ACTIVITIES

1. In the USA, the most common diet is the corn-soybean meal diet supplemented with vitamins and minerals. In other parts of the country and of the world, very different diets are fed. Consider how the vitamin and mineral nutrient supplementation were to change if corn and soybean meal were replaced with:
   a. Barley and Peas (as in the Pacific Northwestern USA)
   b. Alfalfa and liquid whey (as in parts of Mexico)
   c. Bakery waste from a cookie plant as the major energy source
   d. Wheat and fish meal (as in parts of Latin America)
2. What is the energy value of liquid potato slurry? If the potato product is 90% water, can the pig eat enough potato slurry to meet its energy needs?
3. Meat and bone meal has been eliminated as a feed source when fed to ruminants from ruminants. At the same time, there is pressure based on a fear (perhaps unreasonable) of BSE to not feed meat and bone meal to pigs. What can be used as a substitute for meat and bone meal in pig diets? (this could be a combination of feedstuffs).
4. Which molds and fungi can contaminate feedstuffs? Which molds are of greatest concerns in corn?
5. Sometimes sows, particularly the peri-parturient sow, can become constipated. Which feedstuffs or feed additives can be used to relieve constipation among sows?
6. Some feeds are pelleted to improve feeding value. Based on the scientific literature, how much can a producer expect to improve feed efficiency when the growing pig diet is pelleted? Based on this improvement in feed utilization, what would be the break-even price a producer could pay to have his pig feed commercially pelleted, in dollars per ton?

LITERATURE CITED

Association of American Feed Control Officials. 1987. Plant By-product Nomenclature. Association of American Feed Control Officials, Washington, DC.

Chiba, L. I. 2001. Protein Supplements. In: A. J. Lewis and L. L. Southern (eds.) Swine Nutrition, 2nd ed., pp. 803–838. CRC Press, Boca Raton, FL.

FAO United Nations. 2001. Animal Feed Resources Information System (AFRIS) Available at: *http://www.fao.org/ag/guides/resource/data.htm*. Accessed May 16, 2001.

Lewis, A. J. and L. L. Southern. 2001. Swine Nutrition. 2nd ed. CRC Press, Boca Raton, FL.

Miller, E. R., D. E. Ullrey, and A. J. Lewis. 1991. Swine Nutrition. Butterworth-Heinemann, Boston, MA.

Morrison, F. B. 1956. Feeds and Feeding. 22nd ed. Morrison Publishing Co., Ithaca, NY.

Myer, R. O. and J. H. Brendemuhl. 2001. Miscellaneous Feedstuffs. In: A. J. Lewis and L. L. Southern (eds.) Swine Nutrition, 2nd ed., pp. 839–864. CRC Press, Boca Raton, FL.

National Research Council. 1998. Nutrient Requirements of Swine. 10th ed. National Academy Press, Washington, DC.

Patience, J. F. and P. A. Thacker. 1989. Swine Nutrition Guide. Prairie View Swine Center, University of Saskatchewan, Saskatoon, Canada.

Plunknett, D. L. 1979. Small-scale Processing and Storage of Tropical Root Crops. Westview Press, Boulder, CO.

Pond, W. G. and J. H. Maner. 1974. Swine Production in Temperate and Tropical Environments. W. H. Freeman, San Francisco.

Pond, W. G. and J. H. Maner. 1984. Swine Production and Nutrition. AVI, Westport, CT.

Pond, W. G., J. H. Maner, and D. L. Harris. 1991. Pork Production Systems, pp. 289-308. van Nostrard Reinhold, New York, NY.

Sauber, T. E. and F. N. Owens. 2001. Cereal Grains and By-Products for Swine. In: A. J. Lewis and L. L., Southern (eds.) Swine Nutrition, 2nd ed., pp. 785–802. CRC Press, Boca Raton, FL.

INTERNET RESOURCES

Tri-State Swine Nutrition Guide, Bulletin 869-98, Major Factors Influencing Nutrient Recommendations:
http://ohioline.ag.ohio-state.edu/b869/b869_3.html

Swine Nutrition:
http://www.ansi.okstate.edu/course/4643/nutr.htm

Nutrient Requirements of Swine, Tenth Revised Edition, 1998:
http://www.nap.edu/catalog/6016.html

Amino Acids in Pig Diets:
http://www.dpi.qld.gov.au/pigs/4374.html

Evaluating Vitamin Premixes for Swine:
http://muextension.missouri.edu/xplor/agguides/ansci/g02351

Full-Fat Soybeans for Pigs:
http://www.ianr.unl.edu/pubs/Swine/g994.htm

12
FORMULATING DIETS FOR PIGS

INTRODUCTION

Pig performance (growth, reproduction, and lactation) is greatly influenced by nutrition. Formulating diets that meet the nutrient requirements of pigs can be done, as a first approximation, in a simple manner with long-held empirical methods. However, to more precisely meet all nutrient requirements of pigs of specialized genotypes, computerized software programs are needed.

Pigs diets can be simple, consisting of a few ingredients, or complex. A common, simple diet used in the United States for growing and finishing pigs may contain corn, soybean meal, and a mineral-vitamin supplement. In contrast, European pigs of the same age may be fed diets containing 20 or 30 ingredients to provide the same nutrient requirements. The primary difference is the cost of feedstuffs available for use in a particular locale; local feedstuffs tend to be available at lower cost than imported feedstuffs.

This chapter briefly describes the principles of diet formulation for pigs and introduces some of the methods currently used by nutritionists in the feed industry and by pig producers who prepare their own pig feeds from ingredients produced on the farm or purchased for use in feed formulation. The National Research Council's (1998) Nutrient Requirements of Swine, 10th rev. ed., presented a brief introduction to pig diet formulation. This chapter uses the NRC data as a guide for application of computers to diet formulation.

IMPORTANT CONSIDERATIONS IN DIET FORMULATION

Diet formulation requires the consideration of four primary factors:

- The nutrient requirements to be met (discussed in Chapter 10)
- The available ingredients or feed resources (discussed in Chapter 11)
- The cost of supplying the required nutrients from available ingredients
- Feed processing as related to diet intake, nutritive value, and pig performance

MEETING NUTRIENT REQUIREMENTS

Quantitative requirements for specific nutrients for various stages of the life cycle were presented in Chapter 10. After the producer knows the quantitative nutrient requirements during a specific stage of the life cycle, the next step is to consider all of the available feedstuffs and their composition and price. Tables 11–1 through 11–4 in Chapter 11 of Nutrient Requirements of Swine (National Research Council, 1998) list most of the common feedstuffs used in swine diet formulation and present the dry matter; digestible, metabolizable, and net energy; crude protein; crude fat, linoleic acid; neutral and acid detergent fiber (NDF and ADF, respectively); and calcium, phosphorus (total and bio-available), mineral, vitamin, and amino acid contents of each ingredient. Tables 11–1 and 11–4 of that publication were reproduced in Chapter 11 of this book as Table 11–2a (chemical composition) and Table 11–2b (essential amino acid composition), respectively. These tabular values are used by nutritionists in diet formulation, along with values for vitamins and other required mineral elements (not included in this book, but presented in Tables 11–2 and 11–3 of Nutrient Requirements of Swine, National Research Council, 1998).

DETERMINATION OF DIGESTIBLE, METABOLIZABLE, AND NET ENERGY VALUES OF FEEDSTUFFS

Knowledge of the methods commonly used to determine nutritive value of feedstuffs is essential as a basis for adequate diet formulation. These methods are based on analyses that accurately reflect the nutrient composition of the available ingredients. The analyses measure the energy value (measured in calories and expressed as digestible or metabolizable energy/kg or lb) as well as the protein (N × 6.25) and ash concentrations/kg or lb. The traditional proximate analysis provides information on the concentrations of the major classes of nutrients, which include protein (N × 6.25), ether extract, crude fiber, well-digested carbohydrates, and ash. All constituents, except ash, contribute energy to the diet, as discussed in Chapter 10. The following constituents are included in the proximate analyses of feedstuffs for each of the major classes of nutrients:

- Nitrogenous compounds (mainly protein, but also include amino acids, polypeptides, purines, pyrimidines, urea, nucleotides, nucleic acids, ammonia, and other nonprotein substances). Nitrogenous compounds are measured as N in the standard Kjeldahl procedure (N × 6.25 = crude protein; most proteins contain about 16% N).

- Fat-soluble compounds (ether extract) include mono-, di-, and triglycerides, fatty acids, waxes, fat-soluble vitamins, and sterols.
- Poorly digested carbohydrates (analyzed as crude fiber or as neutral and acid detergent fiber based on the methods of Van Soest, 1967).
- Ash (determined by weight loss on ignition of the fresh sample). Individual mineral elements are not measured; their concentrations are measured by specialized methods peculiar to each group of minerals.
- Well-digested carbohydrates (calculated by difference after accounting for protein, fat, ash, and crude fiber).
- Vitamins are not measured in the proximate analysis; their concentration in feedstuffs is determined by specialized methods peculiar to each vitamin.

A more direct approach to energy utilization is to consider calorie absorption and utilization. A calorie (as defined in Chapter 10) is the amount of heat needed to raise the temperature of 1 g of water 1°C. The calories contained in any material can be determined by measurement of the heat produced by complete combustion in a bomb calorimeter. This gross energy (GE) can be expressed as calories/g of material. Feedstuffs widely different in utilizable energy (such as corn versus corn cobs) will yield similar values for gross energy. Digestible energy (DE) is calculated by the difference between the energy in a given amount of feed consumed and the energy lost in the feces after completely combusting both. Although DE provides useful information about the energy value of feedstuffs, it does not take into account losses of energy through the urine, expired air, and other metabolic processes. More refined techniques take these losses into account. The energy remaining for use by the body after urinary and expired air losses is accounted for in metabolizable energy (ME). After a pig eats a meal, extra heat is produced associated with assimilation of the meal—this is called the *heat increment.* The energy remaining for use after the heat increment has been deducted is the net energy (NE) of the feed. NE represents the most refined expression of the energy value of a feedstuff. Figure 12–1 provides a flowsheet of these relationships in describing the energy value of a feedstuff.

FIGURE 12–1

Relationships between Gross Energy (GE), Digestible Energy (DE), Metabolizable Energy (ME), and Net Energy (NE).

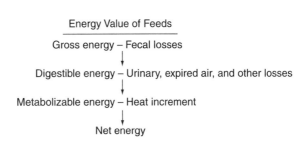

COST OF DIETARY INGREDIENTS

The most nutritionally adequate feedstuff may be too expensive to justify its use for profitable pork production. At the other end of the scale, the lowest priced feedstuff may be so nutritionally inadequate for a particular stage of the life cycle that its use cannot be justified for profitable pork production. Adequate diet formulation requires attention to both nutritional value and the price of the ingredients. No two feedstuffs are identical in nutrient composition. The concept of combining feedstuffs to create a nutritionally well-balanced diet is the basis of economical diet formulation. By selecting appropriate amounts of ingredients, deficiencies in one ingredient are offset by an abundance of nutrients in one or more other ingredients.

Computer technology introduced least-cost formulation of diets by linear programming and has brought computer programs for least-cost diet formulation to the farm. Individual pork producers and commercial feed manufacturers use computers to capture optimum combinations of available feedstuffs at least cost.

MECHANICS OF DIET FORMULATION

Generally, diets are first formulated for one nutrient, followed by checking the ingredients to determine if other feedstuffs need to be included. Protein or lysine (the first limiting amino acid in corn and most cereal grains) is often balanced first, followed by energy. The National Research Council (1998) described a simple approach to diet formulation for swine:

> From a nutritional standpoint, there is no "best formula" in terms of the ingredients that are used in the diet. Therefore, ingredients should be selected on the basis of availability, price, and quantity and quality of the nutrients that they contain. Corn, grain sorghum, barley, and wheat are the primary energy-supplying ingredients in diets for swine weighing 10 kg or more. These cereal grains are severely deficient in several essential amino acids, minerals, and vitamins. Soybean meal, other oilseed meals, and animal-protein meals are generally added as sources of supplemental amino acids to the grain, but they too are deficient in many of the essential minerals and vitamins.
>
> Swine diets can be formulated using rather simple mathematical procedures with a hand-held or desk calculator when a few ingredients are used in the diet. However, more sophisticated formulation procedures are needed to more precisely meet the dietary requirements on a bio-available nutrient basis and when using larger numbers of ingredients that differ in their nutrient bio-availability. These formulation procedures often require computer programs and the expertise of a professional nutritionist.

The National Research Council (1998), using standard feed composition tables (some of these tables are reproduced in Tables 11–2a and 11–2b in Chapter 11) and information on nutrient requirements for each productive function (e.g., growth, reproduction, lactation), presented procedures to formulate diets based on total nutrients and, in the case of amino acids and phosphorus, on a bio-available nutrient basis. The National Research Council (1998) also provided examples of the calculation procedures. The procedures address formulation of a diet using corn and soybean meal as the primary ingredients. The following sections summarize the NRC calculation procedures. Diets can be formulated on a total nutrient basis or on an available nutrient basis. For the following example, the formulation is on a total nutrient basis.

In swine diets formulated with corn and soybean meal, the two ingredients contribute about 97.5% of the total diet. The remaining 2.5% consists of mineral supplements and carrier mixes containing vitamins, trace minerals, and additives. Both corn and soybean meal are high in digestible energy (DE) concentration. Any combination of these two ingredients will result in a relatively high-energy diet.

STEPS IN DIET FORMULATION

The steps in diet formulation are quoted directly from National Research Council (1998). The first step in diet formulation is presented in Equation 9-1, where C is the percentage of corn and S is the percentage of dehulled soybean meal in the diet.

$$C + S = 97.5 \quad (9\text{-}1a)$$

or

$$S + 97.5 - C \quad (9\text{-}1b)$$

Lysine is the first limiting amino acid in corn-soybean meal diets. Because of this, producers can manipulate the proportions of corn and dehulled soybean meal to meet the required concentrations of this amino acid and be reasonably sure that the requirements for all other essential amino acids will be met and that the amount of nonessential amino acid nitrogen will be adequate. Equation 9-2 formulates a corn-soybean meal diet for a 40-kg pig:

$$(A + C) + (B \times (97.5 - C)) = (L \times 100) \quad (9\text{-}2)$$

where A is the percentage of lysine in corn, C is the percentage of corn in the diet, B is the percentage of lysine in soybean meal, 97.5 − C is the percentage of soybean meal in the diet, and L is the lysine requirement of the 40-kg pig, expressed as a percentage of the diet.

Values for A, B, and L are then inserted into Equation 9-1, leaving only one unknown (C). The percentages of corn and soybean meal in the diet can then be solved as follows:

$$0.26C = 3.02(97.5 - C) = (0.90 \times 100)$$

where C is 74.1% corn in the diet. Because S is 97.5 − C, S is 23.41% soybean meal in the diet.

The next step is to add an ingredient to supply inorganic phosphorus to complete the requirement (0.50%) for total phosphorus. If dicalcium phosphate, which contains 18.5% phosphorus, is selected, Equation 9-3 will show the percentage of dicalcium phosphate (DP) to include in the diet.

$$(18.5 \times DP) = (0.50 \times 100) - (74.1 \times \% \text{ P in corn}) - (23.4 \times \% \text{ P in soybean meal})$$

$$(18.5 \times DP) = (0.50 \times 100) - (74.1 \times 0.28) - (23.4 \times 0.69) \quad (9\text{-}3)$$

$$DP = 0.71\% \text{ dicalcium phosphate in diet}$$

The next step is to add an ingredient to supply calcium to complete the requirement for calcium (0.60%). If ground limestone, which contains 38% calcium, is selected, Equation 9-4 will show the percentage of ground limestone (GL) to include in the diet.

$$(38 \times GL) = (0.60 \times 100) - (74.1 \times \% \text{ Ca in corn}) - (23.4 \times \% \text{ Ca in soybean meal}) - (0.71 \times \% \text{ Ca in dicalcium phosphate})$$

$$(38 \times GL) = (0.60 \times 100) - (74.1 \times 0.03) - (23.4 \times 0.34) - (0.71 \times 22)$$

$$GL = 0.90\% \text{ ground limestone in diet}$$

One can completely fortify the swine diet by adding 0.25% sodium chloride; a vitamin premix that supplies the vitamins deficient in the corn-soybean meal mixture (vitamins A, D, E, K, B_{12}, riboflavin, niacin, and pantothenic acid); a trace mineral premix that supplies the trace minerals that may be deficient (iron, zinc, copper, manganese, iodine, and selenium); and, if desired, a premix that contains one or more antimicrobial agents. The fortified diet is shown in Table 9–2 of National Research Council (1998) and in Table 10–6 in Chapter 10 of this book. Refer to Chapter 10 to study the composition of the diet formulated in the exercise from the National Research Council (1998). The diet is made to total 100% by increasing the amount of corn to 74.44%.

Formulation on a true or apparent digestible lysine basis is essentially the same as just described except that the true or apparent digestible lysine values for corn and soybean meal are used in the calculations.

In essence, the mechanics of least-cost formulation of diets for pigs require that the following information be supplied:

1. Tolerance (upper and lower limits) of specific ingredients
2. Accurate feed composition data
3. Current ingredient costs

If these three pieces of information are used, diets that meet all mathematical requirements can be formulated, but the final measure of acceptability is the performance of the pigs fed the diet. Computerized diet formulation does not account for factors such as palatability, which may have an important effect on pig response. In the final analysis, the most sophisticated computer cannot take the place of the pig for which the diet was formulated in determining the acceptability of the diet.

EFFECTS OF FEED PROCESSING ON NUTRITIVE VALUE, PALATABILITY, AND FEED CONSUMPTION

Hancock and Behnke (2001) and Hogberg et al. (1980) wrote excellent reviews of the effects of ingredient and diet processing on pig performance. Many potentially useful feedstuffs for pigs would remain unused if appropriate processing were not possible. The feeding value of some feedstuffs is low because a chemical property reduces the bio-availability of one or more nutrients (for example, heat-labile inhibitors in raw soybeans). Several processing methods, including grinding, fineness of particle size, pelleting, extrusion/expansion, heating, drying, and oil extraction have been applied in the feed industry to improve a variety of feedstuffs for pig feeding. These methods and the types of mills and mixer designs and procedures and their effects on pig performance have been summarized by Hancock and Behnke (2001).

Recent evidence (Hancock and Behnke, 2001) indicates that the high incidence of esophagogastric ulcers in pigs fed finely ground diets is associated with development of low pH in the stomach contents. The problem of stomach lesions appears to be reduced by the addition of alkaline salts (buffers) to neutralize the acidity of the stomach. This exemplifies the unforeseen problems sometimes encountered by feed processing techniques designed to improve pig performance.

Processing can be a physical, chemical, thermal, bacterial, or other alteration of a feed or feed ingredient. Feeds are processed for several reasons:

1. To alter the physical form or particle size
2. To isolate specific parts
3. To preserve
4. To improve palatability
5. To improve digestibility
6. To alter nutrient composition
7. To detoxify

Grinding, pelleting, and extrusion/expansion generally improve growth and feed utilization of pigs. The improvement observed in pigs fed pelleted diets is generally greater with barley and other higher fiber cereal grains than with corn or grain sorghum. Milling and distillery by-products available for pig diet formulation are examples of feedstuffs produced by isolation of specific parts—in this case, to obtain starch and other valuable products for human consumption. Soybean meal and other seed legume protein supplements represent another example of feedstuffs available as a result of the isolation of specific parts of the seed for another purpose (in this case, vegetable oil).

A common example of feed preservation is the ensiling of high-moisture grain or forage. However, drying of newly harvested corn grain to reduce its moisture content to 15% or 16% also preserves it from mold or fungus damage. Heating can be a two-edged sword. It improves the utilization of some feed ingredients if properly applied, but it may reduce efficiency of utilization if excessive. Over-heating can affect the nutritive value of protein and carbohydrate fractions and may destroy vitamins. When damage occurs, it is generally because of inadequate control of the processing conditions.

Hogberg et al. (1980) summarized the effects of processing feedstuffs on growth rate and efficiency of feed utilization of growing pigs (see Table 12–1). The direction and magnitude of response to a specific set of processing conditions is never completely predictable due to variables such as genotype and environment of the pigs, plant growing conditions, crop harvesting times and methods, and plant variety. The development of genetically modified (GM) crops introduces an additional variable of unknown consequence to feedstuff processing responses.

PRACTICAL SWINE DIETS

As a practical matter, swine diets can be formulated to precisely meet the needs of the growing pig for maximum rate of body weight gain and efficiency of lean growth, or

TABLE 12–1
Effects of Processing Feedstuffs on Pig Performance.

Processing Method	Type of Grain	Growth Rate	Change in Feed Intake	Feed Conversion	Comments
Grinding	Corn	Improved 3–5%	No effect	Improved 3–5%	Medium screen (1/4–3/8 in.). Generally most acceptable.
	Grain sorghum	Improved	No effect	Improved	Too fine of grind can reduce palatability.
Pelleting	Corn, grain sorghum	Improved 3–6%	Reduced 1–3%	Improved 5–8%	
	Barley, oats, alfalfa, wheat, bran	Improved 3–6%	Reduced 1–3%	Improved 7–10%	Greater improvement with higher fibrous materials.
Paste	Corn	Improved 10–15%	Increased 10–15%	No change	Feed:Water ratio of 1.2–1.5:1. Advantage with G-F pigs but not weanlings.
Liquid	Corn	Improved	Increased	Improved	Water:Feed ratio of 2.1. Advantages with limit-fed systems. Full-fed levels fail to improve performance.
Roasting	Corn	No change	No change	Slightly improved	Processing cost greater than improvement return.
	Soybeans	No change	Decreased	Improved 4–6%	Necessary for adequate performance. Comparisons made with SBM.
Steam flaking	Corn, grain sorghum	No change	No change	No change	
Micronizing	Corn, grain sorghum	No change	No change	Variable	Availability of amino acids may be reduced.
Extruding	Wheat, grain sorghum	No change	No change	No change	
	Soybeans	No change	Decreased	Improved 4–6%	Comparison made with SBM—necessary for adequate performance.

the needs of the gestating and lactating sow for maximum reproductive and lactation performance. Alternatively, diets can merely meet the basic needs of the pig or sow for modest growth and reproductive performance. The pork producer should determine the degree of complexity and cost of the diet based on the level of production desired and on an economic analysis of the particular short- and long-term production goals and limitations. Over time, the best diets based on economic analysis of the particular production unit may be in continuous change.

Many examples of practical swine diets are available in publications from extension personnel in university departments of animal science in the United States and in other countries. Diets recommended by Kansas State University swine nutritionists (Tokach et al., 1997a and 1997b) have been selected to represent examples of nutritionally adequate practical complete formulas for grower, finisher, gestating, and lactating swine.

Producers must consider how much feed pigs will consume during each phase of post-weaning growth. Table 12–2 (Dritz et al., 1997) provides a feed budget chart for pigs from 9 to 299 lb body weight. Efficiency of feed utilization over the entire period is 3:1 (875 lb of feed/pig from 9 to 299 lb = 875/290 = 3.02 lb of feed/lb of gain). A higher feed-to-gain ratio would result in a higher feed requirement to reach market

TABLE 12–2
Standard Feed Budget Chart Based on a Feed Efficiency of 3.0 from 50 to 250 lb.

Pig Weight	Total Feed	Pig Weight	Total Feed	Pig Weight	Total Feed	Pig Weight	Total Feed	Pig Weight	Total Feed
9	0	68	110	126	258	184	434	242	640
10	1	69	112	127	261	185	437	243	644
11	2	70	115	128	263	186	440	244	647
12	3	71	117	129	266	187	444	245	651
14	6	72	119	130	269	188	447	246	655
15	7	73	122	131	272	189	450	247	659
16	8	74	124	132	275	190	454	248	663
17	10	75	126	133	277	191	457	249	667
18	11	76	129	134	280	192	460	250	671
19	12	77	131	135	283	193	464	251	675
20	14	78	133	136	286	194	467	252	678
21	16	79	136	137	289	195	470	253	682
22	17	80	138	138	292	196	474	254	686
23	19	81	141	139	295	197	477	255	690
24	20	82	143	140	298	198	481	256	694
25	22	83	145	141	300	199	484	257	698
26	24	84	148	142	303	200	487	258	702
27	25	85	150	143	306	201	491	259	706
28	27	86	153	144	309	202	494	260	710
29	29	87	155	145	312	203	498	261	714
30	31	88	158	146	315	204	501	262	718
31	33	89	160	147	318	205	505	263	722
32	34	90	163	148	321	206	508	264	726
33	36	91	165	149	324	207	512	265	730
34	38	92	168	150	327	208	515	266	734
35	40	93	170	151	330	209	519	267	739
36	42	94	173	152	333	210	522	268	743
37	44	95	175	153	336	211	526	269	747
38	46	96	178	154	339	212	529	270	751
39	48	97	180	155	342	213	533	271	755
40	50	98	183	156	345	214	536	272	759
41	52	99	185	157	348	215	540	273	763
42	54	100	188	158	351	216	543	274	767
43	56	101	191	159	354	217	547	275	772
44	58	102	193	160	357	218	551	276	776
45	60	103	196	161	360	219	554	277	780
46	62	104	198	162	364	220	558	278	784
47	64	105	201	163	367	221	561	279	788
48	66	106	204	164	370	222	565	280	793
49	68	107	206	165	373	223	569	281	797
50	70	108	209	166	376	224	572	282	801

(continued)

TABLE 12–2 (continued)
Standard Feed Budget Chart Based on a Feed Efficiency of 3.0 from 50 to 250 lb.

Pig Weight	Total Feed	Pig Weight	Total Feed	Pig Weight	Total Feed	Pig Weight	Total Feed	Pig Weight	Total Feed
51	72	109	212	167	379	225	576	283	805
52	75	110	214	168	382	226	580	284	809
53	77	111	217	169	385	227	583	285	814
54	79	112	220	170	389	228	587	286	818
55	81	113	222	171	392	229	591	287	822
56	83	114	225	172	395	230	594	288	827
57	85	115	228	173	398	231	598	289	831
58	88	116	230	174	401	232	602	290	835
59	90	117	233	175	404	233	606	291	840
60	92	118	236	176	408	234	609	292	844
61	94	119	238	177	411	235	613	293	848
62	96	120	241	178	414	236	617	294	853
63	99	121	244	179	417	237	621	295	857
64	101	122	247	180	421	238	624	296	861
65	103	123	249	181	424	239	628	297	866
66	105	124	252	182	427	240	632	298	870
67	108	125	255	183	430	241	636	299	875

Source: Adapted from Dritz et al. (1997), Reproduced with permission from S. S. Dritz, R .D. Goodband, J. N. Nelssen, and M. D. Tokach.

weight, whereas a lower feed-to-gain ratio would result in a lower feed requirement. To aid in planning, feed intake patterns should be determined for each genetic line in a given production unit.

When pigs are fed in a life-cycle feeding scheme, the number of diet changes may vary from three or four to dozens. In reality, pig nutrient requirements change every day, but it may not be practical to change diets every day. Therefore, the number of diets chosen is dependent on factors such as mill capacity, feed storage space, labor supply, and level of stockmanship on the farm.

Table 12–3 lists the dietary and management phases for two scenarios: when segregated early weaning (SEW) is practiced and when pigs are weaned conventionally after 3 wk of age (Tokach et al., 1997b). When SEW is practiced, pigs are fed a complex starter diet from 5 to 11 lb body weight (see composition of an acceptable starter diet in Chapter 10), followed by a transition diet or phase 1 diet from 11 to 15 lb, a phase 2 diet from 15 to 25 lb, and a phase 3 diet from 25 to 50 lb. Phase 2 and phase 3 diets are shown in Table 12–4 and Table 12–5 (Tokach et al., 1997a) respectively. Under either scenario, the diet change schedule is identical from 15 to 50 lb body weight.

After these nursery phases, a grower diet is normally fed from 50 lb to midway through the grower-finisher period, followed by the finisher diet to market weight at 240 to 260 lb (see Tables 12–6 and 12–7, respectively). A range of dietary lysine levels is shown in both tables and a continuum of values extends through the finishing period. This relationship is evident by comparing the values for dietary lysine content

TABLE 12–3
Sequence of Phase-Feeding Programs for Early and Conventionally Weaned Pigs.

EARLY WEANING 5- TO 21-D WEANING		CONVENTIONAL WEANING GREATER THAN 21-D WEANING	
5–11 lb	SEW		
11–15 lb	Transition	11–15 lb	Phase 1
15–25 lb	Phase 2	15–25 lb	Phase 2
25–50 lb	Phase 3	25–50 lb	Phase 3

Source: Adapted from Tokach et al. (1997b), Breeding Herd Recommendations for Swine, MF2302, Kansas State University Agricultural Experiment Station and Cooperative Extension Service. Reproduced with permission from S. S. Dritz, R. D. Goodband, J. L. Nelssen, and M. D. Tokach.

TABLE 12–4
Acceptable Phase 2 Diets for Pigs Weighing 15 to 25 lb.

	NO FAT		ADDED FAT	
INGREDIENT, LB/TON	FISH MEAL	BLOOD MEAL	FISH MEAL	BLOOD MEAL
Corn or milo	1102.8	1117	962.6	976.7
Soybean meal, 46.5% CP	540	540	580	580
Select menhaden fish meal	80	—	80	—
Spray-dried blood meal	—	50	—	50
Spray-dried whey	200	200	200	200
Choice white grease	—	—	100	100
Monocalcium phosphate, 21% P	27	37	27	37
Limestone	14	19	14	19
Salt	5	5	5	5
Vitamin premix	5	5	5	5
Trace mineral premix	3	3	3	3
Lysine HCl	3	3	3	3
DL-Methionine	0.2	1	0.4	1.3
Antibiotic	20	20	20	20
Total	2000	2000	2000	2000
Calculated Analysis				
Lysine, %	1.35	1.35	1.40	1.40
Met:lysine ratio, %	28	28	28	28
Met & Cys:lysine ratio, %	56	56	55	55
Threonine:lysine ratio, %	66	66	65	65
Tryptophan;lysine ratio, %	20	20	19	20
ME, kcal/lb	1,476	1,473	1,574	1,570
Protein, %	21.1	20.8	21.4	21.1
Calcium, %	.90	.89	.90	.89
Phosphorus, %	.80	.80	.80	.80
Available phosphorus, %	.54	.54	.54	.54
Lysine:calorie ratio, g/Mcal ME	4.15	4.16	4.03	4.04

Source: Tokach et al. (1997a), Starter Pig Recommendations, Table MF2300, Kansas State University Agricultural Experiment Station and Cooperative Extension. Reproduced with permission from S. S. Dritz, R. D. Goodband, J. L. Nelssen, and M. D. Tokach.

TABLE 12–5
Acceptable Phase 3 Diets for Pigs Weighing 25 to 50 lb.

	No Fat		Added Fat	
Ingredient, lb/ton	**1.25**	**1.30**	**1.30**	**1.35**
Corn or milo	1276.6	1242.5	1131.3	1096.2
Soybean meal, 46.5% CP	645	680	690	725
Choice white grease	—	—	100	100
Monocalcium phosphate, 21% P	30	29	30	30
Limestone	20	20	20	20
Salt	7	7	7	7
Vitamin premix	5	5	5	5
Trace mineral premix	3	3	3	3
Lysine HCl	3	3	3	3
DL-Methionine	0.4	0.5	0.7	0.8
Antibiotic premix	10	10	10	10
Total	2000	2000	2000	2000
Calculated Analysis				
Lysine, %	1.25	1.30	1.30	1.35
Met:lysine ratio, %	28	28	28	28
Met & Cys:lysine ratio, %	58	58	57	57
Threonine:lysine ratio, %	67	67	66	65
Tryptophan:lysine ratio, %	21	21	21	21
Me, kcal/lb	1,485	1,486	1,583	1,583
Protein, %	20.4	21.1	20.9	21.5
Calcium, %	.76	.76	.76	.77
Phosphorus, %	.70	.70	.70	.70
Available phosphorus, %	.39	.38	.39	.40
Lysine:calorie ratio, g/Mcal ME	3.82	3.97	3.73	3.87

Source: Tokach et al. (1997a), Starter Pig Recommendations, Table 7, MF 2300, Kansas State University Agricultural Experiment Station and Cooperative Extension. Reproduced with permission from S. S. Dritz, R. D. Goodband, J. L. Nelssen, and M. D. Tokach.

shown in Table 12–6 (grower diets) and Table 12–7 (finisher diets). Tables 12–6 and 12–7, which were adapted from Dritz et al. (1997), do not show the changes in ingredient composition that result when energy density of the diet is increased by adding fat. Dritz et al. (1997), as shown in Tables 12–5 and 12–6, illustrated these changes by adding 5% choice white grease to a corn-soybean meal-based diet. The lysine-to-calorie ratio was kept constant by reducing the corn and increasing the soybean meal concentrations. When ingredient price relationships favor fat as an energy source, it is then possible to utilize fat economically as a substitute for some of the common plant energy sources in the diet.

Table 12–8 (Tokach et al., 1997b) shows the composition of practical gestation diets. Note that composition does not change during gestation, but it must be recognized that the daily feed allowance does change as illustrated in Figure 12–2. Feed intake during lactation must be maximized for adequate milk production. Therefore,

TABLE 12–6
Acceptable Grower Diets (Phase 4) for Pigs Weighing 50 to 150 lb.*

	GROWER DIETS WITHOUT FAT LYSINE, %					
INGREDIENT, LB/TON	.80	.90	1.00	1.10	1.20	1.30
Corn or milo	1620.0	1551.0	1476.0	1405.9	1336.7	1261.5
Soybean meal, 46.5%	320	390	465	535	605	680
Choice white grease	0	0	0	0	0	0
Monocalcium phosphate, 21% P	25	24	24	24	23	23
Limestone	19	19	19	19	19	19
Salt	7	7	7	7	7	7
Vitamin premix	3	3	3	3	3	3
Trace mineral premix	3	3	3	3	3	3
Lysine HCl	3	3	3	3	3	3
DL-Methionine	0	0	0	0.1	0.3	0.5
Total	2000.0	2000.0	2000.0	2000.0	2000.0	2000.0
CALCULATED ANALYSIS						
Lysine, %	.80	.90	1.00	1.10	1.20	1.30
Methionine:lysine ratio, %	31%	30%	28%	28%	28%	28%
Met & Cys:lysine ratio, %	68%	65%	62%	60%	59%	58%
Threonine:lysine ratio, %	74%	72%	70%	69%	68%	67%
Tryptophan:lysine ratio, %	22%	22%	21%	21%	21%	21%
ME, kcal/lb	1,502	1,502	1,502	1,501	1,501	1,501
Protein, %	14.3	15.7	17.1	18.4	19.7	21.2
Calcium, %	.66	.66	.67	.67	.67	.68
Phosphorus, %	.59	.59	.61	.62	.62	.64
Available phosphorus, %	.32	.32	.32	.32	.32	.32
Lysine:calorie ratio, g/Mcal ME	2.42	2.71	3.03	3.32	3.61	3.93

*Note the continuum in lysine level and changes in concentrations of corn or milo and soybean meal throughout the grower and finisher periods in Tables 12–6 and 12–7.

Source: Dritz et al. (1997), Growing-Finishing Pig Recommendations, Table 6, MF2301, Kansas State University Agricultural Experiment Station and Cooperative Extension. Reproduced with permission from S. S. Dritz, R. D. Goodband, J. L. Nelssen, and M. D. Tokach.

the diets shown in Table 12–9 (Tokach et al., 1997b) are intended to be fed ad libitum through lactation. Note that the lysine content of the diet in the example in Table 12–9 is altered by increasing the soybean meal and decreasing the grain in the diet. The level of lysine (and protein) selected is determined by a knowledge of the productivity of the sow herd. Table 12–10 (Tokach et al., 1997b) shows the relationship between adjusted 21-d weaning weight, sow lactation feed intake, and daily lysine intake needed to accommodate the productivity level expected.

Boars can be fed diets that are identical to sow gestation diets. Intake must be restricted to prevent overfattening, as with gestating sows.

TABLE 12–7
Acceptable Finisher Diets (Phase 5) for Pigs Weighing 150 to 250 lb.*

INGREDIENT, LB/TON	FINISHER DIETS WITHOUT FAT LYSINE, %					
	.50	.60	.70	.80	.90	1.00
Corn or milo	1820.5	1779.5	1704.5	1634.5	1559.5	1485.5
Soybean meal, 46.5%	130	170	245	315	390	465
Choice white grease	0	0	0	0	0	0
Monocalcium phosphate, 21% P	19	19	19	18	18	17
Limestone	17	17	17	18	18	18
Salt	7	7	7	7	7	7
Vitamin premix	2.5	2.5	2.5	2.5	2.5	2.5
Trace mineral premix	2	2	2	2	2	2
Lysine HCl	2	3	3	3	3	3
DL-Methionine	0	0	0	0	0	0
Total	2000.0	2000.0	2000.0	2000.0	2000.0	2000.0
CALCULATED ANALYSIS						
Lysine, %	0.50	0.60	0.70	0.80	0.90	1.00
Methionine:lysine ratio, %	41%	36%	33%	31%	30%	28%
Met & Cys:lysine ratio, %	91%	80%	73%	69%	65%	62%
Threonine:lysine ratio, %	90%	81%	77%	74%	72%	70%
Tryptophan:lysine ratio, %	25%	22%	22%	22%	22%	21%
ME, kcal/lb	1,511	1,510	1,509	1,508	1,508	1,508
Protein, %	10.8	11.5	12.9	14.3	15.7	17.1
Calcium, %	.54	.55	.56	.57	.58	.58
Phosphorus, %	.50	.50	.52	.52	.53	.54
Available phosphorus, %	.25	.25	.25	.25	.25	.25
Lysine:calorie ratio, g/Mcal ME	1.51	1.79	2.10	2.39	2.71	3.02

*Note the continuum in lysine level and changes in concentrations of corn or milo and soybean meal throughout the grower and finisher phases in Tables 12–6 and 12–7.

Source: Dritz et al. (1997), Growing-Finishing Pig Recommendations, Table 7, MF2301, Kansas State University Agricultural Experiment Station and Cooperative Extension. Reproduced with permission from S. S. Dritz, R. D. Goodband, J. L. Nelssen, and M. D. Tokach.

TABLE 12–8
Acceptable Gestation Diets.

		LYSINE, %	
INGREDIENT, LB/TON	.55	.60	.65
Corn or milo	1684	1665	1611
Soybean meal, 46.5% CP	225	260	300
Choice white grease	—	—	—
Monocalcium phosphate, 21% P	47	46	45
Limestone	21	21	21
Salt	10	10	10
Vitamin premix	5	5	5
Sow add pack	5	5	5
Trace mineral premix	3	3	3
Total	2000	2000	2000
CALCULATED ANALYSIS			
Lysine, %	.55	.60	.65
Met:lysine ratio, %	41	39	37
Met & Cys:lysine ratio, %	90	85	81
Threonine:lysine ratio, %	94	91	88
Tryptophan:lysine ratio, %	27	26	26
ME, kcal/lb	1,480	1,480	1,480
Protein, %	12.8	13.5	14.2
Calcium, %	.90	.90	.90
Phosphorus, %	.80	.80	.80
Available phosphorus, %	.55	.54	.53

Source: Tokach et al. (1997b), Breeding Herd Recommendations for Swine, MF2302, Kansas State University Agricultural Experiment Station and Cooperative Extension. Reproduced with permission from S. S. Dritz, R. D. Goodband, J. L. Nelssen, and M. D. Tokach.

FIGURE 12–2
Gestation Feed Intake Pattern. Dashed lines are options.

Source: Adapted from Tokach et al. (1997).

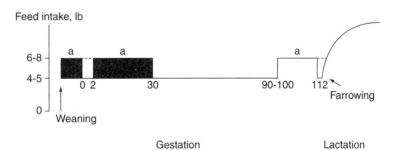

TABLE 12-9
Acceptable Lactation Diets.

	Lysine, %					
Ingredient, lb/ton	.70	.80	.90	1.0	1.1	1.2
Corn or milo	1581	1507	1438	1364	1295	1227
Soybean meal, 46.5% CP	330	405	475	550	620	690
Choice white grease[a]	0 to 5%	0 to 5%	0 to 5%	0 to 5%	0 to 5%	0 to 5%
Monocalcium phosphate, 21% P	45	44	42	41	40	38
Limestone	21	21	22	22	22	22
Salt	10	10	10	10	10	10
Vitamin premix	5	5	5	5	5	5
Sow add pack	5	5	5	5	5	5
Trace mineral premix	3	3	3	3	3	3
Total	2000	2000	2000	2000	2000	2000
Calculated Analysis						
Lysine, %	.70	.80	.90	1.00	1.10	1.20
Met:lysine ratio, %	36	33	32	30	29	28
Met & Cys:lysine ratio, %	79	73	69	66	63	61
Threonine:lysine ratio, %	86	82	79	76	74	73
Tryptophan:lysine ratio, %	25	25	24	24	23	23
ME, kcal/lb	1,480	1,480	1,480	1,480	1,480	1,480
Protein, %	14.8	16.5	17.5	18.9	20.2	21.6
Calcium, %	.90	.90	.90	.90	.90	.90
Phosphorus, %	.80	.80	.80	.80	.80	.80
Available phosphorus, %	.53	.53	.51	.50	.50	.48

[a]If adding fat, substitute for grain on an equal weight basis.
Source: Tokach et al. (1997b), Breeding Herd Recommendations for Swine, MF2302, Kansas State University Agricultural Experiment Station and Cooperative Extension. Reproduced with permission from S. S. Dritz, R. D. Goodband, J. L. Nelssen, and M. D. Tokach.

TABLE 12-10
Dietary Lysine Level Based on Litter Weaning Weight and Sow Feed Intake.

Adj. 21-Day Litter Weaning Weight, lb	Lactation Feed Intake, lb/d								Lysine, grams/d
	8	9	10	11	12	13	14	15	
100	1.0	.90	.80	.70	.70				36
110	1.1	1.0	.90	.80	.75	.70			40
120	1.2	1.1	1.0	.90	.80	.75	.70		45
130		1.2	1.1	1.0	.90	.85	.80	.75	50
140			1.2	1.1	1.0	.95	.90	.80	55
150				1.2	1.1	1.0	.95	.90	60

Source: Tokach et al. (1997b), Breeding Herd Recommendations for Swine, MF2302, Kansas State University Agricultural Experiment Station and Cooperative Extension. Reproduced with permission from S. S. Dritz, R. D. Goodband, J. L. Nelssen, and M. D. Tokach.

SUMMARY

The judicious formulation of swine diets to provide adequate nutrition during all phases of the life cycle ensures high total efficiency of feed utilization for high productivity in a particular swine production enterprise and for the swine industry as a whole. The careful selection of an appropriate blend of economically and nutritionally adequate feed ingredients in diet formulation is fundamental to a flourishing swine industry.

This chapter describes the principles of diet formulation for pigs and introduces the methods used by nutritionists and pig producers in formulating diets in the feed industry and on the farm, based on the use of feed ingredients produced on the farm along with purchased supplements. Diet formulation requires consideration of the nutrient requirements to be satisfied, the available feed resources, the cost of supplying the required nutrients from available ingredients and, finally, feed processing needed to optimize diet intake, nutritive value, and pig performance. The National Research Council's Nutrient Requirements of Swine, 10th rev. ed. (1998), provides the foundation for the application of computer technology to diet formulation for pigs of all stages of the life cycle. Examples of acceptable diets for each phase of the life cycle, based on commonly available feed ingredients in the United States, are provided.

QUESTIONS AND ACTIVITIES

1. If pigs are in a cold environment, they will eat feed to keep their body warm. Thus, body heat may be provided by use of a heat source (such as natural gas) or by allowing the pigs to eat more feed. As the air temperature gets colder, pigs will eat progressively more until they reach a very cold temperature below which eating more feed is not possible (the level we call gut fill). If a pig could double its feed intake and that increase in feed were available for heat energy at a rate of 50% of the gross energy of the feed, how many calories would be able to be burned by this added feed intake for a 100 kg pig?

2. In the USA, most pigs are fed corn-soybean meal diets. By standard nomenclature, this means the diet is primarily corn and then the next most abundant nutrient is soybean meal. A given diet may be 75% corn and 20% soybean meal and 5% vitamin-mineral supplement. In Europe, at times, the price of soybean is actually cheaper than the price of corn. In this case, the feed manufacturer might feed a diet that is 75% soybean meal and 20% corn. Which amino acid would be most limiting in a soy-corn diet?

3. In Europe, due to high feed costs, many pigs are limit fed energy to 95% of ad libitum intake in an effort to improve feed efficiency. Would the diet composition need to be adjusted in such limit-fed pigs? In what way?

4. Listed in Tables 12–4 and 12–5 are diets recommended by Kansas State University swine nutritionists. Considering an example nutrient such as lysine; how do the Kansas State recommendations differ from the NRC (1998) recommendations? Why do you think these different diets/nutrients are recommended?

5. Several universities have swine diet information on the Internet, through the Animal Science Extension Service. How do dietary recommendations compare for swine diets from the University of Illinois, University of Nebraska, Iowa State University and Kansas State University?
6. The primary criterion for setting nutrient requirements in growing pigs is the level of lean gain and the feed efficiency that a given diet and set of nutrients will attain. What would the nutrient requirements look like if other criteria were used? Among the other criteria are: health or immune system optimization, body fat composition, litter size, waste nutrient minimization and others.

LITERATURE CITED

Dritz, S. S., M. D. Tokach, R. D. Goodband, and J. L. Nelssen, 1997. Growing-Finishing Pig Recommendations, MF2301, Kansas State University Agricultural Experiment Station and Cooperative Extension Service, Manhattan, KS.

Hancock, J. D. and K. C. Behnke. 2001. Use of ingredient and diet processing technologies (grinding, mixing, pelleting, and extruding) to produce quality feeds for pigs. In: A. J. Lewis, and L. L. Southern (eds.) Swine Nutrition. 2nd ed., pp. 469–497. CRC Press, Boca Raton, FL.

Hogberg, M., D. Mahan, and R. Seerley. 1980. Physical Forms of Feed—Feed Processing for Swine. National Pork Handbook PIH-71. Cooperative Extension Service, Purdue University, West Lafayette, IN.

National Research Council. 1998. Nutrient Requirements of Swine. 10th rev. ed. National Academy Press, Washington, DC.

Tokach, M. D., S. S. Dritz, R. D. Goodband, and J. L. Nelssen. 1997a. Starter Pig Recommendations, MF2300, Kansas State University Agricultural Experiment Station and Cooperative Extension Service, Manhattan, KS.

Tokach, M. D., S. S. Dritz, R. D. Goodband, and J. L. Nelssen. 1997b. Breeding Herd Recommendations for Swine, MF2302, Kansas State University Agricultural Experiment Station and Cooperative Extension Service, Manhattan, KS.

Van Soest, P. J. 1967. Development of a comprehensive system of feed analyses and its application to forages. J. Anim. Sci. 26:119–128.

INTERNET RESOURCES

Nutrient Requirements of Swine, Tenth Revised Edition, 1998:
http://www.nap.edu/catalog/6016.html

Tri-State Swine Nutrition Guide, Bulletin 869-98, Feed Ingredients:
http://www.ag.ohio-state.edu/,ohioline//b869_51.html

Tri-State Swine Nutrition Guide, Bulletin 869-98, The Starter Pig:
http://ohioline.ag.ohio-state.edu/b869/b869_25.html

Amino Acids in Pig Diets:
http://www.dpi.qld.gov.au/pigs/4374.html

NIH Open Formula Laboratory Animal Diets and Feed Specifications:
http://vrp.odnih.gov/nutrition.htm

Aspects of Feed Formulation and Preparation:
http://www.aps.uoguelph.ca/faculty/delange.html

Formulating Swine Rations/Feed Additives:
http://web.utk.edu/~amathew/4831/ed19.html

SECTION IV

HOUSING, ENVIRONMENT, AND NUTRIENT MANAGEMENT

13

CREATING A COMFORTABLE MICROENVIRONMENT FOR PIGS

INTRODUCTION

In pig production, the term "environment" has recently come to mean the area surrounding the pig facility. People talk about environmental pollution, for example, in terms of air and water pollution. In this case, the term "environment" is used in the broader sense as understood by the public. The area surrounding each pig is its microenvironment. The macroenvironment, or the natural environment in the broad sense, is the area outside the pig facility.

Pigs can be kept extensively or intensively. Extensively kept pigs are in outdoor or open-air facilities (open on at least one side). The traditional contrast is for intensively kept pigs to be inside buildings. However, production intensity actually means that the output of animals is geared toward the pig's biological maximum. Thus, it is possible to keep pigs in extensive environments on an intensive production schedule. Likewise, it is possible, however unlikely, that pigs could be kept indoors on a nonintense production schedule (economic forces usually prevent this model).

Regardless of whether pigs are kept indoors or outdoors, the pork producer must attend to the needs and comfort of the pig by providing an appropriate microenvironment. The natural environment, if poorly managed, can be unkind to otherwise healthy and productive animals. On a cold day, however, deep, dry bedding can often be as comfortable as a heated indoor space.

The microenvironment inside a building can be modified to suit the animals' needs. Generally, animals are confined to improve productivity, as evidenced in improved weight gain and milk production, and the reduction of feed required to yield animal products. This chapter describes the factors that affect the microenvironment and addresses how these factors can be modified or managed to enhance pig comfort.

COMPONENTS OF THE MICROENVIRONMENT

What is the microenvironment? Biology teaches us that the environment is all factors other than genetics. Most people see animal traits and responses, such as rate of weight gain, as functions of both genetics and environment.

Another viewpoint considers the microenvironment as all factors external to the animal. So, the microenvironment is composed of the thermal environment (air temperature, wind speed, and moisture); the physical environment (pens, walls, and floors); the social environment (the animals that interact with each individual); and the disease and microbial environment. Diet can also be considered part of the microenvironment.

A building's physical environment is often chosen on the basis of cost and availability. Producers who are making a decision regarding pen and wall materials should also consider the materials' thermal and sanitary properties (for example, concrete is colder to lie against than plastic, and metal can be disinfected more effectively than wood).

The social environment is not often given enough thought during livestock facility construction. In social groups (many animals in a pen), the level of social stress (fighting) is high and productivity may decline. A large group may be more adversely affected in warm weather. In cold weather, groups of animals conserve heat better than individually penned animals. However, animals are often penned individually to closely regulate nutrition and prevent spread of disease. When the social environment is considered (directly or indirectly), it is possible to optimize group size, space allowance, and feeder, waterer, and resting space.

The microbial environment is controlled by sanitation, ventilation, and management of the social environment. An effective waste management program will go a long way toward reducing disease problems. Microbes are carried on water and dust particles in the air, so high relative humidity (above 80%) or dusty houses are undesirable. In addition, excessive washing of pens may cause microbes to be aerosolized in the spraying water.

THERMAL ENVIRONMENT

Managing the thermal environment is an important objective of animal house managers.

Heating Goals

Animal buildings are heated for three reasons: (1) Young animals (particularly piglets) require a temperature that is warmer than that outdoors; (2) a heated building may be designed to be comfortable for human workers; and (3) animal buildings may be heated to maintain high animal productivity.

Two strategies are used to provide supplemental heating: heating the entire building or heating zones within the building.

A furnace or unit heater is commonly used to heat a building. In most cases, a unit heater cannot warm every location within the building uniformly because the temperature is often warmer near the ceiling. The thermostat should be located at animal level. If it is placed at human eye level, it must be set a few degrees higher than the target temperature to meet the animals' thermal requirements at the floor level. The smallest pigs are best penned nearest the heater.

Zone heating can be used by placing a heat lamp or small space heater within an animal pen. Animals can provide zone heating by huddling or manipulating bedding.

The supplemental heat required depends on the required inside temperature. Other factors that affect supplemental heat are building insulation, ventilation rate, building size, the amount of feed animals are fed, and animal density.

Building Heat Loss

Heat always moves in a warm to a cold direction. The heat loss from an empty building in winter and the heat gain in summer depends on three factors: air temperature (inside and outside), building size, and type or amounts of insulation in the walls, ceiling, and floor. Two additional factors influence heat loss or gain in an animal building: ventilation rate and animal heat production. The greater the ventilation rate, the more heat will be removed or lost. Animals produce substantial amounts of heat. A well-insulated, heavily populated building may require little supplemental heat.

Animal Heat Loss

In most cases, animals are warmer than their environment. Heat moves from warm to cold, so animal heat should be conserved in cold weather and removed in warm weather. Heat is transferred in four major ways between animals and their environment:

1. *Radiation* is heat transferred from one body to another across space. If a poorly insulated building is held at 70°F (21.1°C), the wall temperature could be 50°F (10°C) or less on a cold day. Although the building's air temperature is adequate, animals lose heat through radiation to their cold surroundings.
2. *Conduction* is heat transfer through a solid medium. If an animal lies on a concrete floor, it experiences greater heat loss than if it lies on warmer materials.
3. *Convection* is heat transfer through air currents. A very slight draft causes heat to be moved more quickly from the animal's skin than when air is still. Low-speed drafts are nearly as effective as moderate-speed drafts at chilling young animals. Heat loss from a draft is greater if the animal is wet. Windchill index values indicate the more-than-additive effect that wind has on the perception of cold.
4. *Evaporation* is heat transferred or dissipated when water is evaporated. Evaporation (from liquid water to gaseous water) takes heat energy. If water is applied to the skin, animal heat is used to evaporate the water. An opposite phenomenon, condensation, occurs when water goes from a gas to liquid state and, hence, heat is

transferred to the condensed surface. The effective environmental temperature feels much warmer when the air is progressively more humid.

Effective Environmental Temperature

Heat is exchanged in many ways. The temperature an animal experiences is a function of radiant, convective, conductive, and evaporative heat transfer. Humans often assess the animal environment by evaluating air temperature only. This is not accurate because heat moves by all of the ways just described. Animal environmental managers need to know the animal's effective environmental temperature (EET)—the temperature the animal experiences (Curtis, 1983).

Researchers have demonstrated that animals have a lower critical temperature for tolerance of cold. If a young animal requires 80°F (26.7°C) air temperature, but there is a draft in its pen, it may be experiencing a 70°F (21.1°C) effective environmental temperature. Air temperature should never be the sole measure used to assess the temperature the animal is experiencing.

Animals adapt as they experience colder effective environmental temperatures. They conserve body heat by restricting blood flow to their skin and they cope behaviorally, possibly by huddling. In a progressively colder building, animals change their tissue insulation and their behavior until these strategies alone are not sufficient to fight the cold. At this point—the lower critical temperature—animals must increase their metabolic rate to keep warm. When metabolic rate increases in the cold, productivity decreases because feed energy is needed for body heat instead of growth.

When animals experience a warm effective environmental temperature, they try to limit heat production. Eating increases heat production, so warm environments cause animals to eat less, resulting in lower rate of gain or milk production.

Managers of the microenvironment should try to maintain their animals at a comfortable environmental temperature. Animals are most productive within the zone of thermal neutrality. The temperatures at the upper and lower end of this zone are different for each species (see Table 13–1). Other factors that influence the EET are feeding level, hair coat, physical environment (bedding, floor, and wall temperature), age, and group size.

TABLE 13–1
Acceptable Temperature Limits for Productivity of Pigs. Bedding, Wind, Direct Sunlight, Haircoat, Moisture, and Type of Feedstuff Influence the Acceptable Temperature for Each Animal. Below the Lower Temperature, Animals will Use Feed Energy to Keep Warm and Feed Efficiency will Suffer. Above the Upper Temperature, Feed Intake will be Reduced and Weight Gain will Suffer.

Type and Weight	Preferred Range		Lower Extreme		Upper Extreme	
	°F	°C	°F	°C	°F	°C
Lactating sow	59–79	15–26	60	15	90	32
Nursing piglets	90+	32+	77	25	No practical upper limit	
3–15 kg (7–33 lb)	79–90	26–32	59	15	95	32
15–35 kg (33–77 lb)	64–79	16–26	41	5	95	32
35–70 kg (77–154 lb)	59–77	59–77	23	−5	95	35
70–100 kg (154–220 lb)	50–77	10–25	4	−20	95	35
Over 100 kg (220 lb)	50–77	10–25	4	−20	90	32

Source: Adapted from Curtis (1983) and FASS (1999).

BUILDING DESIGN AND PHYSICAL ENVIRONMENT

Many building designs are suitable for animal production. Choice of design depends on climate, production level, local codes, existing facilities, cost, and personal taste. The type of building can vary from a wind and rain break to an enclosed, automated building.

When choosing a building design, producers must first consider the age of the pigs, the level of production, and their personal tastes for building design. They can choose natural or mechanical ventilation and decide if they want supplemental heat. Other considerations include waste management, light management, pen design, and provision of work and office areas. If the building will be designed to house animals, the needs of the animals should be the first priority. The next priorities should be worker comfort and labor efficiency.

The simplest building, a hut or shelter, can be as simple as a shade structure. The hut can be open on at least one side to provide protection from the wind and rain. Huts that are bedded can provide a comfortable microenvironment in most climates. In some warm weather climates, bedding is not needed in huts.

The next simplest animal building is a cold house. A cold animal house is typically lightly insulated, naturally ventilated, and has no supplemental heat. One or more walls may be open. Cold animal houses are inexpensive to build and maintain, but there is little chance of controlling the inside environment. The temperature and the ventilation rate are determined by wind conditions (see Table 13–2). Some cold animal houses provide animals the opportunity to change their microenvironments—for example, finishing pigs

TABLE 13–2
Table of Windchill Index Values.

Wind (MPH)	Temperature (°F)																	
Calm	40	35	30	25	20	15	10	5	0	−5	−10	−15	−20	−25	−30	−35	−40	−45
5	36	31	25	19	13	7	1	−5	−11	−16	−22	−28	−34	−40	−46	−52	−57	−63
10	34	27	21	15	9	3	−4	−10	−16	−22	−28	−35	−41	−47	−53	−59	−66	−72
15	32	25	19	13	6	0	−7	−13	−19	−26	−32	−39	−45	−51	−58	−64	−71	−77
20	30	24	17	11	4	−2	−9	−15	−22	−29	−35	−42	−48	−55	−61	−68	−74	−81
25	29	23	16	9	3	−4	−11	−17	−24	−31	−37	−44	−51	−58	−64	−71	−78	−84
30	28	22	15	8	1	−5	−12	−19	−26	−33	−39	−46	−53	−60	−67	−73	−80	−87
35	28	21	14	7	0	−7	−14	−21	−27	−34	−41	−48	−55	−62	−69	−76	−82	−89
40	27	20	13	6	−1	−8	−15	−22	−29	−36	−43	−50	−57	−64	−71	−78	−84	−91
45	26	19	12	5	−2	−9	−16	−23	−30	−37	−44	−51	−58	−65	−72	−79	−86	−93
50	26	19	12	4	−3	−10	−17	−24	−31	−38	−45	−52	−60	−67	−74	−81	−88	−95
55	25	18	11	4	−3	−11	−18	−25	−32	−39	−46	−54	−61	−68	−75	−82	−89	−97
60	25	17	10	3	−4	−11	−19	−26	−33	−40	−48	−55	−62	−69	−76	−84	−91	−98

FROSTBITE OCCURS IN 15 MINUTES OR LESS

$$\text{Windchill (°F)} = 35.74 + 0.621\,T - 35.75(V^{0.16}) + 0.4275\,T(V^{0.16})$$

Where, T = Air Temperature (°F)

V = Wind Speed (mph)

Source: Adapted from NOAA (2002).

housed in a modified, open-front building in the winter could make a straw bed to provide a warm, effective environmental temperature.

A warm animal house is well-insulated, heated, and mechanically ventilated. It provides a constant thermal environment. The energy costs are greater, but a uniform, high level of animal productivity can be achieved.

There are many types of buildings in modern animal production. Buildings may have components of both the warm and cold animal houses. Each building is unique because weather of conditions, the age of pigs, and manager expertise.

Ventilation Goals

Ventilation objectives differ with changing seasons. The manager conserves energy in cold weather by reducing the ventilation rate. However, the moisture produced by animals and their waste must be removed, so a minimal winter ventilation rate would be the lowest rate of air exchange that still removes moisture. Excessive moisture (above 80% relative humidity) provides a vehicle for microorganisms, damages insulation, and wets animals.

If condensation forms inside the animal house, the humidity is too high, the ventilation rate is too low, or the building lacks sufficient insulation. Water on the walls or ceiling can pose animal health risks.

Calculations can determine the minimum ventilation rate for removing animal moisture (see Curtis, 1983). For proper ventilation, air must move from air inlet to exhaust fan. If air moves too slowly, or if the distance from inlet to fan is too far, cold air may enter the inlet and drop on the animals, causing a cold draft. To avoid this, the ventilation rate must never fall much below 6 air changes/hr. This is only a rule of thumb and particular buildings may have different requirements. An important criterion is to ensure that ventilation is uniformly distributed throughout the building.

A properly ventilated building is free of drafts and provides clean, fresh air without chilling the pigs. A draft is a stream of air that has a chilling effect. Ventilation air, in contrast, has a refreshing, temperature-neutral effect. The first air current that enters the building is called the primary air current. As the primary air current moves through the building, it warms and generates secondary air currents (see Figure 13–1). Secondary air currents, which ventilate the animal, are warmer because they have had time to equilibrate with the inside air temperature.

In warm weather, the objective is to keep air moving to remove animal heat. If this is accomplished, the ventilation rate in summer is greater than in winter. Moisture is removed when the ventilation rate is high.

A mechanical ventilation system will not provide an inside temperature lower than the outside temperature. Two things happen when air enters a building: The air is warmed and moisture is added. For example, if outside air is 90°F (32.2°C) with a 60% relative humidity, the animals may warm the inside air to 95°F (35°C) with a 70% relative humidity. If producers want an inside air temperature of 90°F (32.2°C) or less, they must utilize artificial cooling. Appropriately managed, mechanically ventilated barns can be maintained within 5°F of outside temperatures during hot weather (see Table 13–3).

FIGURE 13-1
Diagram of a Pig Building Showing Modes of Heat Flux (arrows) and Streams of Air Representing the Primary and Secondary Air Currents. Heat Is Lost through the Walls, Foundation, Perimeter, (through the floor and out the wall-level of the foundation) and Ceiling.

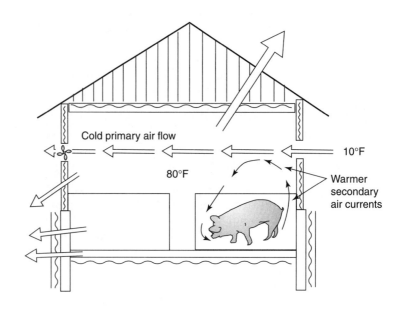

TABLE 13-3
Heat Index Chart (temperature and relative humidity).[a]

RH %	Temperature (°F)															
	90	91	92	93	94	95	96	97	98	99	100	101	102	103	104	105
90	119	123	128	132	137	141	146	152	157	163	168	174	180	186	193	199
85	115	119	123	127	132	136	141	145	150	155	161	166	172	178	184	190
80	112	115	119	123	127	131	135	140	144	149	154	159	164	169	175	180
75	109	112	115	119	122	126	130	134	138	143	147	152	156	161	166	171
70	106	109	112	115	118	122	125	129	133	137	141	145	149	154	158	163
65	103	106	108	111	114	117	121	124	127	131	135	139	143	147	151	155
60	100	103	105	108	111	114	116	120	123	126	129	133	136	140	144	148
55	98	100	103	105	107	110	113	115	118	121	124	127	131	134	137	141
50	96	98	100	102	104	107	109	112	114	117	119	122	125	128	131	135
45	94	96	98	100	102	104	106	108	110	113	115	118	120	123	126	129
40	92	94	96	97	99	101	103	105	107	109	111	113	116	118	121	123
35	91	92	94	95	97	98	100	102	104	106	107	109	112	114	116	118
30	89	90	92	93	95	96	98	99	101	102	104	106	108	110	112	114

[a]Exposure to full sunshine can increase HI values by up to 15°F.
Source: NOAA web page (2002).

Heat is consumed when water evaporates. The environmental manager may use the cooling properties of water in two ways: (1) spraying the animals with water and (2) using a mechanical system.

The simplest cooling method is to spray the animals with water during warm weather. This system, most common in cold animal houses and feedlots, can be automated with thermostatically controlled foggers.

When holding pigs in warm houses during the summer months, producers may use a mechanical evaporative cooling system. This system places evaporative cooling pads over the air inlets; water trickles over these pads. As air is drawn into the building through the pad, water evaporates, removing heat from the air, which is then cooled. The amount of cooling depends on the moisture present in the outside air. In most dry climates, air coming off the evaporative cooling pad is 10 or more degrees lower than the outside air temperature. Air-conditioning is occasionally used in animal buildings because it is less dependent on the moisture content of the outside air. However, such units are expensive to buy and maintain.

SPACE NEEDS OF PIGS

Pigs need enough space to meet one of three types of space requirements: static space, dynamic space, and social space.

The static space needs of pigs is represented by the physical space of the pig's body. Static space is measured by the width, length, and height of the pigs. Regression equations are available (Baxter, 1987) to define these space needs (see Chapter 8).

When pigs make normal postural adjustments, like standing up and lying down, they require more space than their static space needs. The space needed to make normal postural adjustments is called the dynamic space requirement and is, by definition, larger than static space needs. The dynamic space needs of pigs can be met with or without social housing.

Social space allowance is the space needed to allow pigs to interact socially. Interestingly, the social space allowance and the dynamic space allowance occupy about the same floor area. Table 13–4 gives recommended space allowances for individually penned and group-housed pigs. Table 13–5 compares the spaces needed for a sow to meet her static, dynamic, and social space needs.

Group size has an effect on the space needs of pigs (McGlone and Newby, 1994). Pigs have two sorts of floor space in their pen: used space and free space. Used space is the area occupied by the pig's body. Free space is the unoccupied space in a pen at any given time. Free space changes with time of day. Pigs that are lying down laterally take up more space than pigs that are standing or lying on their sternum. If some pigs are walking, some pigs are feeding, and some are drinking (all done from a standing posture), the pigs in a pen have less total space needs. If all the pigs are lying down, the most space is needed. Pigs show a diurnal pattern of general activity, so their space needs vary with time of day. Most pigs are less active at night and, thus, the most pen space is needed when people are generally not present.

Figure 13–2 graphs the diurnal cycle of pig behavior, using free space as the variable. Pigs have more available free space around midday because they are up and moving around.

SECTION IV HOUSING, ENVIRONMENT, AND NUTRIENT MANAGEMENT

TABLE 13–4
Space Allowances for Pigs of All Ages.[a]

STAGE OF LIFE	INDIVIDUAL PIGS		GROUPS OF PIGS[a]	
	FT²	M²	FT²	M²
Litter and lactating sow, pen	35	3.15	—	—
Litter and lactating sow, sow portion of crate[b]	14	1.26	—	—
Nursery, 3–27 kg (7–60 lb)	6	.54	1.7–4.0	.16–.37
Growing, 27–57 kg (60–125 lb)	10	.90	4–6	.37–.56
Finishing, 57–104 kg (125–230 lb)	14	1.26	6–8	.56–.74
Late finishing, 105–125 kg (231–275 lb)	14	1.26	8–9	.74–.84
Mature adults[b]	14	1.26	16	1.49

[a]The amount of space per pig decreases as group size increases.
[b]Minimum stall width should be 22 in (56 cm). The minimum length should be 2.2 m (7 ft). Young adult females require only 2 m (6.5 ft) of stall length.

TABLE 13–5
Space Needs for an Adult Sow to Meet Her Static, Dynamic, and Social Space Needs. This Concept Is Based on a Paper by Stan Curtis and His Students. Table Values Are for a Sow That Weighs about 250 kg (550 lb).

SOW'S BODY PART	STATIC SPACE		DYNAMIC SPACE		SOCIAL SPACE[a]	
	M	FT	M	FT	M	FT
Length	1.61	5.2	2.20	7.2	~3	~9
Width or breadth at shoulders	.43	1.4	.86	2.8	~3	~9
Height	.89	2.9	.99	3.2	NA	NA
Total floor area	0.7 m²	7.3 ft²	1.9 m²	20 ft²	1.9 m²	20 ft²

[a]Assumes at least 4 sows per pen.
Source: Adapted from Curtis et al. (1989).

FIGURE 13–2
Graph of Free Space as Time of Day Varies. Note That Where the Arrow Points Is the Time of Day (around midnight) When the Least Free Space Is Available. This Is because the Pigs Are Sleeping. Solid Line Is the Mean. Dashed Line Is the 95% Confidence Interval.

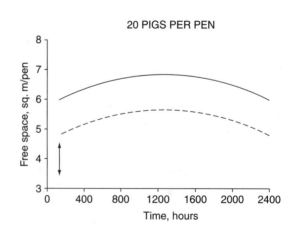

Pigs that weigh 220 lb (100 kg) take up around 1.0 sq m (10 sq ft) each when they are lying down (see Figures 13–3 and 13–4). The space that is in contact with the floor, however, is only about 6 ft^2 (0.56 m^2) per pig. The published space allowance is around 8 ft^2 (0.74 sq m)/pig. The free space in a pen of pigs is around 2 ft^2 (0.19 m^2)/pig or about 25% of the total space.

As group size increases, half of the free space can be removed without compromising pig performance. When producers remove half of the free space, they can add another 10% to 15% more pigs. This difference becomes significant when group sizes are very large (500 or more pigs/pen).

FIGURE 13–3
Overhead View of Pig (100kg or 220 lb) Size When Resting in Lateral Recumbency. Total Area Is about 1 sq m (10 sq ft) per Pig. The Area in Contact with the Floor Is about 0.6 sq m (6 sq ft) per Pig.

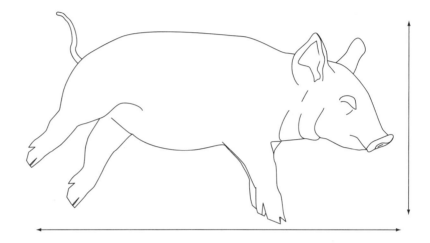

FIGURE 13–4
Graph Showing How the Numbers of "Extra" Pigs per Pen can Be Added as Group Size Increases from 10 to 1,000 Pigs per Pen. The Y-axis Is the Number of Extra Pigs That can Be Added when Assuming 1 sq ft per Pig Is not Used ("free space"). Note That about 10% to 15% More Pigs can Fit in the Same Space when Very Large Group Sizes Are Used. This Model Assumes Totally Slotted Flooring Good Waste Management and Ventilation.

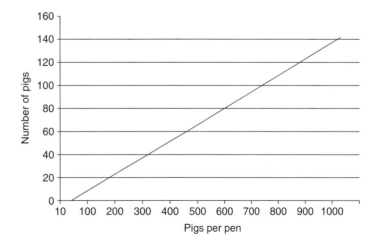

> As group size or the numbers of pigs per pen increases, the space needs per pig decrease slightly.

The same principles apply to pigs on bedding, except they require more space than pigs on slotted floors. This is because pigs on any flooring other than total slats must have separate areas for eating, drinking, dunging, and resting. Without separate functional areas, the pigs will be a mess with manure covering their skin. Risk of disease increases in dirty pens. If pigs are covered with manure, the building's waste and/or ventilation system is not working properly.

The economics of space utilization are well described. When the performance of individual pigs is considered, pigs should be given a generous amount of space. However, space is costly, and the greater the cost of space (depreciation or building cost per year), the higher the number of units that need to be produced from that space. When the performance of the building is considered, economic forces provide advantages to crowding the pigs—giving them less than the recommended space allowance, but getting more value in pig numbers/building. Space utilization based on weight of pig produced/unit of space is a common measure of efficiency.

It is not just floor space that must be managed. Pigs experience space constraints on feeder, waterer, and social spaces in the pen. Equipment space, such as the feeder, can be occupied at an economically efficient level, just like floor space. Figure 13–5 shows that with two feeder spaces/20 pigs, the feeding behavior follows a diurnal pattern (McGlone et al., 1992). As feeder space becomes limited, pigs tend to eat at a faster rate. When there is only one feeder space/20 pigs, the pigs cannot eat fast enough to satisfy their ad libitum feed intake drive. Thus, not only is feed intake limited, but the feeder

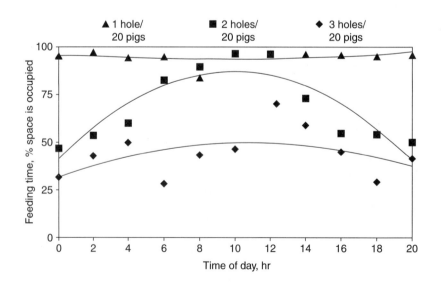

FIGURE 13–5
Diurnal Pattern of Feeding in Pigs Given 1, 2, or 3 Feeder Holes per 20 Pigs in Each Pen. SEM = 8.4 % Units. N = 60 Pigs or 3 Pens/Mean.

is fully occupied for 24 hr/d (see Figure 13–5—one feeder space for 20 pigs). For a detailed discussion of space needs for growing pigs, see Chapters 15 and 19.

THE DISEASE AND MICROBIAL ENVIRONMENT

The microbial environment is less well-studied than other components of the environment. Producers know that early-weaned pigs kept in very clean environments grow very rapidly. They also know that germ-free pigs do not respond to antibiotics very well, suggesting that the microbes in a typical production environment limit pig productivity.

Two general methods can be used to keep pigs in a low-microbe environment: (1) Increase the standard of sanitation by cleaning pens and (2) give the pigs more space, especially by housing them outdoors and thereby lowering the air and surface microbe counts.

By increasing the level of sanitation, producers can improve pig growth and decrease mortality. In a Danish study, (reported by John Gadd) scientists examined 26 batches of pigs whose pens were either partially or totally cleaned between batches. Total cleaning improved weight gain over 7% and decreased mortality 40% (for more discussion, see Chapter 20).

SUMMARY

A comfortable environment for pigs is important to their overall well-being and productivity, just as it is for other domestic animals and, indeed, for humans. This chapter describes the components of the microenvironment of the pig (all factors external to the animal). The components include thermal environment, physical environment, social environment, disease and microbial environment, and nutritional (diet) environment. Each component is described in terms of management factors needed to optimize animal comfort and productivity. The various components of the microenvironment are interrelated to one another, such that a change in one may affect the animal response to another (e.g., a change in building design [physical environment] may affect decisions about managing the thermal environment, social environment, and/or disease and microbial environments). Consolidation of all of these factors is crucial to the development of a comfortable environment for pigs in both extensive and intensive pork production systems.

QUESTIONS AND ACTIVITIES

1. Compare each mode of heat flux for a pregnant sow in (a) an outdoor system with (b) a gestation stall in a building.
2. Compare each mode of heat flux for a pregnant sow in (a) a group pen on indoor bedding with (b) a gestation stall in a building.

3. Conduct a survey of classmates and see the percentage of students (or other groups) that believes sows should be given enough space to accommodate their static, dynamic, or social space needs.
4. Consider a pig at a given stage of production. Describe each component of its environment. What can be improved? What is the cost of each improvement relative to the economic pay-off (if any)?
5. Asset utilization is a term that applies to buildings and equipment within the building. How would you determine rules to follow to best utilize a feeder in a finishing barn?

LITERATURE CITED

Baxter, S. 1987. Intensive Pig Production: Environmental Management and Design. Granada, London.

Curtis, S. E. 1983. Environmental Management in Animal Agriculture. Iowa State University Press, Ames.

Curtis, S. E., R. J. Hurst, H. W. Gonyou, A. H. Jensen, and A. J. Muehling. 1989. The physical space requirement of the sow. J. Anim. Sci. 67:1242–1248.

Fass. 1999. Guide for the care and use of agricultural animal teaching and research. Fass. Savoy, IL.

McGlone, J. J. and B. Newby. 1994. Space requirements for finishing pigs in confinement: Behavior and performance while group size and space vary. Appl. Anim. Behav. Sci. 39:331–338.

McGlone, J. J., R. I. Nicholson, and T. Hicks. 1992. Performance of Pigs with "Limited" Feeder Space. Texas Tech Univ. Agric. Sci. Tech. Rep. T-5-317, pp. 41–43.

INTERNET RESOURCES

For Wind Chill Index information:
http://lwf.ncdc.noaa.gov/oa/climate/conversion/windchillchart.html

For Heat Index Chart:
http://lwf.ncdc.noaa.gov/oa/climate/conversion/heatindexchart.html

For Swine housing and environment publications:
http://pasture.ecn.purdue.edu/~epados/swine/house/pubs.htm

14

PRODUCTION SYSTEMS FOR ADULT PIGS

INTRODUCTION

The pig was a good choice for domestication largely because of its plasticity and outstanding biological efficiency. For an animal to be relatively plastic, it must bend, rather than break, when faced with environmental challenges. The pig does quite well at bending and adapting in a wide variety of environments. Signs of lack of adaptation in the adult pig focus on lower—but usually not zero—reproductive rates. Thus, during prolonged times of stress in the wild, conception rates and litter size might be reduced in the pig, while reproduction may completely stop in other wild species. When ovulation rate is reduced in a species that has only one offspring/pregnancy, reproduction may be delayed until the stress is relieved.

Around the world, pigs reproduce in cold and warm climates, in dry and humid climates, and at sea level and high altitude. Because of the extraordinary plasticity of the domestic sow, it is a challenge to define her specific environmental needs. Simply knowing the bare, essential needs for the sow leads to low reproductive performance. Instead, pork producers should seek the highest output possible within economic constraints. However, when economics is added to the equation, lower output may be accepted (pigs per sow per year) if input costs are low enough. On the other hand, in a high-investment facility, producers must have high output. The optimal situation is one with low-cost input and sustained, high output.

This chapter addresses the problem of finding the right system by addressing the two extremes: the high- and low-investment facilities. Once producers commit to a confinement building, their objective *must* be to maximize output from that investment. With a progressively lower investment, producers might be willing to accept progressively lower reproductive output. In all cases, the objective is to maximize output regardless of the investment in the production unit.

PRODUCTION SYSTEMS: REASONABLE CHOICES

The choices discussed here refer mainly to gestation systems. Several other systems are in place in the breeding and farrowing areas, but typically, once a given system is decided upon in gestation, the breeding and farrowing choices are easier to make. On a weekly production schedule and with 3-wk weaning, the sow inventory falls in the following areas, as a percentage of the population:

- 14.3% in farrowing (3/21 sow groups)
- 28.5% in breeding and heat check (6/21 groups)
- 57.2% in gestation (12/21 groups)

Thus, the gestation unit will drive the rest of the system.

What system should be used in gestation and farrowing? The European magazine, *Pigs* (1994), recently concluded that the options to pork producers are:

- Outdoor production—Use of this system is increasing rapidly in Europe and involves housing sows in intensively managed pastures in a manner similar to how they might be kept indoors. Outdoor stock people have a philosophy that centers on animal care.
- Pens or yards with individual feeders—This system can be used indoors on slotted floors or bedded floors. Sow feed intake can be regulated by use of individual feeders.
- Electronic sow feeders—The electronic sow feeder business seems to rise and fall in the United States, but it has reasonable representation in Europe. Evidence is insufficient in the United States to recommend use of these systems, especially if the equipment is made in Europe and replacement parts and technical service are not readily available.
- Automated floor feeding—In this system, feed is provided on the floor, usually by use of an auger or by hand feeding. Because of the lack of control over individual feed intake, sows must be sorted by stage of gestation and condition. This means there must be fairly small social groups, which adds to the cost. Sow condition tends to be variable in this system.
- Trickle feeding—Also known as "Biofix," this system supplies feed at a very slow rate. Sows learn to be still and wait for feed because if they steal their neighbors' feed, their spot opens. After an acclimation period, sows are very quiet during feeding. This system uses specialized equipment, but is fairly simple. Developed in Sweden, the system really has not gotten started in the United States.

There are more system choices in the United States because legislation does not prevent the use of any system for pork production. The following system choices are available to U.S. producers in addition to the European systems:

- Stalls or crates—The crate is the most common system in the United States based on the number of sows. The overwhelming majority of new U.S. units use gestation crates.

Europeans do not consider some of the fairly common systems in use in the United States. These systems are not typically recommended:

- Tethers
- Group pens (indoor or outdoor) without control of feed intake
- Post-harvest foraging grain and other crops
- "Fend-for-themselves systems" in forests, swamps, or other uncontrolled areas. These are clearly not modern production systems.

Producers must first decide if gestating sows will be kept outdoors or indoors. Climate is a factor, but outdoor breeding in cold, northern climates does occur. Productive outdoor breeding and farrowing units are now found all the way from the tropics to Sweden, Denmark, and England. Confinement units are found in an equally wide range of climates.

The British seem to have revived outdoor production. Although encouraged earlier, a 1991 UK law forced producers to phase out and eventually abandon tethers and crates. Thus, they began to re-invent the outdoor system. As UK pork producers considered new facilities, they discovered that some producers were very efficient in using intensive outdoor units. Field data and some controlled studies showed outdoor productivity to be quite good.

An outdoor, low-intensity system is probably a bad idea. The word *probably* is used because if the inputs are low enough (say the pigs live in a forest or in a recently harvested field), it may still be cost effective. However, if sows are outside and not managed closely, herd productivity is low. Modern-day pork production—the business—does not reward units with poor records, poor health management, and poor facilities. Some people assume that outdoor, intensive breeding/gestation/farrowing means providing a few fences and letting the pigs fend for themselves in dirt lots. This is clearly not the intent of intensive, outdoor production.

If producers decide to bring the breeding herd inside, they must first decide if the hogs will be kept in groups or in individual pens. If they are in groups indoors, they could either be on slotted floors or on solid floors, or on solid floors with bedding. If they are housed individually, they are most likely to be on slotted floors. With these generalizations understood, the field data become a bit more difficult to interpret. When field data compare group pens versus individual stalls, the group-housed sows typically have bedding and solid floors while the individually housed sows have slotted floors. The comparison of the performance of indoor, crated sow herds on slats with that of sows in group-housed, outdoor herds with natural vegetation and bedded huts is clearly confounded—it is difficult to know which of the individual elements (floor type, social housing, space) might be responsible for production differences. In addition, units with sows in stalls typically have more complete records and they individually hand-mate or use artificial insemination, while many units that group-house sows may pen-mate and have less precise records.

The bottom line is that field data are always suspected of being inaccurate or unreliable, but they often have very large sample sizes. However, when field data support controlled studies, producers can begin to make important generalizations.

To make sense out of field and controlled studies, producers need an understanding of pig behavioral biology. With such an understanding, labor and facility decisions become more clear. Although producers may not fully understand everything about boar and sow sexual behavior, they can improve the process of designing facilities by making use of the present base of knowledge.

THE BREEDING AREA

Among the many types of breeding areas, the "Lubbock" system is in common use. This system was first widely used by the original Lubbock Swine Breeders (then called DeKalb Swine Breeders, now a subsidiary of Monsanto). In the Lubbock system, recently weaned sows are placed in stalls and a boar is in a pen behind them (Figure 14–1).

The major disadvantage of the Lubbock breeding system is that the boars constantly stimulate the estrous sows and, as a result, sows may get tired or refractory to mating when put with a boar. One solution to the problem is to have a completely separate mating area where boars and sows meet for breeding. Another common alternative is to breed sows in the boar's pen or in the aisle. Diagrams of well-thought-out breeding designs for pen-mating, hand-mating, and AI facilities are available from Dr. Don Levis (1995), long-time faculty member at the Department of Animal Science University of Nebraska, Lincoln (now at Ohio State University).

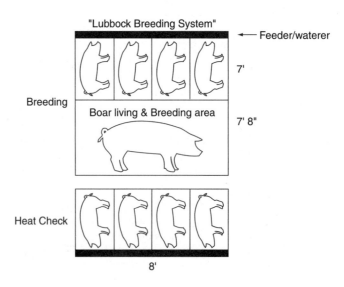

FIGURE 14–1
Common Confinement Breeding Area Known as the "Lubbock" System.

PREGNANCY MAINTENANCE

At first thought, producers might expect pregnancy to be easy to maintain. However, the average farrowing rate in surveys of herds is often less than 80%—even the best farms have trouble keeping the farrowing rate above 90%. The typical 10% to 30% of pregnancies lost on commercial farms represents a huge economic loss. It is not clearly understood why farrowing rates are not near-perfect, but there are some clues.

It is known that in most species, including the pig, reproduction is very sensitive to stress. Stress can have profound negative effects on reproductive rates of pigs (stimulating puberty in gilts is the exception). Poor handling, heat stress, and social stress have significant effects on reproduction (the effects of social stress on reproduction are discussed later).

FARROWING AND LACTATION

Sows seek isolation when they prepare for farrowing. Among outdoor sows, the behavior to seek isolation is quite strong. The only problem with isolation-seeking is that prefarrowing sows will "bother" the peri-parturient sows. For these reasons, individual housing is a reasonable biological choice during the peri-parturient and lactating phases. The important question becomes how to reduce piglet mortality, which was 15% in a recent national survey (Tubbs et al., 1993). On the best farms with the best-trained people and equipment, pre-weaning mortality can be less than 10%.

Reduced mortality can be achieved in two ways: (1) give the sow less room (with a farrowing crate) or (2) give her a great amount of room (in certain pen configurations). These options are discussed in the section on controlled studies.

FIELD DATA ON GESTATION AND FARROWING SYSTEMS

INDOOR VERSUS OUTDOOR SYSTEMS

The use of the outdoor system seems to be increasing in parts of Europe and North America but, in fact, is not a recent development. Some producers have been producing higher numbers of pigs per sow per year for many years in low-investment, high-output units. In most years in the UK record system, the indoor and outdoor units have similar productivity.

The output in terms of pigs per sow per year is similar in the indoor and outdoor units in the United Kingdom, even in 1988 (see Table 14–1). In many data sets, the outdoor system is slightly less productive on average. There is also similar productivity between indoor and outdoor herds in France. However, there is a trend that is quite annoying to the supporters of intensive, outdoor production for the outdoor system to have a greater pre-weaning mortality and, therefore, fewer pigs weaned. In the UK data presented, the outdoor unit had 0.2 fewer pigs weaned and in the French report, the outdoor units showed 1.4 fewer pigs weaned than the indoor unit. The increase in piglet mortality is thought to be due to a greater rate of crushing outdoors

TABLE 14-1

Comparison of Outdoor and Indoor Herds in the United Kingdom in 1988.

Measure	Outdoor Herds	Indoor Herds
Litters per sow per year	2.2	2.3
Pigs born alive/litter	10.4	10.7
Pigs weaned/litter^a	9.4	9.6
Mortality, % of born alive	10.2	10.9
Pigs weaned per sow per year	20.9	21.8

^{a}This measure was not in the original paper; it was obtained by calculation.
Source: Adapted from MLC Pig Year Book (1989).

TABLE 14-2

Comparison of Outdoor and Indoor Herds in France in 1993.

Measure	All Outdoor Herds	Best One-Third Outdoor Herds	National Average of All Herds
Litters per sow/yr^a	2.3	2.4	2.4
Pigs born alive/litter	10.8	11.3	10.9
Pigs weaned/litter	9.1	9.7	9.5
Mortality, % of born alive^a	10.8	9.3	6.9
Pigs weaned per sow per year	21.3	23.2	22.7

^{a}This measure not in the original paper; it was obtained by calculation.
Source: Adapted from Le Denmat et al. (1994) and Berger et al. (1997).

than indoors, on average. Once again, some producers can make the outdoor system a highly productive system. Note the excellent records for the one-third better producers using the outdoor system in France (Table 14–2).

There are very few intensive, outdoor units in the United States. The intensive, outdoor herd comprises less than 2% of all sows, even if the percentage of low-investment sows is a much greater percentage. The U.S. data represent a mixture of low-investment systems that includes only a few European-style, intensive outdoor herds.

Two sources of information with very large databases were examined for effects of indoor versus outdoor productivity: (1) The PigChamp database, which includes herds of a variety of genotypes and housing systems (see Table 14–3); and (2) the Pig-Tales (1994) database from the Pig Improvement Company (PIC). The PIC® database is a much more narrow genetic base, but it includes a variety of housing systems.

The PigChamp® database was surveyed by Polson (1994) and includes a sample of closely scrutinized herds. Among the PigChamp database of over 143,000 sows, outdoor herds weaned about 1.7 fewer pigs per sow per year than did indoor herds.

Companies are sensitive to providing actual means for housing systems. They report odds ratios (odds of being among the 20% of poor herds). Indoor herds in the PIC database are eight times less likely to be in the bottom 20% of herds than are herds that have any sows outdoors.

The observations from both the PigChamp and PIC databases fit conventional, U.S. wisdom that indoor systems are better, on average, in terms of reproductive performance.

TABLE 14–3
Comparison of Outdoor and Indoor Herds in the United States. From a Survey of 143,695 Sows Participating in the PigChamp Recordkeeping System. Data Represent Herds that Hand-Mated and Were on a Weekly Farrowing Schedule.

MEASURE	OUTDOOR HERDS[a]	INDOOR HERDS[a]
Litters per sow per year	2.1	2.3
Pigs weaned/litter	8.6	8.7
Pigs weaned per sow per year	18.1	19.8

[a]Rounding causes numbers to not precisely multiply and agree.
Source: Adapted from D. Polson (1994).

However, it is important to remember that the conventional U.S. outdoor system is not typically the European-style intensive, outdoor system.

One outdoor hog enterprise was developed on the Southern High Plains as a 2,400-sow breeding and farrowing complex with off-site nurseries and growing-finishing buildings. Several 2,400-sow-intensive, outdoor herds were also established. While company records are confidential, it was revealed that output (pigs weaned per sow per year) was actually equal or better in the outdoor unit than in the indoor units for its first 3 yr of operation. The successful Plains unit had a few key features: European-style layout, European-style equipment, and a British herdsman. Two key features are clear—good equipment and highly qualified people.

INDOOR SYSTEMS: GROUP VERSUS INDIVIDUAL SYSTEMS

Field data are available in both Europe and the United States to compare certain systems for indoor sows. The indoor systems discussed in this section are systems other than intensive, outdoor systems. Some field data on group pens might use a system that has sows indoors some of the time and outdoors some of the time.

Table 14–4 presents data from the United Kingdom on sow performance when pigs are housed with electronic feeders, group pens, and stalls and tethers. One problem for this type of summary is although both stalls and tethers house sows individually, they are not similar in the sows response to the system.

The British data in Table 14–4 clearly show that the three systems have similar reproductive performance—all systems produce within 0.7 pigs per sow per year. The only system with a hint of lower productivity is the group pen system. Experiences in the United States indicate that stalls and tethers should not be combined because they can result in strikingly different reproductive rates.

Field data from the United States do not entirely agree with UK data presented in Table 14–4 (but they may agree if stalls and tethers are separated). The PigChamp and PIC databases show trends in performance of sows in different systems.

Polson (1994), in a survey of PigChamp participants in the United States, found much better reproductive performance for sows in crates and tethers than for sow herds in pens or on pasture/dirt. The data clearly show that reproductive performance was better by 2.0 to 3.0 pigs per sow per year for sows kept in individual systems than for sows kept in social groups. Combining crates and tethers in the data set may not be such a good idea if the sow's reproductive performance was different in the two systems. However, in

TABLE 14–4
Comparison of Indoor Systems in the United Kingdom in 1993.

MEASURE	ELECTRONIC SOW FEEDERS	GROUP PENS	STALLS AND TETHERS
Litters per sow per year	2.31	2.25	2.27
Pigs born alive/litter	10.7	10.9	10.7
Pigs weaned/litter	9.4	9.5	9.5
Mortality, % of born alive	12.0	12.6	11.3
Pigs weaned per sow per year[a]	21.7	21.0	21.7

[a]Numbers of pigs weaned per sow per year do not agree with other data in the table, but are presented as published. Rounding errors probably explain the discrepancy.
Source: Adapted from MLC Pig Year Book (1994) and MLC pigplan data (adapted from *Pigs,* October, 1994).

TABLE 14–5
Comparison of Reproductive Performance for Sows in Three Common Systems in the United States.

MEASURE	CRATE OR TETHER	PEN	PASTURE OR DIRT[a]
Farrowing rate, %	83.8	78.2	78.2
Pigs born alive/litter	10.2	10.1	9.9
Pigs weaned/litter[b]	8.98	8.42	8.01
Mortality, % of born alive	12.0	16.6	19.1
Pigs weaned per sow per year[c]	20.6	18.5	17.6

[a]This system is not the intensive, outdoor system.
[b]Calculated from other data. Rounding effects probably explain the discrepancy.
[c]Calculated assuming 2.3 litters per sow per year for crates/tethers and 2.2 for others.
Source: Adapted from Polson (1994).

Polson's data set, which contained only a few herds with tethers, the pigs per sow per year were similar for the crate system (20.4 pigs weaned per sow per year from 112 herds) and for the tether system (20.7 pigs weaned per sow per year from six herds).

In the PigTales records, data comparing indoor and outdoor systems agree in general with the Polson (1994) data (see Table 14–5). PigTales estimates that herds with any degree of pasture/dirt outdoor production were eight times more likely to be among the bottom 20% of herds in its database compared with indoor herds. Herds with indoor (confinement) pens were five times more likely to be among the worst 20% of herds than were herds with gestation crates. PigTale's March 1994 report stated that herds with tethers had similar performance as herds with crates.

CONTROLLED STUDIES ON GESTATION AND FARROWING SYSTEMS

INDOOR VERSUS OUTDOOR SYSTEMS

Modern publications comparing herd performance indoors with intensive, outdoor systems are rare. An investigation with nearly 300 gilts was recently completed at Texas Tech University, in which the confinement herd was restocked simultaneously with the establishment of an intensive outdoor unit. Fifty gilts of each of three genotypes started either indoor or outdoors. Genotypes were PIC Camborough-15 (C-15) or PIC Camborough-Blue (C-Blue) or a cross of Yorkshire by Landrace (YL). The C-15 is a typical confinement line and the C-Blue was a line developed for outdoor production; it contained a portion of the British outdoor breed the Wessex Saddleback.

The outdoor herd was on pasture or dirt, depending on rainfall and season. Sows were kept in groups, fed on the ground, and contained by electric fence. The American-style porta-hut, bedded with wheat straw, was used to farrow the gilts and sows. Sows always had shade and during warm weather had a man-made mud wallow.

The indoor unit was on slotted floors in a mechanically ventilated building. Buildings were heated by gas furnace and cooled by water spray or drip in the summer. The farrowing rooms also were evaporatively cooled.

The interaction between environment and genotype was not significant. Each genotype responded about the same both indoors and outdoors. Other work has shown that genotypes by environmental interactions are found but, in this case, for example, the C-Blue did not do much better outdoors than the confinement line (the C-15).

Reproductive performance was much better indoors than outdoors during this first phase (Table 14–6). Indoor sows over the first two parities weaned 8.6 pigs/litter while outdoor sows weaned only 7.5/litter. Assuming equal litters per sow per year, indoor sows approached 20 pigs per sow per year, while outdoor sows weaned just over 17 pigs per sow per year.

The results from this controlled study fit nicely with the field data in the United States, suggesting indoor herds have higher reproductive rates than outdoor herds. But, the experiences on local farms and in Europe seem to be different than the U.S. field data and the results from the first phase of the Texas Tech University investigation. In examining the difference, the focus is on one key element—good equipment.

A major difference between a local farm and the Texas Tech University unit was that the local farm used an English-arc type of farrowing hut. The data in Table 14–7

TABLE 14–6

Farrowing Environment and Sow and Litter Performance. Data in Table Are Least Squares Means-Averaged Over the First and Second Parities.

MEASURE	INDOORS	OUTDOORS	SE[a]	P-VALUE[b]
Number of litters	148	210	—	—
Number born/litter	11.0	11.0	.27	.97
Number born live/litter	9.9	10.6	.30	.08
Number found dead/litter	1.47	0.38	.15	.0001
Birth weight alive, lb/piglet	4.1	4.2	.06	.17
Number of pigs weaned/litter	8.6	7.5	.28	.007
Pre-weaning mortality, %	12.3	27.5	2.17	.0001
Weaning weight, lb/pig	14.3	15.3	.48	.106
Total weaning weight, lb/litter	120.5	114.6	4.45	.30
Sow farrowing weight, lb	454.8	469.0	5.44	.11
Sow weaning weight, lb	381.7	400.7	6.59	.079
Sow lactation weight loss, lb	72.0	66.6	3.37	.21
Lactation length, days	28.1	25.7	.59	.007

[a]SE is the standard error of the mean.
[b]P-value (probability) for environment effect.
Source: Adapted from Nicholson (1994) Texas Tech University.

show that litters farrowing in the English-style hut with 31% more floor area (9 ft wide and 5 ft deep, or 2.8 × 1.5 m or 4.2 m²) had lower mortality and weaned more pigs than did sows in the American-style hut (5 ft wide and 7 ft deep, or 1.5 × 2.15 m or 3.2 m²). In the American-style hut, sows weaned 8.3 pigs/litter, while in the English-style arc, sows weaned 10.3 pigs/litter. This is an amazing difference of over 4.5 more pigs per sow per year in the English-style arc than in the American-style hut. This finding may help explain the phenomenal productivity of the local outdoor unit. The results also give encouragement to pork producers who prefer the outdoor system and have the land and climate suitable for such a system.

INDOOR SYSTEMS: GROUP VERSUS INDIVIDUAL SYSTEMS

Sows housed in groups often, but not always, have lower reproductive rates than sows housed in crates (see previous field data). Texas Tech University initiated a study in 1994 to compare reproductive performance and other measures for sows in groups of three or housed in crates. Group-housed sows were placed in a pen and videotaped. Initial fights and a competitive feed test determined socially dominant, intermediate, and submissive sows. The results were striking (see Table 14–8).

TABLE 14–7
Performance of Outdoor Sows Farrowing in American-Style Porta-Huts or English-Style Arc Huts. Huts Were Made From Galvanized Steel by the Porta Hut Company.

MEASURE	ENGLISH STYLE	AMERICAN STYLE	P-VALUE[a]
Number of litters	11	64	—
Number born alive/litter	11.0 ± 0.88	9.9 ± 0.38	0.220
Number found dead/litter	0.18 ± 0.25	0.45 ± 0.11	0.320
Birth weight, lb/piglet	4.36 ± 0.24	4.11 ± 0.11	0.400
Piglets weaned/litter	10.32 ± 0.86	8.21 ± 0.38	0.028
Litter weaning weight, lb/litter	128.19 ± 12.23	110.76 ± 5.47	0.198
Number of piglets that died	0.73 ± 0.50	1.68 ± 0.22	0.087
Mortality, %	6.83 ± 5.94	18.97 ± 2.61	0.065
Sow farrowing weight, lb	407.30 ± 17.24	406.19 ± 7.90	0.950
Sow weaning weight, lb	359.58 ± 18.37	358.03 ± 8.28	0.940
Days of lactation	27.92 ± 1.71	27.95 ± 0.77	0.980

[a]Treatment effect.
Source: Adapted from McGlone and Hicks (2000).

TABLE 14–8
Comparison of Crated (individually) Housed Sows and Group-Housed Sows in Terms of Reproductive Performance.

MEASURE	CRATED	GROUP	DOMINANT	INTERMEDIATE	SUBMISSIVE
Farrowing rate, %	100	90.4	100	85.7	85.7
Pigs born/litter	13.0	11.1	12.1	10.2	10.9
Pigs weaned/litter	9.1	9.3	10.2	7.9	9.7
Pigs weaned per sow per year[a]	20.9	20.5	23.4	16.6	20.4

[a]This measure was not in the original paper; it was obtained by calculation, assuming 2.3 litters per sow per year for crated and dominant sows and 2.1 litters per sow per year for intermediate and submissive sows.
Source: Adapted from Nicholson (1994) Texas Tech University.

In comparing the average fertility of group-housed sows and sows in individual crates, the means are quite similar. However, not all group-housed sows have similar reproductive performance. Socially subdominant sows had clearly suppressed reproductive rates. Socially intermediate sows had lower farrowing rates, fewer pigs born, and fewer pigs weaned/litter. Socially dominant sows had about as good a reproductive rate as sows kept in gestation crates.

NOT ALL INDIVIDUAL SOW-HOUSING SYSTEMS ARE CREATED EQUAL

Many reports of field data and summaries of controlled studies group individual housing systems. However, not all individual systems produce similar sow behavior, physiology, and performance. Results from the study described previously clearly showed that one type of outdoor hut (that provided more room) produced many more weaned piglets than did other systems. The same might be true for indoor gestation and farrowing systems.

The two most common indoor systems for gestating and farrowing sows are the crate and tether. In some cases, countries have legislated these systems as being similar and in others as possibly being different. As of 1991, UK pork producers cannot install either crates or tethers in new facilities. In the European Union, tethers are now banned—crates will be banned on all farms by 2013.

Table 14–9 presents data from McGlone et al. (1994) in which behavior, physiology, and reproductive performance for gilts and sows in crates or girth tethers during gestation and lactation were compared. Sows (but not gilts) in gestation/lactation crates had more pigs born and weaned than the sows in the girth tether. The suppressed reproduction among tethered sows was an effect during gestation because the stillbirth and pre-weaning mortality data were statistically similar for the two systems. The girth tether either suppressed ovulation rate or caused more early embryonic mortality. The gestation crate would be clearly favored in terms of pork producer economics, based on this study.

TABLE 14–9
Reproductive Performance of Sows Kept During Gestation and Lactation in Either a Crate or a Girth Tether. Data for Gilt Litters Were not Different among Systems, but for Sows (N = 112 gilts and N = 59 sows), the Number of Pigs Born and Weaned Were Fewer (P < .05) for the Girth Tether System than for the Crate System.

MEASURE	CRATED	GIRTH TETHER
Farrowing rate, %	89.6	83.3
Pigs born/litter	10.6 ± .5	9.1 ± .5
Pigs weaned/litter	9.4 ± .4	8.1 ± .4
Pigs weaned per sow per year[a]	21.6	18.6

[a]Calculations assume 2.3 litters per sow per year for each system (farrowing rates were not significantly different between systems).
Source: Adapted from McGlone et al. (1994).

SOW AND LITTER PRODUCTIVITY

The McGlone et al. (1994) data show that outdoor huts can strongly influence sow and litter productivity. In addition, by providing more room in an outdoor hut, producers can save baby pigs and wean more pigs/sow. What if more room is provided for sows indoors?

Many older pork producers remember that a greater piglet mortality was sustained when sows were farrowed in pens. Many farrowing pen designs have come and gone, but in today's high-sanitation confinement units where there is great control over waste, producers cannot easily use straw or bedding of any kind. Therefore, one important question is, can producers improve the quantity and quality of space without compromising piglet mortality?

A 1990 Texas Tech University study examined sows that were farrowed in level or sloped crates or pens (McGlone and Morrow-Tesch, 1990). The crates were standard 5 × 7 ft crates (1.5 × 2.15 m, with a 0.6 × 2.15 m sow area). The pens had an area of 7 × 7 ft (2.15 × 2.15 m) with a creep area of 7 × 2 ft (2.15 × .55 m).

Sows in the level pen crushed more piglets than did sows in the level crate, confirming why the industry moved to the farrowing crate (Table 14–10). However, when the sow was provided the increased area and when the pen was sloped 8%, the crushing rate was reduced in the pen. The sloped pen gave the same high weaning rate as the level crate. Interestingly, there was a classic interaction in that the sloped crate actually increased crushing of piglets. The crushing of piglets in the sloped crate occurred primarily during farrowing (as pigs were born, they were pressed against the back of the farrowing crate). The crushing in the level crate was throughout the 28-d lactation period.

The sloped pen provided more room, but it also changed the quality of space. Sows had to stand and lie down much more carefully in the sloped pen. They rested either with their udder pointed downhill or with their head uphill most of the time. Thus, producers can provide more room, without bedding, and obtain the low mortality rates without use of the crate—but with a cost. The sloped farrowing pen requires 40% more room; this added room would add to construction costs.

TABLE 14–10
Comparison of Sow and Litter Performance for Level and Sloped Crates and Pens (N = 40 sows and litters).

Measure	Level Crate	Level Pen	Sloped Crate	Sloped Pen	Pooled SE
Pigs born alive/litter	8.3	9.1	'10.4	9.6	0.71
Crushed piglets/litter[a]	0.51	1.53	1.29	0.27	1.0
Mortality, %[a]	10.8	27.1	17.2	9.1	4.5
Pigs weaned per litter[a]	8.2	6.6	7.6	8.4	0.51
Pigs weaned per sow per year[b]	18.9	15.2	17.5	19.3	—

[a]Significant interaction, P < .05.
[b]Calculations assume 2.3 litters per sow per year for each system (farrowing rates were not significantly different between systems).
Source: Adapted from McGlone and Morrow-Tesch (1990).

ANIMAL WELFARE CONSIDERATIONS

Producers must have an understanding of pig behavior and physiology to understand pig welfare. While other physiological systems are changed in stressful environments, reproductive performance is considered primarily as a sensor of physiology. The premise includes two main points: (1) If an environment is stressful to the sow, she will have reduced fertility (or reduced pigs per sow per year); and (2) if an environment supports a high rate of reproduction, the environment cannot be said to be stressful. Data from field studies and controlled studies provide some generalizations about how sows perceive their environments (in terms of reproductive responses). Taken with behavioral measures, producers should be able to design facilities that are comfortable for the breeding herd, promote sound well-being, provide relatively high economic returns, and are acceptable to consumers.

Human perception can be the only measure taken into consideration in legislating appropriate systems. Even if decisions are not based on science, producers may still be able to have high productivity in alternative systems. This is possible because the sow is highly plastic and because tools are available to design highly productive systems both indoors and outdoors.

The two most important elements to successful pork production are to have good pig people (Hemsworth and Barnett, 1990; Hemsworth et al., 1994) and good equipment. Good people understand, through intuition or training, that sows, boars, and growing pigs have behavioral needs. Good equipment is designed to accommodate the behavioral biology of the pig. Field data and controlled studies show trends in reproductive rates under common systems. The pig is amazingly plastic and can perform well in an plethora of systems. When pig behavioral needs are met and when the environment accommodates the appropriate physiology, producers find a high level of productivity that rewards workers and owners.

BUILDING DESIGNS

In spite of the large number of countries around the world and an apparent wide variation in views about pig biology and management, a few common building layouts are found around the world. Figures 14–2 through 14–6 show some examples.

SUMMARY

This chapter addresses the choices of production systems for sows in response to major shifts in the structure of the swine industry. The discussion focuses on the selection of appropriate housing and husbandry conditions during gestation and lactation. Information is presented from research data and actual production data, primarily from Europe and the United Kingdom, on reproductive performance in conventional indoor confinement systems at one extreme versus outdoor intensive

FIGURE 14–2
Lubbock System: Four-Row Hand-Breeding Facility.
Source: Courtesy Donald J. Levis.

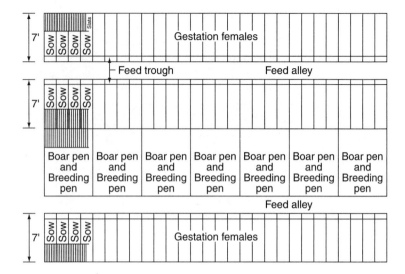

FIGURE 14–3
LEVIS System: Four-Row Hand-Breeding Facility (combination of boar stalls and boar pens).
Source: Courtesy Donald J. Levis.

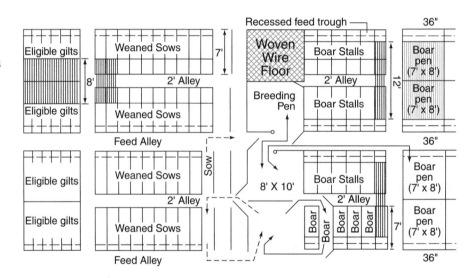

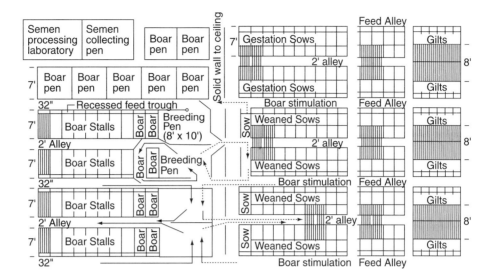

FIGURE 14–4
LEVIS System: Six-Row Hand-Breeding/AI Facility.
Source: Courtesy Donald J. Levis.

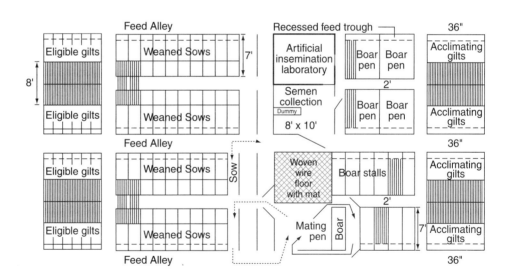

FIGURE 14–5
LEVIS System: Hand-Breeding/Artificial Insemination Facility (converting an existing four-row system to utilize an AI program) (on-farm semen collection and AI laboratory).
Source: Courtesy Donald J. Levis.

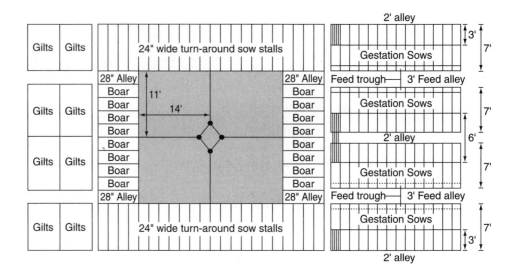

FIGURE 14–6
Breeding Square for Hand-mating and Artificial Insemination (also known as Connor Square).
Source: Courtesy Donald J. Levis.

production systems at the other extreme. Behavior and physiology are affected by the level of stress associated with the environment in which sows are kept, which in turn affect reproduction performance. Based on the information obtained from comparing different production systems for gestating and lactating sows, facilities may be designed that provide comfort and well-being to the animal and relatively high economic returns to the producer. The appropriate facility design depends on many factors, including geographic location, climate, market infrastructure and population density, and environmental concerns.

QUESTIONS AND ACTIVITIES

1. What is meant by "confounding of variables" when farm records are compared? Consider the example of typical indoor versus outdoor systems. How many variables are different when one farm's records are compared with another farm's records?
2. Historically, a low-input production unit, like an outdoor unit, also is low output. Is it possible to have a lower-input farm with a high output?
3. Compare the outdoor system of 50 years ago with a modern, intensive outdoor system.
4. Why do you think the gestation crate was not banned in most of Europe in early legislation, but the tether was (both are a method of individually housing sows)?
5. From an economic point of view, why is the gestation crate favored for indoor pigs compared with group housing? Is economics the only criterion that should guide a producer's decision about a housing system?

6. Ergonomics is the study of human time and motion. In the pig industry, particularly in the breeding area, ergonomic evaluations become important. Do you think it takes longer to pen-breed, hand-breed, or breed by AI? You might find literature by Dr. William (Billy) Flowers from North Carolina State University of value in addressing this question.

LITERATURE CITED

Berger, F., J. Dagorn, M. le Denmat, J. P. Quillien, Vaudelet, J. C., and J. P. Signoret 1997. Perinatal losses in outdoor pig breeding. A survey of factors influencing piglet mortality. Annales de Zootechnie. 46:321–329.

Hemsworth, P. H. and J. L. Barnett. 1990. Behavioural responses affecting gilt and sow reproduction. J. Reprod. Fert. Suppl. 40:343–354.

Hemsworth, P. H., G. J. Coleman, and J. L. Barnett. 1994. Improving the attitude and behaviour of stockpersons towards pigs and the consequences on the behaviour and reproductive performance of commercial pigs. Appl. Anim. Behav. Sci. 39:349–362.

Le Denmat, M., J. Dagorn, A. Aumaitre, and J. C. Vaudelet. 1994. Outdoor pig breeding in France. Pig News Info.

Levis, D. G. 1995. "LEVIS" swine breeding facilities. Extension Report. University of Nebraska, Lincoln.

McGlone, J. J. and J. Morrow-Tesch. 1990. Productivity and behavior of sows in level vs sloped farrowing pens and crates. J. Anim. Sci. 68:82–87.

McGlone, J. J. and T. A., Hicks, 2000. Farrowing hut design and sow genotype (Camborough-15 vs 25% Meishan) effects on outdoor sow and litter productivity. Journal of Animal Science. 78:2832–2835.

McGlone, J. J., J. L. Salak-Johnson, R. I. Nicholson, and T. Hicks. 1994. Evaluation of crates and girth tethers for sows: Reproductive performance, immunity, behavior and ergonomic measures. Appl. Anim. Behav. Sci. 39:297–311.

Nicholson, R. I. 1994. Gestating sows: Influence of housing, rearing environment, gut-fill, and pharmacological manipulations on stereotyped behavior. Ph.D. dissertation, Texas Tech University, Lubbock.

Pigs. 1994, October. European housing laws mean simpler systems for pigs. Pig-Misset.

PigTales. 1994, March. Pig topics. Series 11, Number 2. Pig Improvement Company, Inc., Franklin, KY.

Pig Year Book. Various years. Meat and Livestock Commission, Milton Keynes, UK.

Polson, D. 1994. Problem-solving in swine breeding herds: Methods for analyzing production problems, and assessing herd-level risk factors. Ph.D. dissertation, University of Minnesota.

Tubbs, R. C., S. Hurd, D. Dargatz, and G. Hill. 1993. Preweaning morbidity and mortality in the United States swine herd. Swine Health and Production. 1:21–28.

INTERNET RESOURCES

Texas Tech University has a Pork Industry Institute home page at:
http://www.pii.ttu.edu

Temple Grandin has a home page that provides information on handling pigs and provides some diagrams:
http://grandin.com/

Grandin's design of pig chutes is found at:
http://grandin.com/design/blueprint/pigrace.html

The University of Nebraska has a home page that includes information on pork production. One site, designed by Don Levis et al., contains information on a spring-loaded gate latch for swine breeding facilities:
http://ianrwww.unl.edu/ianr/pubs/extnpubs/farmbuil/g1287.htm

The Midwest Plan Service has many publications available at a reasonable cost:
http://gaia.ageng.umn.edu/extens/mwps.html

15

PRODUCTION SYSTEMS FOR GROWING PIGS

INTRODUCTION

Over the past decades, pork production traditionally consisted of three types of enterprises: farrow-to-finish, feeder pig production, and finishing feeder pigs to slaughter weight. Many traditional pork production enterprises included one of two phases in which pork producers either produced feeder pigs or finished feeder pigs for slaughter. The feeder pig producer bred sows, cared for nursing sows and their piglets, and produced weaned pigs that were kept in nurseries until the pigs were 8 to 12 wk old. The second phase of these production enterprises was the feeder pig finisher. These units took the feeder pigs from 18 to 23 kg (40 to 50 lb) body weight up to a market weight of 100 to 114 kg (220 to 250 lb). These long-held standards have changed in some markets during the past decade. A major change is the gradual increase in the market weights over the last 50 years. The *average* market weight is now about 118 kg (260 lb)—and rising! The weights and ages of pigs in each stage are evolving in U.S. modern pig production.

The traditional stages of weaned, growing pigs are nursery, grower, and finisher. Farrow-to-finish producers have all stages under their direct management. A pig producer who has just nursery, just grower-finisher, or both nursery and grower finisher is often called a "grower."

MODERN-DAY CHANGES IN GROWING PIG STANDARDS

Certain production practices may be found in modern, growing pig production systems. These include segregated early weaning (SEW) and all-in-all-out systems of disease control; use of superior genetics from a healthy source; use of diets formulated to meet pig nutrient needs (probably on a least-cost basis); and a willingness to supply capital, equipment, or labor to allow the production system to succeed.

Producers must consider several factors in determining which finishing system will work best for a given operation:

1. Available capital
2. Through-put of pigs required
3. Environmental situation (laws, regulations, farm situation)
4. Nearest neighbors and the need to reduce offensive odors
5. Need to capture manure nutrients for crops
6. Skills of the stockpeople
7. Availability of bedding
8. Availability and cost of power
9. Market location and market requirements
10. Labor

MARKET WEIGHTS

Average market weights have increased because of packer requirements for heavier hogs. Packers discovered that for some genetic lines, heavier pigs were still lean. Economics dictate that the labor to process a carcass is quite similar for a 300-lb (136 kg) pig and a 220-lb (100-kg) pig. Thus, heightened labor efficiency and greater plant through-put of salable pork are achieved with heavier pork carcasses as long as the heavier carcasses are relatively lean. Packers' demand for heavier carcasses is achieved by the grower through two general methods: (1) holding pigs in finishing longer and/or (2) genetically selecting for increased rate of lean weight gain.

If a packer demands (on a short time schedule) that the best price will be paid for pigs with a heavier body weight, the grower has no choice but to increase the time pigs spend in the finishing phase. These added days are very costly, however. The pig's feed efficiency gets progressively worse as it gets heavier.

Breeders are currently selecting not just for greater average daily gain (ADG), but for greater average daily lean gain (ADLG). If the production facilities are scheduled with a fixed time in finishing, the grower will increase the final market weight as selection for increased ADLG progresses.

While the present market weight average is 250 to 260 lb (114 to 118 kg), the U.S. industry is moving toward the day when market weights average closer to 300 lb (136 kg).

WEANING AGE

Sows in the wild wean (separation of mother and piglets) their piglets at a highly variable age from a few days to 6 mo of age. Most wild sows gradually wean the litter over several weeks or months, with the average weaning age between 3 and 6 mo of age. In the last century, people who weaned pigs at 8 wk of age were considered as practicing "early weaning." In the 1950s through the early 1970s, it was considered "early weaning" to wean pigs at 4 wk of age. Economic models and biological studies then began to investigate very early weaning, starting at birth. Machines were built to provide hourly milk to the piglets. Most of these early efforts to wean piglets at 1 to 21 d of age were not widely adopted due to piglet health problems, problems with feeding liquid diets, and an incomplete understanding of piglet nutrient requirements.

The 1990s combined better diets with economic models and drove weaning age down to 14 to 19 d. Piglet diets improved even more in the 1990s and a discovery was made about piglet health: Piglets that were taken from their mothers at a very early age (5 to 15 d) had superb health. This technique is called segregated early weaning (SEW). The conventional wisdom is that SEW pigs have a high level of passive immunity, but they do not have enough time on the sow to be exposed to pathogens. Maternal antibodies are transmitted from the sow to the piglets through colostrum; these maternal antibodies last at least 21 d. Removing the piglets from the sow before she infects them and while they still have passive immunity makes sense. SEW can be used to produce healthy pigs. SEW is less effective for certain diseases, but works quite well for others.

Growers can now buy SEW piglets. This is a new concept because the traditional feeder pigs are 40 to 50 lb (18 to 23 kg) and the SEW pigs are 5 to 15 lb (2 to 7 kg) and could be 5 to 19 d of age. Furthermore, they may be taken off the sows and transported to a distant grower, a relatively new practice.

CONTRACTING

The concept of contracting, while familiar to the broiler industry, is relatively new to the pig industry. The general concept and the most common model is that the grower owns the land and buildings and supplies the labor to care for the pigs. Another individual or corporation owns the pigs and supplies the feed and veterinary care. In essence, the grower has minimal risk and is paid a flat fee, perhaps with a bonus for good pig performance. The owner of the pigs assumes more market risk, but while there is a greater risk of losing money, there also is a much greater chance of making a profit. The owner of the pigs can put more money in sow units as a result of the contract grower sharing the total capital expense of production.

Contracting changes people's motivation. Contract growers may show less care of the pigs because they do not own them. Or, alternatively, they may show greater care for the animals because of the contractual arrangement.

The design of facilities and management of the animals in a contract facility is a concern to both the owner of the pigs and the contract grower.

WEAN-TO-FINISH

Wean-to-finish buildings are becoming more popular on new farms in the United States. In the 1980s, some farms had three buildings from weaning to market: nursery, grower, and finisher. The grower and finishing buildings were then combined to make grow-finish buildings. In the wean-to-finish barn, one building is used from weaning to market for about 26 wk. In the old models, nurseries housed pigs for 4 to 9 wk and the pigs spent the remaining weeks in the growing-finishing building (17 to 22 wk).

An analysis of the economics of the nursery-growing-finishing period favors the newer wean-to-finish system over conventional two-stage buildings (Stein, 1999). Wean-to-finish systems have lower transportation costs, lower animal health costs, better pig survival, lower water costs, and lower waste management costs (see Figure 15–1). The ADG tends to be faster, but is not significantly better for wean-to-finish pigs (see Table 15–1) because, unlike conventional systems, there is no set-back in growth associated with moving and mixing pigs with the wean-to-finish buildings. The inefficient space utilization during the early weeks is a disadvantage of the wean-finish concept but may be offset by improved productivity.

FIGURE 15–1
Three Models of Facilities for Growing Pigs. Model 1 (three buildings) Was Popular in the Early 1980s in the United States. Model 2 Was Popular in the Early 1990s. In the Late 1990s, the Third Model became a Popular Scheme for Taking Pigs from Weaning (generally at less than 21 d) to Finishing (slaughter weight in the United States of over 260 lb) in a Single Building.

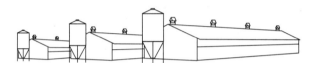

Model 1. Nursery + Grower + Finisher

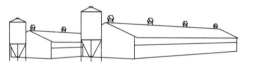

Model 2. Nursery + Grower − Finisher

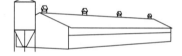

Model 3. Wean-to-finish

TABLE 15–1
Comparison of Pig Performance and Economic Returns for Conventional Two-Stage Buildings (nursery building plus a growing-finishing building).

	FACILITY TYPE	
	CONVENTIONAL NURSERY–GROWING/FINISHING	WEAN-TO-FINISH
Number of sites	68	52
Starting weight, lb	9.5	9.5
Ending weight, lb	255	270
ADG, lb/d	1.37	1.45
Feed intake, lb/d	3.62	3.8
Feed:gain ratio	2.64	2.62
Mortality, %	5.5	3.0
Building cost/pig space, $	$162	$187
Comparative profit, $/pig	Set to 0	+$9.67

Source: Adapted from the Knowledgeworks, Inc. database (Stein, 1999).

BUILDING STYLES

Baxter (1984) reported on the care of growing pigs in medieval times in a system that involved a night-time containment of pigs in a fenced area in combination with daytime foraging of tree mast (acorns, especially). In this system, the herder blew a horn to signal feeding in the evening. Pigs would come running out of the forest and gather in the fenced-in area for a feeding of grains or food scraps. The pigs would bed-down at night in this protected enclosure and were released in the morning to forage again. This system conditioned the pigs to a temporary containment that provided a convenient collection point for the occasional human dinner-time harvest.

Pig housing then moved to full-time containment in pastures, lots, or buildings. In modern pork production, the greater the building investment, the greater control the owner has over pig performance and through-put. From an economic point of view, buildings and pens are depreciated over different time scales, depending upon their expected useful life. If one building costs $100,000 and lasts 10 yr, its per-year depreciation is $10,000. A $200,000 building that lasts 20 yr has the same annual charge for depreciation.

Some facilities that seem inexpensive are actually very expensive. If a low-investment facility costs $50,000 and lasts only 4 yr ($12,500/yr charge), it is more expensive annually than a $100,000 building that lasts 10 yr ($10,000/yr). A facility's cost is determined by its per-year cost, not the total cost of the facility.

Building intensity is a term used to describe the relative concentration of production in the facility. Generally, but not always, buildings with a more intense production through-put are more expensive to build and operate.

The U.S. pig industry has almost come full circle in relative intensity of growing pig facilities. In the 1950s through the 1960s, the U.S. pig industry saw a clear movement to indoor systems for containing growing pigs. Pigs moved from the old-style sty with a simple wooden shelter to open-front buildings on concrete to entirely enclosed buildings. Briefly, in the late 1970s, buildings for finishing pigs were totally enclosed with a heater and mechanical ventilation. This building style is still in use in the northern Midwest and in Canada. In the core of the Midwest and southern parts of the country, however, a fully enclosed, mechanically ventilated, heated building was not needed (nor is such a building cost effective in southern climates) (see Figure 15–2).

From the totally enclosed building, the industry moved to naturally ventilated buildings. These buildings were operated more as a shaded structure in the summer and a poorly insulated building in the winter. Pig performance, air quality, and worker comfort were very good in these buildings. Some producers skipped the mechanically operated building phase and have used one of several styles of curtain-sided buildings ever since the 1970s.

The tunnel-ventilated building is a newer style of mechanically ventilated building. In this style, large fans operate on the ends of the building. One end uses either an air inlet or an evaporative cooling pad. The longer building sides have curtains. These curtains do not serve as ventilation inlets, but rather serve three purposes: (1) They let light in and add to pig and worker comfort; (2) they drop in the event of a power outage, thereby preventing pig loss due to lack of ventilation; and (3) they provide a lower cost for materials—a curtain costs less than a solid wall.

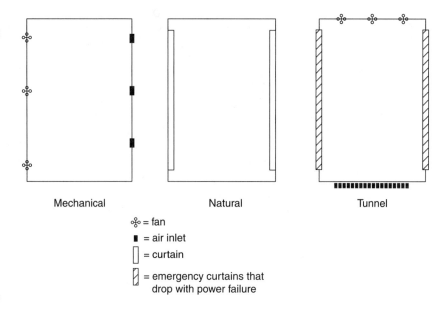

FIGURE 15–2
Three Types of Common Pig Buildings. The Mechanically Ventilated Building Has Fresh Air That Is Brought in by Use of Electric Fans. The Naturally Ventilated Building Can Have One Curtain (sometimes called a modified-open front building), Two Curtains, or Three Curtains (the third curtain is in the center of the roof). The Tunnel-Ventilated Building Is Ventilated from the Long Side of the Building Rather than from the Sides. The Air Inlets Often Have Evaporative Cooling Pads to Cool Incoming, Warm Air. The Long Sides of the Tunnel-Ventilated Building Have Emergency Curtains.

A typical modern building houses 1,000 pigs (± 250). All pigs that arrive at the building are of a common age group and from a single genetic source. The building is emptied over a 3- or 4-wk period in five semi-truckloads (one truck holds about 200 pigs). The production schedule of the farm dictates how long the pigs are there and how many days or weeks they may be on feed. Final market weights are higher with more days on feed.

How does pig performance compare in different facilities? Table 15–2 shows that the pig performance was much better in the low-investment facility than in the high-investment building. Nicholson et al. (1995) conducted the study in Texas in the fall of the year when pig performance was high because the weather was relatively mild.

When the same sort of comparison was done in Iowa, a different outcome was found. Iowa pig performance was essentially equal in the indoor and the hoop-style buildings (see Table 15–3). Pigs in hoop-style buildings tend to grow at a slower rate and with a lower feed efficiency in cooler climates. Colder climates place a greater demand on pigs to use part of their feed for body warmth. In warm climates, productivity in hoops is likely to be at least equal to that in warm housing. Tables 15–2 and 15–3 illustrate that pigs can be raised successfully in a wide variety of housing systems.

Controlled data comparing pig performance in many systems are not available. Sometimes field data are not as reliable as data obtained in controlled studies, but field data may indicate general trends that might help producers make informed decisions.

The Cover-All Company in Canada reported field data on pig performance under different housing systems. In this report, all pigs were of similar genetics and they were fed similar diets. The modified open-front building—the standard high-performance building of the 1970s in the midwestern United States, produced the

TABLE 15–2
Comparison of Pig Performance and Economic Returns for Mechanically Ventilated and Sheltered Lots on Dirt in Texas.

	FACILITY TYPE		
	DIRT FLOOR, OUTDOOR RUN, WITH SHELTER	INDOOR, FANS AND HEATED BUILDING	SE
Number of pigs	119	120	—
Starting weight, lb	52.8	52.8	—
Ending weight, lb	243.9	223.7	4.72
ADG, lb/d[a]	1.7	1.5	0.02
Feed intake, lb/d[a]	6.1	5.5	0.11
Feed:gain ratio	3.5	3.7	0.10
Mortality, %[b]	3.4	2.5	—
Cost of production, $/lb	$0.42	$0.45	—

[a] Housing systems differ, $P < .05$.
[b] Four pigs died in the low-investment facility; three pigs died in the high-investment facility.
Source: Adapted from Nicholson, McGlone, and Ervin (1995).

TABLE 15–3
Means ($\pm$SE) for Pigs Grown in a Hoop-Style or Conventional Facility in Iowa. Total Pigs Evaluated Was 583.

	FACILITY TYPE	
MEASURE	HOOP-STYLE, BEDDED	CONFINEMENT
Start wt, lb*	12.6 $\pm$ 0.17	11.9 $\pm$ 0.12
Ending wt, lb	259.6 $\pm$ 1.6	260.0 $\pm$ 1.1
ADFI, lb/d	4.43 $\pm$ 0.11	4.35 $\pm$ 0.08
ADG, lb/d*	1.63 $\pm$ 0.03	1.53 $\pm$ 0.02
F:G ratio*	2.71 $\pm$ 0.03	2.83 $\pm$ 0.03

* $P < 0.05$ (a statistically-significant difference).
Source: Larson and Honeyman (2002).

lowest pig performance. Pigs in a double-curtain building (curtains on two side walls) and in the hoop-style building appeared to have the best overall performance.

ALTERNATIVE FINISHING SYSTEMS

Alternative finishing becomes more attractive when the following conditions are present:

1. Less capital is available
2. Bedding is available
3. Labor and heavy equipment are available

Alternative production systems are being developed, in part, because every pork producer should want to build facilities with lower capital costs. Dealing with the large mass of dry or damp bedding is the most significant obstacle to development of bedded facilities. Additional advantages to bedded facilities compared with concrete slat buildings include:

1. Pigs on bedding show less tail biting than pigs on slats.
2. Pigs on bedding have fewer foot pad lesions than pigs on slats.
3. Pigs on bedding have fewer leg problems than pigs on slats.
4. Pigs on bedding in naturally ventilated structures tend to have fewer respiratory problems.

MAJOR TYPES OF ALTERNATIVE FINISHING SYSTEMS

The most common type of finishing building in west Texas and western Oklahoma (present number of pigs finished is over 5 million/yr) and among newer U.S. buildings in other regions is a tunnel-ventilated building with total concrete slats. This building houses about 25 pigs/pen and is sized for 1,000 pigs ($\pm$ 200 head). Older (1 to 8 yr old) buildings in the region have a separate nursery and newer buildings are built as wean-to-finish buildings.

The most common, newer or alternative finishing system is the bedded, naturally ventilated, open-air building. Two styles showing some success are hoop buildings and turkey buildings. The hoop building is designed for 200 to 250 pigs; the turkey building is designed for 1,000 to 2,000 pigs. Systems of finishing pigs that have lost favor for various reasons include slatted-floor, open-air buildings; totally enclosed, mechanically ventilated buildings; partially slatted, partially solid-floored pens; and true pasture rearing of finishing pigs. It seems that the most successful buildings are either the totally slatted (often tunnel-ventilated) or bedded and open-air (but not really outdoors).

The newer alternative production systems for finishing pigs are best characterized by the hoop buildings or the scaled-up, former poultry house versions (see Table 15–4). Major differences in these newer, more successful systems include:

1. Use of large group sizes (one pen/building)
2. Use of corn stalks, wheat, or barley straw as bedding
3. Use of dry or wet-dry feeders
4. All-in-all-out systems
5. Use of superior genetics and healthy pigs
6. Utilization of dry bedding as a nutrient for crops

Concrete, Metal, and Wood Structures

Buildings in this category are referred to as total confinement buildings or indoor systems. Such buildings have fans that operate to draw in fresh air and exhaust stale air. They also have heaters in most climates and cooling systems in many buildings. Many of these buildings use total or partial slats for waste containment and removal.

Disadvantages of this type of building include:

- These buildings use the most energy of the alternatives.
- They are the most expensive to build.
- Pigs and workers are subjected to artificial light.

TABLE 15–4
Relative Pig Performance and Costs of Construction for Different Styles of Buildings for Growing-Finishing Pigs.

Measure	Double-Curtain, Total Slats	Modified, Open-Front	Cargill-Style, Indoor-Outdoor	Hoop-Style Bedded Building
Construction cost, $/pig capacity	$180	$120	$80	$55
Mortality, %	3.6	10.6	1.12	2.2
Feed intake, lb/d	4.8	4.6	5.7	5.6
ADG, lb/d	1.66	1.41	1.57	1.77
F:G ratio	2.90	3.25	3.61	3.14

Source: Adapted from a compilation of field data from Canadian company Cover-All Shelter Systems.

- The air environment is often poor, especially in northern areas in the winter months.
- Floors are often slotted, which is a challenge for pigs' feet and legs.
- Pig health is at risk during power failures.
- Once disease organisms enter, they spread quickly.
- Buildings emit odors through exhaust fans.

Advantages of the total confinement building include:

- They promote uniform pig growth.
- In some seasons, especially winter, the best feed conversion is obtained.
- Disease organisms can be kept out through strict biosecurity.
- Seasonal delays or performance problems are minimized.
- Waste can be contained and managed to recycle nutrients and reduce offensive odors.
- Buildings have a long life.

Indoor-Outdoor Lots

Structures in this category include facilities with an indoor, sheltered area and an outdoor run. The sheltered area is often constructed of wood posts or poles. The unit is sometimes called a pole shed.

Floors in these units may be earthen or concrete. On occasion, a slotted area is provided, more often on the outside run. One classic building style is referred to as the Cargill-style building. In this structure, the floor is concrete and is sloped from the shed to the outside concrete run. The sheltered area is usually bedded, but the outside run is typically not bedded.

The open-front building with an indoor and outdoor run is considered a facility that compromises cost and, therefore, productivity. Very few of the newly constructed buildings use this design, yet many thousands of pigs are raised in these buildings today.

Disadvantages of this type of building include:

- Producers are unable to regulate the environment.
- They promote poor pig health.

- They are intermediate in cost to build, but are more costly than the deep-bedded, hoop-style buildings.
- Pig performance is intermediate to total confinement and the better, bedded facilities.
- Most open-front facilities emit an offensive odor due to accumulating waste that is infrequently removed.
- Waste recycling is not on-going, but periodic.
- Workers experience the variable outdoor environment (especially the extremes of winter and summer).

Advantages of this type of building include:

- Their construction costs are lower than those of total confinement buildings.
- Older buildings may be available at a reasonable cost.
- During mild weather, workers may enjoy working outdoors.

Hoop-Style Design

The hoop-style building is one of the newer-designed buildings for housing growing and finishing pigs. It is basically a deep-bedded building that is Quonset shaped. The Quonset shape provides a simple and inexpensive building. The hoop-style buildings house small groups of pigs (150 to 250 pigs) in one large pen.

The deep bedding may be a material such as wheat straw or corn stalks. Other local bedding materials may be used (cotton gin trash, rice or grass hulls, etc.). The bedding is first installed to a depth of at least 18 in. Bedding is then added over time to create a dry material on top with an underlying, fermenting material. The manure pack generates a fair amount of heat as it ferments and digests the waste. The dry bedding that covers the manure pack provides an insulation and odor-trapping function.

The construction of the hoop building includes an optional concrete pad in the front of the building that extends out to create a concrete slab on which equipment may be driven or pigs herded.

The walls of the structure are 4 to 6 ft tall and are typically wooden planks with posts sunk deep (below the frost line) for support. Metal pipes or trusses are fastened on top of the walls to make the hoop shape from one long wall to the other. A canvas or plastic tarp is drawn over the metal hoops.

Pork producers who have these hoop structures report great success in both pig performance and worker satisfaction. Because this building design is relatively new, there are no good estimates on the expected life of these structures. Several buildings are fully functional after 4 yr and their expected life is probably well over 10 yr.

Two main groups have utilized the hoop design: One group is from the plains of Canada (Manitoba and Saskatchewan) and the other group is in Iowa and Nebraska. Both groups report a positive outcome for the hoop building in terms of economics and producer satisfaction.

Disadvantages of this type of building include:

- Pig performance in northern climates is slightly worse than that of comparable indoor systems.

- The environment may be humid at some times of the year, causing condensation to drip from the bars.
- Flies may be a problem in the summer months.
- Because the social groups are large, pig fighting may continue for much longer than would occur in smaller, indoor groups.
- Removing manure can be very time-consuming (but probably not much more so than power washing an indoor unit and dealing with the effluent).
- Feed efficiency may be 10% to 20% lower in the hoop structure than indoors.
- Bedding use averages over 200 lb per pig per cycle or turn.
- Feed efficiency worsens in colder climates.
- Weight gain and fatness may increase in warmer climates.

Advantages of this type of building include:

- Very low capital cost is needed to build the structure.
- Well-designed and efficiently operated buildings provide a pleasant environment for pigs and workers.
- Pig mortality is reported to be low in the hoop building.
- Weight gains are comparable to those of pigs in indoor systems.
- The added bedding costs and poor feed efficiency are reported to be more than offset by the lower building costs.

Tent Design

The tent design is a variation on the hoop design. The tent system is a European design and is not in common use at this time in the United States. The side walls of the tent are made of stacked-up straw or bedding materials. Large, square straw bales are the most common material. The thick wall of bedding provides significant insulation from temperature extremes.

The top of the tent is often a triangular-shaped metal frame that is covered with canvas or plastic tent material. The facility is easily moved.

The floor is earth, with a healthy cover of bedding (more than 2 ft deep to start). Feeder and waterer numbers and space needs for the tent structure are just as those for the hoop structure.

BUILDING AND LOT LAYOUTS

GENERAL ORGANIZATION

Buildings and lots for growing pigs need to accommodate the needs and comfort of the pigs and the workers. The most common design has a center aisle with pens on each side. In wider buildings, having pens on just one side of the building and an aisle along one side is less efficient space utilization than to have a center aisle that serves two sets of pens (see Figure 15–3).

To overcome the inefficiency of the aisle, buildings have been built without an aisle. These general building layouts are shown in Figure 15–4.

FIGURE 15–3

Diagram of a Hoop-Style Building. The 200+ Pigs Are in One Large Pen. Bedding Is Provided and About 12 sq ft/Pig of Floor Space.

Source: Adapted from Brumm et al. (1997).

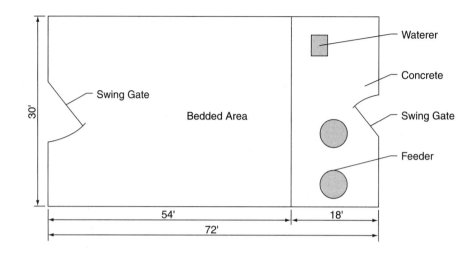

FIGURE 15–4

General Configurations for Buildings.

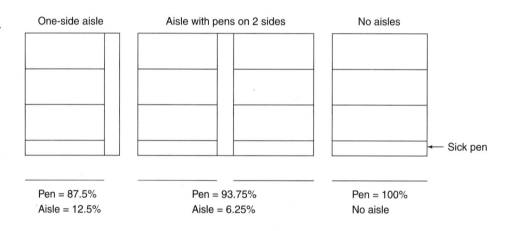

For each of these buildings, calculations are based on a pen that is 6 × 3 m (20 ft × 10 ft) that provides 0.8 m² (8 ft²)/pig for 25 pigs. The example building has a 0.8 m (2.5 ft) wide aisle. In the building on the left with an aisle and pens on just one side, the aisle represents 12.5% of the floor area of the building. For the middle building, which in this example is twice as wide, the aisle occupies only half as much space (as a % of total space). Aisles are considered by some people to be a waste of space because pigs do not occupy that space on a full-time basis. Building costs are increased proportionally by added aisle space.

The aisle is not an absolute necessity in the management and care of the pigs. In the building without aisles, workers can walk from pen to pen among the pigs to observe animal health. Alternatively, some buildings have a type of catwalk over the fencing material so people can observe the pigs from above. This requires a taller building and is an added cost.

The building configuration impacts how pigs are observed and handled. Even the fence height influences how the pigs are handled—if the fence material is too tall, it inhibits people from walking from pen to pen. If the fence material is too low, pigs will jump from pen to pen causing injury, excessive fighting, and even pig deaths.

Producers should consider how the pigs will be handled before the building is occupied. Pigs may need to be sorted, moved, and treated for illness. Provisions should also be made to accommodate dead pig removal.

ORGANIZATION OF PENS WITHIN THE BUILDING

Within the growing pig building, a sick pen (pen for congregating sick or injured pigs) is usually placed on the end of each aisle of pens (see Figure 15–5). Some newer farms put the sick pen in the center of the building, which reduces the maximum distance to walk or move sick pigs. Center pens are often warmer and are subjected to fewer drafts from open doors on the ends of buildings. However, placing the sick pigs in pens in the center of the building exposes more pigs to potential pathogens. In the absence of serious diseases, the sick pens should be smaller than an ordinary pen. Fewer than 0.5% of the pigs in a given building should be in the sick pen at any one time. Many modern buildings use the center pens for sick, injured, or smaller pigs; these pens are the same size as ordinary pens.

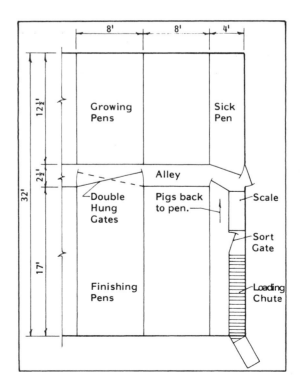

FIGURE 15–5
General Organization of Pens within Growing Pig Buildings. An Alternative Is to Use Center Rather than End Pens for Sick Pens.

Source: Adapted from Midwest Plan Service (MWPS-1).

ORGANIZATION OF OUTDOOR LOTS

Many pork producers still finish pigs in open lots or in low-investment facilities (see Figure 15–6). The capital costs/yr for low- and high-investment facilities are often very similar. Low-investment facilities use equipment that is depreciated over 7 or 8 yr. A building is an asset that can be depreciated over 15 or more yr. Thus, the capital cost/yr may be similar if the low-cost alternative is about one-half the cost of the indoor unit. The general pen and feeder/waterer organization is used for both low- and high-investment facilities.

ORGANIZATION OF HOOP STRUCTURES

Hoop-style structures can be organized in a manner similar to that used in indoor units. Animal handling areas (chutes, scales, etc.) are typically outside the main building, rather than inside the building.

Hoop-style buildings are stocked differently than typical indoor facilities. Most often, the building holds 200 to 250 pigs in one large pen. The pen building has the required amount of feeding and watering spaces. The space requirements/pig have not been fully evaluated; however, the recommendation of 12 sq ft/pig is common; this space allowance supports good animal performance.

When several buildings are lined up on a single site, there should be a 40-ft space between them to allow for air movement and ventilation of each structure.

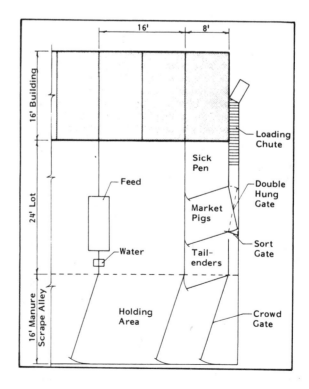

FIGURE 15–6
Design for a Low-Investment Finishing Facility. Note the 16-ft Deep Building or Covered Shed. A 40-ft Outside Run Is Provided with a 24-ft Outside Run along with a 16-ft Area for Manure Collection and Pig Movement. The End Pen Is Used for Sick Pigs and as a Pre-Loading Area.

Source: Adapted from Midwest Plan Service (MWPS-1).

STANDARD PEN LAYOUTS

Standard pen layouts are most often rectangles. Rectangular-shaped pens are easy to construct and equip. If pens were round or odd-shaped, it would be difficult to align a series of pens.

Pigs use their space based on quantity and quality of space provided. A pen is a collection of individual components that come together to form a microenvironment for the inhabitants. Pigs experience the microenvironment in ways producers are only beginning to understand.

Standard equipment within the pen includes flooring, fencing, feeders, watering devices, resting places, and free space. Some farms now use enrichment devices to provide some entertainment for pigs during idle times. How pigs view the quality of space is largely unknown.

FLOORING

Flooring can be solid, solid and bedded, partially slatted, or totally slatted. Most solid floors use bedding of some sort. Baxter (1984) suggested that a good floor should:

1. Not cause injury
2. Not contribute to disease
3. Not cause discomfort or distress
4. Not be inconvenient (to humans)

In addition, a floor should:

5. Firmly support the pig's weight
6. Allow manure to drop through or manure moisture should be absorbed by bedding
7. Be easy to clean or self-cleaning
8. Not be slippery, but not too abrasive either

Bedding

Bedding should provide nonslip footing, warmth, moisture absorbency, and a play or enrichment material. Bedding may also contribute to nutrients eaten. Bedded facilities provide the greatest comfort to pigs in terms of both physical and psychological comforts (bedding provides a substrate on which pigs can root). Pigs, however, can get by quite nicely without bedding if all their other physical and psychological needs are met.

Bedding materials vary in absorbency but should absorb a large amount of water/unit of bedding weight. Table 15–5 lists the relative absorbencies of common bedding materials.

Most modern indoor units use partially or totally slatted floors. If the floor is partially slatted, as is the case in many open-front and curtain-sided buildings, the slatted area should be at least 25%, and preferably more than 33%, slatted. If less slatted area is provided, the pigs will soil the solid areas of the floor. A build-up of manure on the solid floor space is unsightly, unsanitary, and should be avoided.

SECTION IV   HOUSING, ENVIRONMENT, AND NUTRIENT MANAGEMENT

TABLE 15–5
Relative Absorbencies of Common Bedding Materials.

MATERIAL	LB WATER ABSORBED/LB OF BEDDING MATERIAL
Pine wood chips	3.0
Pine sawdust	2.5
Hardwood chips	1.5
Ground corn cobs	2.1
Wheat straw	2.2
Hay, chopped mature	3.0
Peanut or cottonseed shells/hulls	2.5

Source: Values adapted from Midwest Plan Service publication (MWPS-1).

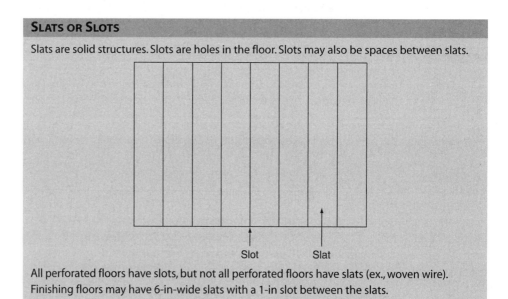

SLATS OR SLOTS

Slats are solid structures. Slots are holes in the floor. Slots may also be spaces between slats.

All perforated floors have slots, but not all perforated floors have slats (ex., woven wire). Finishing floors may have 6-in-wide slats with a 1-in slot between the slats.

Floor Materials

The anatomy and growth of the pig's foot are important features. Pig feet grow in an exponential manner—rapid growth is followed by a leveling off of toe width and length. The pig's inside toes are typically shorter than the outside toes. Toe pad lesions are found on the underside of either toe. Such lesions can be quite painful and should be avoided. A greater % solid area minimizes toe lesions. Rough floor surfaces add to the potential for foot pad lesions. A foot bath of 5% copper sulfate (4 lb/10 gal of water) that pigs walk through three times/wk will help heal toes with lesions. However, the real cause of the excess rate of lesions (such as floor type or condition) should be identified and corrected.

The % solid or the % void (the two measures are perfectly correlated) is an important measure of a slatted floor's features. A floor such as woven wire has a low % solid area and a high % void area. A flooring such as 6-in-wide concrete slats has a high % solid area or a low % void area. Example floors are given in Figure 15–7.

TABLE 15–6
Flooring Characteristics for Common Pig Floors.

Floor Material	% Solid	% Void	ADG, kg/d	Foot Lesion Scores	Thermal Resistance, °C × m²/W
Concrete slat, 6-in wide, 1-in void	86%	14%	NA	NA	0.073
Plastic coated, expanded metal	74%	26%	0.31	1.1	0.15
Woven wire	40%	60%	0.33	1.9	0.12

Source: Adapted from Baxter (1984) and Kornegay and Lindemann (1984).

Younger pigs require a warmer and cleaner floor than do older pigs. This is because younger pigs have a less-developed immune system and a warmer temperature requirement (i.e., a higher lower critical temperature). Thus, younger pigs are more susceptible to disease microorganisms and are more easily chilled.

The trade-off in slatted flooring is between foot support and lameness on the one hand and warmth and cleanliness on the other (see Figure 15–8). The younger the pigs, the greater the need to separate the pigs from their feces and urine for health reasons. A floor with a low % solid area will clean better than a floor with a high % solid area. Larger pigs also have larger feet and they can push the fecal material through the flooring more efficiently than can younger pigs.

Older pigs require greater support for their feet and legs. They will have problems with lameness if their feet are not well supported. The best support is provided by flooring that has a large solid area. Concrete slats have the greatest solid % and woven wire has the least solid area of the common floor types (see Table 15–6).

The most common flooring material in U.S. commercial nurseries is woven wire that covers either the total floor surface or two-thirds of the floor surface. Newer nursery floors use plastic slatted material. Some of the plastic flooring contains a bactericide that retards bacterial growth. Concrete slats or partial concrete slat floors are the most common flooring for growing-finishing pigs in newer U.S. buildings.

In partially slatted buildings, the fencing near the solid flooring is often solid to create a warm, dry area. The fencing over the partial slats is often open materials to allow drafts and wind currents that keep the area cool and encourage elimination.

FENCING MATERIALS

Fencing materials are chosen for their thermal properties, their ability to be sanitized, pen ventilation requirements, and durability. The choice of open or solid fencing is dependent upon the needs for ventilation and the thermal needs of the pigs in the building.

Common fencing materials include solid concrete, poured concrete with open wall slots, painted metal, wood, or plastic. The less porous the fencing and flooring material is, the better it can be sanitized.

What Is the Best Fence Height?

When choosing the height of fencing material, producers must take into account the ability of pigs to climb or jump over fencing, the cost of the fencing, and the ability of

FIGURE 15–7
Materials Used for Slotted Floors in Commercial Swine Operations. Growing-Finishing and Adult Pigs Are More Often Kept on Concrete Slats and Younger Pigs Are Kept on One of Several Floor Types with a Lower % Solids, Enabling Them to Be Kept Warm and Clean. The Top Floor Types Are Often for Farrowing and Nursery Buildings. The Slat Layout (bottom-right) Is the Minimum Area of Slotted Floors Recommended. The Lower-Left Layout Shows a Cross Section of a Concrete Slat for Finishing Pigs.

Source: Midwest Plan Service (MWPS-1).

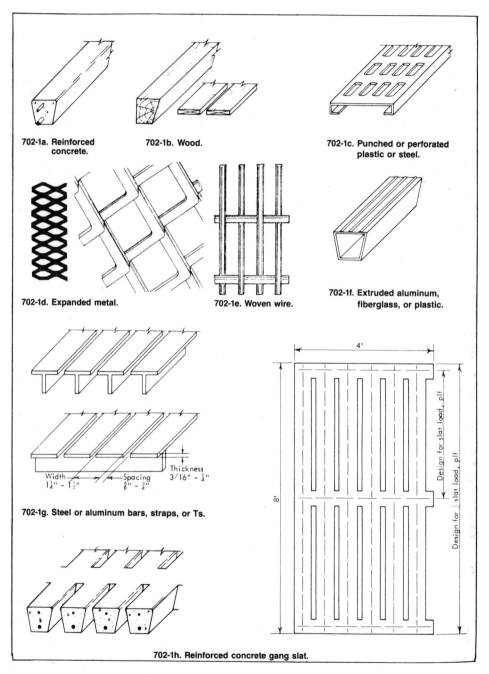

FIGURE 15–8
A Balance Must Be Struck Between Young and Old Pigs in Floor Type. Young Pigs Are More in Need of Warmth and Separation from Their Manure. Older Pigs Are More in Need of Foot Support and Less in Need of Warmth.

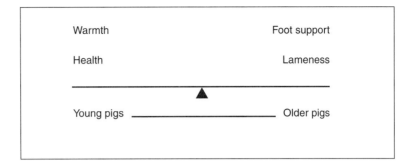

workers to climb over the fencing material as they care for the pigs. Clearly, if the fence height is lower, workers can more easily move from pen to pen. Lower fences generally cost less than taller fences. Pigs are curious animals and they will investigate the fence and may climb or jump from their home pen to the neighbors' pen. When a socially strange pig arrives in a new pen with an established social order, the residents will be aggressive toward the newcomer—they may even kill the "intruder." For this reason, the fence height should prevent pen-to-pen jumping.

In a controlled study, McGlone et al. (1990) examined how high nursery pigs can jump. Pigs were trained to jump over plywood walls cut to different heights. Some nursery pigs could jump 0.77 m (30 in) if they had some training. The naïve nursery pig could not jump 0.61 m (24 in). When weanling pigs were acclimated from weaning to fences 24 in tall (Wadsworth et al., 1990), they did not jump over to the neighboring pens. Based on this work, Wadsworth et al. recommended that nursery fencing be 24 in tall. The researchers are not aware of similar work done for finishing pigs. However, some newer-styled finishing buildings in the United States use 24-in-tall fencing material with success.

FEEDER SPACE AND DESIGN

The choice of an appropriate feeder is a critical decision. The feeder is usually depreciated for 8 yr but it is expected to last at least 10 yr. Because of the need for durable equipment, feeders are more often made from stainless steel and plastic rather than wood or steel. In small swine enterprises, many farmers traditionally construct and repair their own wood and metal feeders.

Pigs have a strong level of curiosity and they will investigate, chew, and root against the feeder and other objects in the pen. The feeder must be durable enough to withstand the constant manipulation.

The simplest way to feed pigs is to feed them on the floor. In this management system, the feed is spread on a concrete, solid floor (indoors or outdoors) or on an earthen floor outdoors. If this method is used, the feed should be spread out so each individual will get enough to eat. This method is used for early weaned pigs to get them to eat dry feed more easily (a little feed is spread on the floor or in a pan on the floor to get them to root and investigate and begin to eat). Outdoor sows or low-investment indoor sows can also be fed on the floor or even on the ground if cubes are used.

Feeders are used (1) to contain the feed in a sanitary manner, (2) to minimize feed wastage, and (3) to conserve space, especially in growing-finishing where effective floor-feeding requires much more space than a feeder occupies.

The amount of feeder space needed by pigs has largely been determined by trial and error on commercial farms with an eye toward providing ample, but not excessive, feeder space. Newer studies indicate that some competition for feeding spaces actually improves performance of the groups of pigs.

As a practical matter, feeder designs are based on the weight and size of pigs in their last days in the building or lot. The feeder space could be too large or too small and each has special problems. If the range of weights is from 15 to 45 lb, the feeder is designed to accommodate the head and shoulder size of the 45-lb pig (Figure 15–9). The farm will experience problems if the range of weights is too large in a given phase. Imagine the problems if the range of weights in a nursery is from 8 to 60 lb and if the feeder is designed for 60-lb pigs. The first problem is that some 8-lb pigs will live in the feed trough. This site, selected as a resting place, prevents other pigs from eating, and could result in soiled feed and injured or trapped pigs. Fixed design equipment is simply not flexible enough to accommodate pigs that change in body weight by eight-fold.

If the feeder is too narrow, pigs will not be able to feed easily. As they approach a weight and size that is too large for the feeder, the caregiver would notice abrasions, cuts, and open sores on the pigs' head and neck. This is a clear sign that the feeder hole is either too small or is improperly designed (perhaps with sharp edges).

The choice of the feeder space width (and other dimensions) is best made based on the ending weight and size of the pigs in a given building. Data based on estimated pig head sizes are presented in Figure 15–10, which shows that a group of 50-lb pigs would have head widths of about 5 in. Thus, one might conclude that the feeder space ought to be about 5 in $\times$ 5 in if it is a square hole; a round hole would need a diameter of at least 5 in.

FIGURE 15–9

Feeder Space Requirements for Pigs Are Based on Shoulder or Head Width Measurements. The Entire Range of Weights for Pigs from Birth to Adult Weights Is Presented. The Formula from Seaton Baxter Is $Y = 61 \times (\text{body weight}^{0.33})$, with Y (shoulder width) in mm and Body Weight in kg. A More Precise Estimate Is Needed for a Narrow Range (see Figure 15–10).

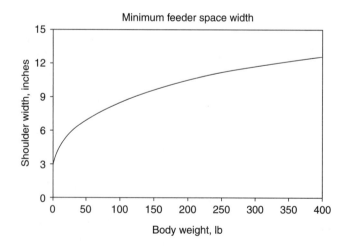

The CV

The coefficient of variation is a percentage figure that is adjusted based on the mean. To calculate the CV, use:

$$100 \times \frac{SD}{Mean} = CV$$

where SD is the standard deviation. Uniformity is associated with a low CV.

FIGURE 15–10

Relationship between Body Weight and Head Width for Nursery Pigs Late in the Growth Phase. The Data Are Based on Individual Pigs. The Regression Equation That Predicts Head Width (Y, cm) Based on Body Weight (X, kg) Is: Y = 11.0 × .05X

Source: Adapted from McGlone et al. (1983).

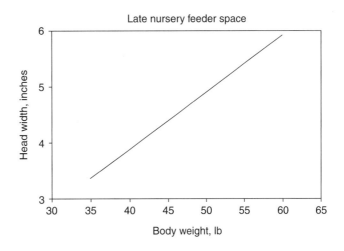

Producers must not only be concerned about the average ending weight, but also the range of weights in the group. Pigs in a given group at a common age have about a 20% coefficient of variation (CV) in body weight. On larger farms, the higher number of pigs available makes it easier to sort pigs by age and weight into uniform pens. Table 15–7 examines the relative variation and the range of pigs' weight. Even under fairly good conditions, the weight range may be from 40 to 60 lb for the average barn full of 50-lb pigs. Thus, the feeder space for the average ending weight of 50 lb should actually be for the upper end of the pigs in the barn, which may weigh 60 lb when they leave the nursery. The consequence for not meeting the needs of pigs for feeder space is injury to pigs and an equipment-induced limit on feed intake and weight gain.

The problem of wide ranges in body weights and head/shoulder sizes is more of a concern in the nursery than in growing-finishing. In growing-finishing, the growth curve levels off above about 100 lb and, therefore, the range of head and shoulder widths is less. The nature of the exponential growth curve is that the rapid growth during the nursery phase expresses itself as a rapid increase in head width. In later stages, however, the head width does not change as much.

To accommodate the wide ranges of pig body weights and head sizes, some feeders have an open trough. The open trough may be long enough to accommodate four or

TABLE 15-7
Variation in Pig Weights for a Group of Pigs That Average 50 lb. SD Is the Standard Deviation. CV Is the Coefficient of Variation (calculated as 100 × (SD/Mean)). With 2 SD on Either Side of the Mean, the Weights Represent 95% of the Pigs in the Group. CI Is the Confidence Interval or the Range of Weights That Contains 95% of the Data.

CV, %	SD	95% CI FOR WEIGHT MINIMUM TO MAXIMUM
5	2.5	45 to 55 lb
10	5.0	40 to 60 lb
15	7.5	35 to 65 lb
20	10.0	40 to 70 lb

TABLE 15-8
Effects of Feeder Space Allowance (feeder spaces/20 pigs) for Finishing Pigs Fed from Dry Feeders.

	FEEDER SPACES				PROB. VALUES[a]	
MEASURE	1	2	3	SE	LINEAR	QUAD.
Number of pigs	200	200	200	—	—	—
Number of pens	10	10	10	—	—	—
Starting weight, lb	134.5	133.8	133.6	3.82	.87	.96
Market weight, lb	228.4	233.9	233.0	4.28	.223	.257
ADG, lb/d	1.46	1.56	1.57	.042	.039	.096
Feed:gain ratio	4.02	3.91	3.80	.136	.137	.914
Feed intake, lb/d	5.68	6.02	5.82	.163	.59	.16

[a]P-values for the linear and quadratic effects of feeder space allowance (spaces/20 pigs).
Source: Adapted from McGlone et al. (1993).

more pigs eating simultaneously when they weigh 15 lb, but only two pigs eating side-by-side when the pigs weigh 45 lb.

Traditional recommendations for feeder space requirements put the requirement at four to five pigs/feeder space, but more feeder spaces are needed shortly after weaning. Newer information challenges this standard.

The normal biology of the pig is disrupted at weaning when the pigs are separated from their mothers. During lactation, pigs of European origin have a very strong attachment to a single teat—they express teat fidelity (interestingly, Asian pigs such as Meishan have much weaker teat fidelity). When pigs are weaned, the caregiver provides fewer than one space/pig. To improve efficiency, pigs take turns eating out of a feeder. Previously, each pig had its own feeder (teat). Producers disrupt pigs' behavior when they move them to a dry feeder.

Former conventional wisdom suggested that producers should not limit feeder space because it would limit feed intake (FI) and average daily weight gain (ADG). Some competition at feeding or social pressure at the feeder actually can stimulate feed intake.

Table 15–8 shows data from a Texas Tech University study of 600 pigs. Growing-finishing pigs were given one, two, or three feeder holes in groups of 20 pigs/pen. With

TABLE 15-9
Feeder Space Recommendations for Pigs of Various Sizes. The Traditional Recommendations Are Based on the Midwest Plan Service and the Pork Industry Handbook. Newer Recommendations Are Based on Controlled Studies and Field Data.

	FEEDER SPACES OR SIZE	
PIG SIZE	TRADITIONAL RECOMMENDATION	NEWER RECOMMENDATION
Early weaned pigs, 7–18 lb	2 pigs/space	2 pigs/space
Nursery, 18–60 lb	3 pigs/space	5 pigs/space
Growing-finishing, 60–300 lb	5 pigs/space	10 pigs/space
Sows and boars	1 ft/sow	19–24 in/sow

Source: Adapted from Midwest Plan Service and the Pork Industry Handbook.

two feeder holes (one space/10 pigs), the pigs actually ate more feed and gained faster than when pigs had fewer feeder spaces. With the old standard of three feeder holes (one hole/six to seven pigs), there was no change in feed intake or ADG. From a statistical perspective, the F:G ratios and ADG for two and three feeder holes/20 pigs were similar.

The data described in Table 15–8 are for pigs fed from a dry, wooden feeder. The results are certainly likely to be different for each model of feeder on the market. The proper pig density/feeder space is very much worth refining in each microenvironment so the greatest economic return can be captured (see Table 15–9).

Wet versus Dry Feeders

A wet feeder refers to one of several styles of feeding equipment. Pigs used to be fed a slop—a suspension of food in water that was poured, shoveled, or otherwise delivered to the pigs. The old-style wet feeder was a trough in which meals were delivered. Wet feeders in the modern sense are feeders that provide the water source and feed source in the same compartment. Many of the modern wet feeders are designed to have the pigs apply the water to the feed or the feed to the water trough. In any case, many argue that "wet feeding" improves F:G ratios and allows more pigs to occupy a feeder space because pigs can eat more quickly from a wet feeder than from a dry feeder (which they have to alternate between feeding and walking over to water). Wet feeders may also conserve water.

The conventional European view is that wet feeding is the most economical approach. In the United States, wet feeding gets mixed reviews but is often recommended for pigs over about 50-lb body weight. A Texas Tech University study showed that in the nursery phase, daily weight gain and feed-to-gain ratios were lower in pigs fed from wet/dry feeders than from dry feeders (see Table 15–10; McGlone and Fumuso, 1993). The resulting ending weight for nursery pigs on the wet-dry feeder was 2 to 3 lb lighter than that for pigs on the dry feeders. The lower final nursery weight would be very costly to the pork producer. An average nursery ending weight that is 1 lb higher should translate into 2 or more lb heavier average market weight, given the same number of days on feed. However, in the finishing phase, results can

TABLE 15–10
Least Squares Means for Pig Performance When Exposed to Three Types of Feeders During a 35-d Nursery Period (weaning at 29 d of age).

MEASURE	FEEDER TYPE			SE	P-VALUE[a]
	RECTANGLE	ROUND	WET/DRY		
Number of pigs	120	120	120	—	—
Number of pens	6	6	6	—	—
Starting weight, lb	15.2	15.0	15.2	.23	.870
End weight, lb	38.0[d]	37.8[d]	35.2[e]	.34	.008
ADG, lb/d	.65[d]	.64[d]	.57[e]	.01	.014
Feed:gain ratio	2.02[b]	1.97[b]	2.32[c]	.06	.035
Feed intake, lb/d	1.30	1.30	1.32	.02	.588

[a] P-value for treatment effect.
[b,c] Treatments in rows with different superscripts differ, $P < .05$.
[d,e] Treatments in rows with different superscripts differ, $P < .01$.
Source: Adapted from McGlone and Fumuso (1993).

be quite different. In Texas Tech University studies (unpublished), wet/dry feeders in finishing caused an increase in ADG of 14% and about the same feed efficiency.

In comparison, a rectangular, stainless steel feeder and a plastic, round feeder yielded similar pig performance. Given equal pig productivity, the decision on feeder model should be based on cost, durability or longevity, and ease of adjustment. Producers prefer feeders that require little adjustment and still deliver adequate feed with little waste.

Feeder models vary considerably. Pig performance must be determined for each feeder. In addition to the variation from one feeder model to another, feeder manufacturers change design features regularly in an attempt to improve pig performance and sales of their feeder.

WATERERS

Pigs require water to sustain their life and to grow. The nature of the watering device can impact animal growth and development. Pigs cannot live without water for very long; they suffer if they are without water for more than a few hours. Water deprivation for 12 to 24 hr has significant negative effects on pig behavior, feed intake, and growth.

In most situations, growing pigs should be given ad libitum access to water. The water should be free from minerals, bacteria, and other contaminants. Gilbert Hollis (1996) of the University of Illinois reports that water intake is at least twice the feed intake. The water intake for growing pigs is given in Table 15–11.

Some systems provide water in set watering bouts. Some systems provide water only when feed is delivered. These meal-fed and interval-watered systems are more common in Europe where pigs are more often limit-fed than in the United States where most pigs are fed ad libitum. When water is available in intervals, it should be left on for at least 30 to 45 min at a given time. Water should always be available when pigs are feeding.

Water flow rate has a strong influence on water consumption. If water pressure is too low, pigs will not be able to consume sufficient water, they will eat less feed, and they will grow slower. Jerry Bodman at the University of Nebraska suggests the minimum flow rate for waterers should increase with pig size:

Nursery pigs 0.23 l/min (0.06 gal/min)
Grower pigs 0.45 l/min (0.12 gal/min)
Finisher pigs 0.72 l/min (0.19 gal/min)

Pedersen (1994) suggests a greater flow rate for growing pigs (from 0.5 to 1.2 l/min). The European systems of limit feeding should require greater water flow rates because pigs can be fed water in meals rather than ad libitum. Water flow rate is clearly proportional to the pressure in the water line. Water pressure in the line should be 10 to 80 psi. Some nursery waterers may limit water intake in the 2 to 5 psi range. At this low pressure, pigs' water intake is limited and feed intake and growth will be proportionately reduced.

Watering devices come in various forms, including streams, a trickling hose, nipples, cups, and troughs. While most waterers are fixed against a wall, a newer style of waterer swings from the center of the pen. Pigs gain some enrichment or entertainment value from the swinging waterers and farmers report that the swinging waterers conserve water. Recent research at the University of Nebraska (Brumm and Dahlquist, 1997b) showed that the Trojan WaterSwing reduced water use by 11.1% and reduced manure volume by 16.2%. This is an advantage in the management of manure because wasted water is a large volume of the effluent volume. Anything that can be done to reduce water waste is welcome. However, the total manure nutrients are not reduced simply because of a reduced water volume—the manure is less dilute but contains the same total mass of nutrients that must be dealt with.

The Midwest Plan Service (1987) requires one waterer/10 nursery pigs and one waterer/15 growing-finishing pigs. If more than one waterer is used/pen, they should be spaced at least 12, 18, and 36 inches apart during nursery, growing, and finishing phases.

TABLE 15–11
Water Intake for Pigs of All Sizes.

	WATER INTAKE	
WEIGHT (LB)/AGE OF PIGS	**GAL/D**	**L/D**
25	0.4	1.5
50	0.6	2.3
75	0.9	3.4
100	1.0	3.8
150	1.3	4.9
200	1.7	6.4
250	2.0	7.6
300	2.1	8.0
400	2.3	8.7
Pregnant sows/gilts	4.5	17
Lactating sows	6.0	22.7
Boars	4.5	17.0

The height of the waterer should vary as the pigs grow. The desired height of the waterer varies depending on whether the waterer has a large or a small angle (90° or 45°). As pigs grow, the waterer should be raised to the pigs' shoulder or head height so that drinking is easy and comfortable (see Table 15–12).

Water can contain contaminants that limit water consumption. McFarlane (1995) gives values for various common contaminants in Table 15–13. He also provided a few alternatives if quality water is in limited supply on some farms. Among the alternatives to conserve water are recycling water from lagoons to the building flush systems; reducing water pressure to a minimal level to reduce water waste; revitalizing the wells on the property if they are older; and installing a filter or water treatment system. Newer swing waterers may also conserve water.

TABLE 15–12
Recommended Nipple Waterer Height (in) above the Floor.

AGE OR STAGE	90° ANGLE	45° ANGLE
Nursing piglets	4–9	6–12
Nursery	9–16	12–18
Growing	16–21	18–25
Finishing	21–26	25–30

Source: Adapted from Pedersen (1994).

TABLE 15–13
Limits of Some Common Water Contaminants for Water Sources for Pigs. Zero Contamination Is Preferred.

ITEM	MAXIMUM ALLOWED
Hardness; calcium carbonate, ppm	180
Total dissolved solids, ppm	< 3,000
pH	5–8
Coliforms, number/100mL	1
Algae	None
Nitrate, ppm	300
Sulfate, ppm	< 3,000
Chlorine, ppm	300
Copper, ppm	0.5
Fluorine, ppm	2
Arsenic, ppm	0.2
Cadmium, ppm	0.5
Lead, ppm	0.1
Mercury, ppm	0.01
Selenium, ppm	0.05
Zinc, ppm	25
Chromium, ppm	1
Cobalt, ppm	1
Nickel, ppm	1
Vanadium, ppm	1

Source: Adapted from McFarlane (1995).

SPACE NEEDS

Pigs grow in an exponential manner, with the exponent of the curve less than 1.0 units. The general nature of pig growth follows a pattern of rapid early growth followed by a leveling-off of the growth rate. Not only does pig weight follow this pattern, but pig body size and shape follow the same pattern (see Chapter 8).

Groups of pigs require different space requirements than those previously discussed for individual breeding animals. The bodies of a group of pigs require a certain amount of space, referred to as occupied space. The space in a pen that remains is called free space. The amount of space pigs occupy depends on their posture and behavior. Pigs require less space when they are standing than when they are resting. Resting on their side (lateral recumbency) requires less space than does lying on their belly/chest (sternal recumbency). At any given time in a group of pigs, individual pigs are standing, walking, lying in lateral recumbency or sternal recumbency, or sitting. At times, pigs interact socially in various manners; these behaviors require even more space, referred to as social space.

Seaton Baxter (1984) described the amount of space pigs occupied in different postures. The equations are plotted in Figure 15–11. Pigs of about 250-lb body weight require about 6 sq ft for their bodies to rest on while lying down. At 250 lb, pigs require about 20% less space while standing or lying on their sternum than when they are lying on their side.

Pigs vary their behavior with time of day. Depending upon the effective environmental temperature and other components of the microenvironment, pigs generally show a diurnal cycle in behavior. A Texas Tech University study showed that pigs were least active around 0100 hours. At this point in the daily cycle, the pigs occupy the greatest amount of space because the greatest numbers of pigs are sleeping or resting (or the greatest number of pigs are assuming the posture that takes up the most amount of space). The amount of free (unoccupied) space is least around midnight and the most space is available shortly after noon when the most pigs are standing.

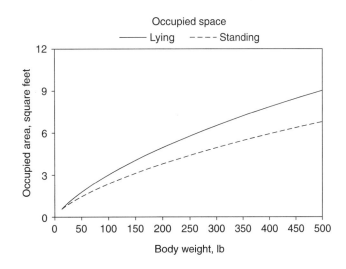

FIGURE 15–11
Space Occupied by Single Pigs of Various Weights While Lying (solid line) or Standing (dashed line). Data Are Quite Similar for Pigs Lying on Their Sternum and Standing. The Formula for Space Occupied while Lying Is: Area (m^2) = .024 × (weight$^{0.66}$). For Standing, the Formula for Occupied Space Is: Area (m^2) = .019 × (weight$^{0.66}$).

The cycle in pig behavior might lead some caregivers to think that pigs are not crowded when, in fact, they are. If the pigs are observed during the day, the caregiver should understand that even less space will be unoccupied or free in the evening when most pigs are lying down.

The amount of unused or free space increases with increases in group size. Figure 15–12 shows data on behavioral measures of space utilization are presented. Research has documented that if all the free space is removed, reduced feed intake and reduced weight gain will result. Attempts to add antibiotics or increasing nutrient density to compensate for the lower feed intake have failed to restore pig performance to the level of the "uncrowded" pigs.

McGlone and Newby (1994) conducted a study in which some or all of the free space was removed. Results showed that removal of 50% of the free space did not cause a pig performance problem. However, removal of too much of the unused or free space did result in a slow-down of ADG, largely due to a reduced feed intake (see Table 15–14).

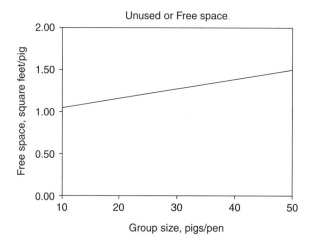

FIGURE 15–12
Space Unused by Group-Housed Pigs (given 8 ft²/pig) Increases with Increased Group Sizes. The Data Here Represent 50% of the Free Space in Pens with Group Sizes from 10 to 50 Pigs/Pen. These Data Suggest That This Amount of Space Could Be Removed without a Performance Problem.

TABLE 15–14
Pig Performance When Given the Traditionally Recommended Space of 8 Sq Ft/Pig or 7 Sq Ft/Pig (removal of 50% of the free space) or 6 Sq Ft/Pig (removal of 100% of the free space).

	SPACE/PIG, SQ FT				
MEASURE	8	7	6	SE	P-VALUE[a]
Number of pigs	80	80	80	—	—
Number of pens	3	3	3	—	—
ADG, lb/d	1.54[a]	1.50[a]	1.32[b]	0.02	<0.05
Gain:feed ratio	0.24	0.24	0.24	0.007	NS
Feed intake, lb/d	6.31[a]	6.25[a,b]	5.57[b]	0.23	<0.05

[a,b]Means with a common superscript do not differ, P < .05.
Source: McGlone and Newby (1994).

The traditional space requirements were established with relatively small group sizes (less than 10 pigs/pen). When larger group sizes are used, there is a greater amount of shared, unused, or free space. Removal of some (up to 50%) of the free space has no negative performance consequence. The findings reported in Table 15–14 were confirmed in work at the University of Nebraska, indicating that 7 sq ft/pig is adequate for maintenance of economical pig growth (Brumm and Dahlquist, 1997a).

The space needs for heavier pigs can be estimated based on the data presented in Figure 15–11. Pigs up to 250-lb body weight and in small group size (less than 20 pigs) require 8 sq ft/pig. Larger group sizes, especially those over 50 pigs/pen, and pigs up to 300 lb-body weight may need only 8 sq ft/pig rather than the 9 sq ft/pig needed in smaller group-size pens. Traditional and newer estimates of space needs for growing pigs are given in Table 15–15.

Space needs for pigs in outdoor lots should be based more on local performance standards than by hard-and-fast numbers. When the weather is cold, less space in outdoor lots is acceptable. When the weather is hot and dry, less space is needed than when the weather is hot and wet. The cleanliness of the pigs and their health will help determine the space needs. Recommendations for space needs for pigs kept in sheds and lots (inside and outside areas) are given in Table 15–16.

TABLE 15–15
Floor Space Needs for Indoor-Housed Growing Pigs. For Partial Slats, Add 10%. For Fully Bedded Barns, Add 50% to the Numbers in This Table. Group Sizes of 40 to 100 Pigs/Pen Are Not Recommended.

		SPACE NEEDS, FT^2/PIG		
WEIGHT OF PIGS, LB	TRADITIONAL	SMALL GROUPS <20/PEN	MEDIUM GROUPS 20–40/PEN	LARGE GROUPS 100+/PEN[a]
5–30	2.5	2.5	2	1.8
30–75	4	4	3	2.8
75–150	6	6	5	4.8
150–250	8	8	7	6.5
250–350	na	9	8	7

[a] These requirements are speculative because they are obtained by extrapolation. See recent studies at the University of Illinois by Mike Ellis on group size effects on growing pig performance
Source: Hyun and Ellis (2001), Wolter et al. (2000), Wolter et al. (2001), and Hyun et al. (1998).

TABLE 15–16
Recommended Space Needs for Pigs in Outdoor Lots.

STAGE	WEIGHT, LB	INSIDE AREA FT^2/PIG	OUTSIDE AREA FT^2/PIG[a]
Nursery	18–75	4	8
Growing-finishing	75–250	6	15

[a] For some outdoor finishing pigs, 50 pigs or less/acre is recommended.
Source: Adapted from the Midwest Plan Service (MWPS-1).

APPROPRIATE GROUP SIZES

The appropriate or best group size is difficult to determine. Every reasonable group size has been tried on farms with varied success. The traditional recommendation is to keep pigs in group sizes of 30 pigs or less. Group sizes of 40 to 100 pigs/pen have more problems with injury and mortality. A study at Texas Tech University indicated that growing-finishing pig mortality was 3.5% or less for group sizes of 10 or 20 pigs/pen. For 40 pigs/pen, however, the mortality was 10%. Similar results have been reported.

Recent work, however, has caused reconsideration of the general recommendation to keep group sizes under 30 pigs/pen. Pigs have been housed in groups of 150, 200, 400, or even over 1,000 pigs/building in a single pen. In these few case studies, pig mortality and productivity were comparable or, in some cases, even better than the traditional 20 to 25 pigs/pen found on many farms. Field trials have been done with 450, 1,000, 1,500 and 1,600 pigs/building in a single pen. In each case, pig productivity was equal to or better than that for pigs in traditional systems with 25 pigs/pen. How could this be? What mechanisms might be at play in large social groups? Further research is needed.

There is a limit to how many pigs a given pig can recognize and remember. This number is not known, but estimates vary that a given pig can recognize from 6 to 40 individuals. If pigs are regularly bumping into strangers, they experience social stress. When pigs live in very large group sizes, they establish smaller social groups that function as gangs of pigs. Each gang has a territory that it roams over. With this behavior, pigs do not often bump into unknown pigs. If this theory is correct, it would argue to spread feeding and watering stations over the large barn to prevent pigs from having to travel great distances for nutrients and to prevent unwanted and negative social interactions. The broiler industry, with thousands of chickens/barn, has utilized this concept for decades.

LOADING CHUTES ON THE FARM

SINGLE OR DOUBLE CHUTES

Loading chutes are needed to move pigs from buildings or lots onto trucks or trailers. If the loading chute is well-designed, pigs should flow like water through the chute. In well-managed, double-wide chutes, pigs can walk at a speed of over 1,000 pigs/hr. Features of a well-designed chute include:

- Fences should be strong and well-constructed.
- Fences can be solid or mesh as long as the mesh does not throw shadows on the floor.
- Lighting should be uniform.
- A straight chute is preferred to one that bends.
- Sharp, angled turns should be avoided.

- If a turn is required, a rounded fence is preferred.
- Cleats on the floor should be spaced as appropriate for the size of the pigs to be moved through the chute.
- If more than 600 pigs/hr are to be moved, a double chute should be provided.
- Chutes should not contain drafts.

Most nursery and growing-finishing barns should be able to have loading chutes that exit in a straight direction out of the building. It is a good idea to have the chute covered to avoid drafts, uneven light distribution, and outside disturbances. Nothing stops indoor-reared pigs more effectively than to be loaded in the early morning with the sun shining through the chute and a draft hitting them in the face. These conditions should be avoided.

Figures 15–13 and 15–14 present two loading chute designs. Figure 15–13 shows a double chute that is nearly straight. Some producers favor the slight bend in the chute so that pigs at the entrance to the chute cannot see the truck and activity at the chute-truck juncture.

Figure 15–14 shows a circular chute design. The circular design for the pre-chute area allows for a great number of pigs to enter the loading chute.

A funnel-shaped design for the pre-loading area does not work very well. As the funnel narrows, pigs get wedged at their shoulders and they cannot move forward. A dog-legged shape actually allows pigs to move more quickly than does a funnel-shaped entrance.

FIGURE 15–13

A Permanent Loading Chute with an Attached, Raised Catwalk for Use by Animal Handlers.

Source: Adapted from the Midwest Plan Service (MWPS-1).

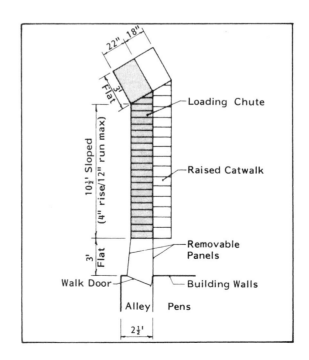

FIGURE 15–14
Diagram of a Round Loading Chute Designed to Efficiently Move Larger Numbers of Finishing Pigs.

Source: Adapted from the Midwest Plan Service (MWPS-1).

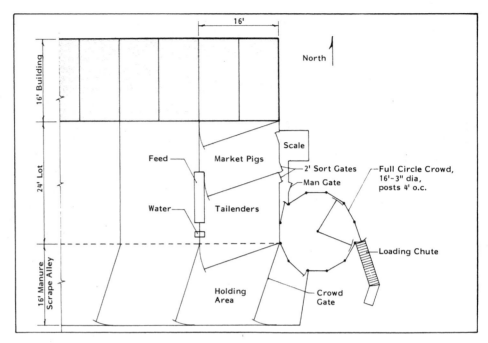

SUMMARY

Pork production systems have changed dramatically over the past decades and continue to do so. This chapter focused on the production practices currently used for grower-finishing pigs. Changes in grower-finisher pig standards, including increased slaughter weights and earlier weaning, have accompanied the evolution of different types of housing, facilities, equipment, and marketing. Choices in building styles, ventilation systems, building and lot layouts, pen layouts, and types of fencing, flooring, bedding, feeder design, and waterers were described. Space needs of pigs and appropriate group size, from the standpoint of animal comfort and optimum performance, were discussed, and the latest research-based recommendations were provided. Finally, the design and characteristics of loading chutes needed for movement of pigs from pens or lots to vehicles for transport was addressed in terms of animal comfort and efficient marketing.

QUESTIONS AND ACTIVITIES

1. What are the trade-offs when producers decide upon low- or high-investment finishing facilities?
2. Why did wean-to-finish systems begin to flourish in the late 1990s?
3. How do the thermal needs of weaned pigs from an SEW unit weaning at 7 d differ from the needs of pigs weaned at 21 d in a conventional system?

4. How do the air temperature and space allowances vary for a wean-to-finish building compared with a conventional nursery? Should the wean-to-finish building be managed differently when nursery-age (weaning through 10 wk of age) pigs are present?
5. What effect does the production system have on pork eating quality, if any? (Search the literature and the Web for answers to this question; note the work of L. L. Hansen from Denmark and J. Gentry from the USA).

LITERATURE CITED

Baxter, S. 1984. Intensive Pig Production. Granada Publishing Ltd., London.

Brumm, H. and J. Dahlquist. 1997a. Effect of floor space allowance on barrow performance to 300 pounds. University of Nebraska Swine Report.

Brumm, H. and J. Dahlquist. 1997b. Impact of feeder and drinker designs on pig performance, water use and manure production. University of Nebraska Swine Report.

Brumm, M., J. D. Harmon, M. C. Honeyman, and J. R. Kliebenstein. 1997. Hoop structures for grow-finish swine. MWPS AED 41.

Hansen, L. L., A. E. Larsen, M. Hammershoj, P. Sorenson, and J. Hansen-Moller. 1999. Influence of aromatic components from pig manure on odour and flavour of cooked chicken meat. Meat Science. 52:325–330.

Hollis, G. 1996. University web page (*http://www.aces.uiuc.edu/~pork/*). Univ. IL.

Hyun, Y., M. Ellis, and R. W. Johnson. 1998. Effects of feeder type, space allowance, and mixing on the growth performance and feed intake patterns of growing pigs. J. Anim. Sci. 76:2771–2778.

Hyun, Y. and M. Ellis. 2001. Effect of group size and feeder type on growth performance and feeding patterns in growing pigs. J. Anim. Sci. 79:803–810.

Kornegay, E. T. and M. D. Lindemann. 1984. Floor surfaces and flooring materials for pigs. Pig News and Information. 5:351–357.

Larson, M. E. and M. S. Honeyman. 2002. Performance of pigs in hoop structures and confinement during summer with a wean-to-finish system. Iowa State University web publication. *http://www.ae.iastate.edu/hoop_structures/animal_performance.htm*

McFarlane, J. 1995, August. How to swell your water supply. Pork '95.

McGlone, J. J., T. E. Held, and S. Hayden. 1983. Physical and behavioral measures of feeding space for nursery-age swine. In: Proc. Western Section of ASAS. 34:66–68.

McGlone, J. J. and C. Fumuso. 1993. Performance of nursery pigs using a rectangle, round or wet/dry feeder. Texas Tech University Agricultural Sciences Technical report No T-5-327. p. 42.

McGlone, J. J., T. Hicks, R. Nicholson, and C. Fumuso. 1993. Feeder space requirement for split-sex or mixed-sex pens. Texas Tech University Agricultural Sciences Technical report No T-5-327. pp. 45–46.

McGlone, J. J. and B. Newby. 1994. Space requirements for finishing pigs in confinement: Behavior and performance while group size and space vary. Appl. Anim. Behav. Sci. 39:331–338.

Midwest Plan Service. 1987. Structure and Environment Handbook. MWPS-1.

Nicholson, R. I., J. J. McGlone, and R. T. Ervin. 1995. Economic comparison of pig feedlot housing facilities in the Southern high plains of Texas. TX J Agric. Nat. Resour. 8:19–26.

Olsson, O. 1981, December. Water supply system for pigs—drinking from nipples.

Pedersen, B. 1994. Water intake and pig performance. Teagasc Pig Conference, Fermoy, Ireland.

Purdue University. 2002. Pork Industry Handbook. Cooperative Extension Service. West Lafayette, IN.

Stein, T. 1999, October 15. The new economics of wean-to-finish production. National Hog Farmer. pp. 38–42.

Wadsworth, J. R., B. Owen, and J. J. McGlone. 1990. Partition height requirement for nursery-age pigs. Texas Tech University Agricultural Sciences Technical report No T-5-283. pp. 67–68.

Wolter, B. F., M., Ellis, S. E. Curtis, E. N. Parr, and D. M. Webel. 2000. Group size and floor-space allowance can affect weaning-pig performance. J. Anim. Sci. 78:2062–2067.

Wolter, B. F., M., Ellis, S. E. Curtis, N. R. Augspurger, D. N. Hamilton, E. N. Parr, and D. M. Webel. 2001. Effect of group size on wean-to-finish production system. J. Anim. Sci. 79:1067–1073.

INTERNET RESOURCES

North Dakota State University provides a list of agricultural engineering publications at: *http://www.ageng.ndsu.nodak.edu/exten/midwest/BOOKS.HTM*

Some swine education and extension sites are found at:
University of Illinois: *http://www.aces.uiuc.edu/~pork/*
University of Nebraska: *http://anr.ces.purdue.edu/anr/anr/swine/porkpage.htm*
Oklahoma State University: *http://www.ansi.okstate.edu/breeds/swine/*
Kansas State University: *http://www.oznet.ksu.edu/dp_ansi/swine/swine.htm*
North Carolina State University: *http://jah.asci.ncsu.edu/*
Purdue University: *http://anr.ces.purdue.edu/anr/anr/swine/porkpage.htm*
Texas Tech University: *http://www.pii.ttu.edu/*
USDA-MARC/Swine: *http://psru.marc.usda.gov/swine.htm*

Plans for handling facilities are found at Temple Grandin's home page: *http://www.grandin.com/*

For information on hoop structures:
http://www.ae.iastate.edu/hoop_structures/animal_performance.htm
http://www.ae.iastate.edu/hoop_structures.htm

16
WASTE AND NUTRIENT MANAGEMENT

INTRODUCTION

In years past, pig waste was typically disposed of by spreading it on the land. In modern times, producers generally recognize that waste is actually valuable nutrients needed for crop production and other purposes. Utilization of manure nutrients is an integral part of a sustainable pork operation. To be sustainable, the manure nutrients must be utilized in a manner that recycles or captures the nutrients. While some manure management programs can reduce the nutrient load by various fermentation or aerosolization processes, in the scheme of the world's shortage of plant nutrients and energy, a more sustainable option is to utilize these manure nutrients for some productive purpose.

Waste and manure management is all about options and choices (Melvin et al., 1979). Producers have many methods at their disposal to accomplish manure management and utilization. This chapter discusses some of the options and choices.

MANURE MANAGEMENT

There is great public concern about potential environmental pollution resulting from run-off of animal manure from feedlots and fields on which it is spread for soil fertilization. This concern has led to federal regulations to control discharges of pollutants from concentrated animal feeding operations (CAFOs) to water supplies. The United States Environmental Protection Agency (EPA) is responsible for monitoring and enforcing the regulation as part of the Clean Water Act of the U.S. Congress. The acronym "CAFO" applies to animal feeding operations that confine or house livestock or poultry prior to the animal being marketed or slaughtered. In the case of swine, any production unit with more than 1,000 animal units (defined as 2,500 or more swine weighing 55 lb (25 kg) or more) is classified as CAFO if the following criteria are present (Concentrated Animal Feeding Operations, 2001):

1. Animals are maintained in confinement for 45 d or more in a 12-mo period.
2. Crops, post-harvest residues, or vegetation forage growth are not sustained in the area of confinement in the normal growing season. This criterion distinguishes confined feedlots from pasture settings.
3. The facility meets a threshold of 2,500 or more swine weighing 55 lb or more. For dual discharges that pass through the area of confinement or indirect discharge via a man-made conveyance, the threshold is 300 animal units (750 or more swine weighing 55 lb or more).

Each producer should consult agricultural engineers and local regulators before deciding on a manure management system. In addition to cost and ease of management, producers should consider the impact of design features on neighbors and the environment. Each farm should have a manure management plan that defines design features of the production system. The manure management plan should also attempt to minimize the risk of environmental pollution.

Manure is the combined feces, urine, and added products such as water, wasted feed, hair, and bedding (if used). Manure can be handled as a liquid slurry, a semi-solid, or a dry form. Manure has characteristics of a liquid when it has 15% solids or less. When manure has over 20% solids, it may have the characteristics of a solid.

The volume of manure produced by pigs of various stages is given in Table 16–1. Among growing pigs, finishing pigs produce the greatest volume of manure compared with other stages of production—clearly, larger pigs produce a greater volume of manure. Among the breeding herd, lactating sows and their piglets produce far more manure than do pregnant sows primarily because the lactating sows are fed ad libitum and may consume up to 9 kg/d (20 lb/d) compared with the pregnant sows who are fed ~2 kg (4 to 5 lb/d). Limit-fed animals produce less manure than do full-fed animals.

Producers can calculate the amount of land required for a pig production unit to fully utilize the nutrients it produces. The exact amount of land required depends, in large part, on the crop to be grown and its nutrient needs as well as what

TABLE 16–1
Manure Production by Pigs of Various Stages. Manure Is Assumed to Be about 91% Water. Consult the Midwest Plan Service for More Complete Data (MWPS-1).

	Pig Weight		Manure Production		
Stage of Production	LB	KG	LB/DAY	CU FT/DAY	GAL/DAY
Nursery pig	35	16	2.3	.038	.27
Finishing pig	150	68	9.8	.16	1.13
Pregnant sow and boar[a]	275	125	8.9	.15	1.1
Sow and litter of 8 pigs	375	170	33	.54	4.0

[a]Assumes sows and boars are the same weight and are limit-fed and eating less than 50% of ad libitum intake.
Source: Adapted from Midwest Plan Service (MWPS-1) (1987).

the pigs are fed and the amount of manure produced. Thus, a sustainable pig farm will require about 1 acre/sow for a farrow-to-finish unit. If the farm contains only pregnant sows, lactating sows, nursing piglets, and nursery pigs, the farm will require about 2.7 acres/sow. Finishing pigs require, on average, a land area of 10.4 pigs/acre (see Table 16–2).

Less land is needed if certain technologies are used to reduce the manure mass. For example, if a methane digester is used to turn manure nutrients into gas and if the gas is utilized for heat production, less land is needed. If a series of anaerobic and aerobic digestion is used, the land requirement for disposal of nutrients is less.

The largest bulk of material in manure falls into a discrete number of categories. Each nutrient must be either released into the air or recycled, often through crops. Categories of manure nutrients are:

- Water (H_2O)—Hydrogen and oxygen
- Carbon—Primarily in the form of partially digested plant materials. Carbon is often overlooked as a manure nutrient. Carbon is utilized by direct spreading of solids or semi-solids or released into the air in various forms such as carbon dioxide (CO_2) or methane (CH_4).
- Nitrogen—The nitrogen content of pig manure varies, but is about 0.5 mg/kg (10 lb/ton) of manure. Nitrogen is found in the forms of elemental nitrogen, ammonia (NH_3), ammonium (NH_4), and urea ($CO(NH_2)_2$) as well as various metabolites of urea. Urea is a product of the breakdown of protein. Overfeeding protein will increase the nitrogen content of the manure. In liquid pits, nitrogen is found at 36 lb/1,000 gal (16 kg/3,785 l) but the concentration is only 4 lb/1,000 gal (1.8 kg/3,785 l) in lagoons. Figure 16–1 shows how nitrogen flows through the environment.
- Phosphorus—Phosphorus is found as a part of partially digested plant material in the form of phytate and in the plant-available form of P_2O_5. Phosphorus (as P_2O_5) is found in manure at about 0.4 mg/kg (9 lb/ton) of manure. In liquid pits, phosphorus (as P_2O_5) is found at about 27 lb/1,000 gal (12.2 kg/3,785 l), but in lagoons its concentration is only about 2 lb/1,000 gal (1.1 kg/3,785 l).

TABLE 16–2
Whole Farm Design Values for Total Manure Production/Animal Unit. See Pork Industry Handbook (PIH-67) for More Details.

Type of Production	Animal	Pig Weight		Manure, GAL/DAY[a]	Annual N Produced, LB/YEAR	Land Needed per Animal[b]
		LB	KG			
Farrow-to-finish	Sow	1,417	644	14.2	248	1 acre/sow
Farrow-to-feeder pig	Sow	522	237	5.2	91	2.7 sows/acre
Finishing only	Pig	135	61	1.4	24	10.4 pigs/acre

[a]Multiply by 1.52 to get tons/year. N content is 11.4 lb/ton on average.
[b]Land needs is a generalization based on a crop that requires 250 lb/acre of N for optimum yield. Specific crops may need more or less N/acre. Other nutrients, particularly P and K, must be considered as well as metals that may accumulate.
Source: Adapted from Melvin et al. (1979), Pork Industry Handbook (PIH-67).

- Potassium—Potassium is found in partially digested plant materials and in the available form of K_2O. In liquid pits, potassium in the form of K_2O is 22 lb/1,000 gal while it is only 4 lb/1,000 gal (1.8 kg/3,785 l) in a lagoon.
- Sulfur—Sulfur compounds are found in proteins of plant, animal, and bacterial matter. As a part of proteins, sulfurs do not have much of an odor. However, in the form of hydrogen sulfide (H_2SO_4), manure has the odor of rotten eggs.
- Other—Many other manure chemicals are produced. Some are harmless and some have a strong odor. A detailed list of components of manure, including the less-concentrated but more volatile (and thus more odorous) compounds, are given by Mackie et al., 1998.

The N:P:K ratio in manure is about 10:9:8, while a plant such as corn needs 10:4:10. As an example, 180 bu/acre corn requires 240 lb/acre N, 100 lb/acre P_2O_5, and 240 lb/acre K_2O. Thus, because plants pick up and incorporate nitrogen and potassium at higher levels than they do phosphorus, it is the phosphorus that accumulates on cropland. If a producer applied enough manure nutrients to meet only the phosphorus needs of the crop, then either supplemental nitrogen and potassium would have to be added, or the crop yield would be reduced.

Pork producers should be careful when they consider which crops to use to utilize manure nutrients. Conventional wisdom is that legumes, which fix nitrogen from the air, require less nitrogen than do nonlegume crops. However, legumes will absorb and utilize nitrogen at a higher rate than will corn or other grains when nitrogen is applied. Table 16–3 lists some values of crop utilization of the three primary nutrients (nitrogen, phosphorus, and potassium).

If the objective is for the selected crop to pick up the most nitrogen, alfalfa is a good choice as a crop. However, if the objective is to recycle the greatest amount of

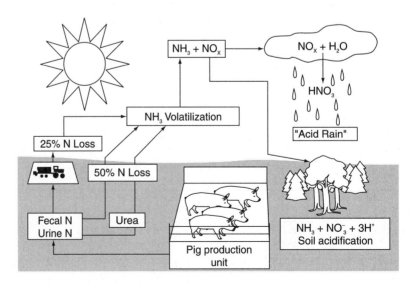

FIGURE 16–1
Movement of Nitrogen Through the Environment.

Source: Adapted from Mackie et al. (1998).

TABLE 16–3
Nutrient Utilization by Selected Crops.

		Application Rate (lb/acre)		
Crop	Yield	N	P_2O_5	K_2O
Corn	180 bu/ac	240	100	240
Sorghum	4 tons/ac	250	90	200
Soybeans	60 bu/ac	336	65	145
Alfalfa	8 tons/ac	450	80	480
Tall fescue	3.5 tons/ac	135	65	185

Source: Adapted from Midwest Plan Service MWPS-1 (1987).

phosphorus, corn is the best choice among the crops listed in Table 16–3. Plant utilization of manure nutrients is an important research topic worldwide. The search for plants that utilize manure nutrients efficiently in a manner that can be utilized by other animals, particularly ruminants, is very important for worldwide development of sustainable pork production systems.

REGULATIONS AND POTENTIAL FOR POLLUTION

The potential for pig manure-related pollution can fall under one of two categories: water or air pollution. Every U.S. pig production unit should strive to minimize the risk of environmental pollution.

Water pollution can kill fish and other wildlife and it can pollute drinking water to the point at which it is not safe for drinking and other uses. Because water is necessary for life and because the federal government claims ownership of the waters of the United States (the same for other countries), water pollution is of concern to national governments and citizens.

Fresh manure does not have much of an odor. However, with just a few minutes of bacterial fermentation, the manure takes on an offensive odor. The key is to contain the odor by a physical barrier, to spread it over a large area in a short period of time, or to allow the animals to spread the manure themselves. Odor pollution is of greatest concern to people who live or work near the source of the odors. Offensive odors have been studied in many industries, including pig farms. People who live near pig farms that have what they report as offensive odors, report symptoms of physical and psychological illness that they associate with the pig farm (Schiffman, 1998).

Pig production units in the United States are nonpoint sources of pollution, by regulation. Pig farms are not allowed, except by an act of God, to discharge into U.S. waterways (streams, lakes, or underground aquifers). Some states have extensive regulations regarding liquid manure handling, but minimal regulations about handling of dry manure. Dry manure poses a lower risk of environmental pollution than liquid manure handling systems. Dry manure is less likely to flow toward a waterway than is liquid manure. Secondly, dry manure, if it is truly dry, has less of an odor than liquid manure.

Extensive safeguards can be put in place to minimize the risk of environmental pollution with liquid manure systems, usually by adding more expensive liquid manure handling systems.

MANURE NUTRIENT COLLECTION OPTIONS

The first choice in manure nutrient collection is to determine if manure is handled in a dry or liquid form (semi-solid is included in the liquid category here). If the manure is handled in dry form, there are two options (with sub-choices): (1) Allow the pigs to self-distribute the manure nutrients or (2) collect and mix the manure with a bedding source (see Chapter 13).

When manure is handled in a dry form mixed with bedding, the largest challenge is to get the manure-bedding mixture out of the building or lot. This is accomplished either by hand or with the use of machines. The dry manure-bedding mix can either be applied directly to croplands or composted.

Some degree of separation is needed when manure is handled in a liquid form. Two methods are common: use of a settling basin or use of a liquid-solid separator. If the liquid-solid separator is used, the solids must be either composted or spread directly on the fields. If a settling pond is used (see Figure 16–2), the settling basin must be emptied at some time.

Liquid manure systems can be anaerobic or aerobic. When a manure pond is deeper than 2 ft (0.6 m), and especially when it is deeper than 4 ft (1.22 m), the lower portion of the pond will be anaerobic. Anaerobic and aerobic ponds have a different odor. An aerobic pond has a more constant odor year-round and has a less-intense odor. Anaerobic ponds may have little odor at some times, but at other times a very strong odor. Deep, anaerobic ponds will "turn over" when the seasons change from warm to cool or cool to warm. This temperature inversion causes the smelly bottom materials to move up toward the surface.

FIGURE 16–2

An Example Manure-Handling System. Note the Four Necessary Components: (1) a Building (could be any number of styles); (2) a Settling Basin (could also be a settling pond); (3) a Holding Pond or Lagoon or Tank; and (4) a Field on which the Manure Nutrients Are Spread.

Source: Midwest Plan Service (1987).

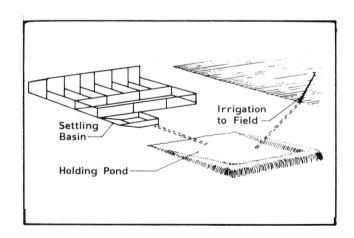

Manure handled in a dry form
Manure is handled in a dry form in different ways for outdoor and indoor production systems.

- Outdoor
  - With a vegetative ground cover and posture rotation (recommended)
  - Without a vegetative ground cover (not recommended unless there is a vegetative or other boundary)
  - On solid concrete or hard ground with slotted floors
- Indoor with bedding
  - Manure handled by hand or machine
  - Manure removal automated

Manure handled in liquid or semi-solid form (indoors with slotted floors)
Manure under slotted floors is best handled in a liquid or semi-solid state. Being liquid facilitates movement of the material out of the facility.

- Under-floor storage; pit pumped out as needed
- Under-floor flush
  - One or more flushes/d
  - Pull-plug system (plug pulled and building flushed as needed)
- Under-floor scraper
  - Scraped each day
  - Scraped as needed
- Other

Liquid manure storage and handling options
After the manure leaves the building, it is stored for short or long durations (hours through years). Storage containers can be equipment on site or can be built into the ground.

- Multi-stage lagoon (anaerobic or aerobic)
- Slurry-storage tank (anaerobic or aerobic)
- Covered lagoon or tank for fermentation and possible methane production and capture
- Under-floor pit
- Other

SUMMARY

Pigs generate manure which in the past was considered a waste product. Today, manure is recognized as a valuable nutrient that can be recycled through one of several means including on crops or generation of an alternative fuel such as methane. Manure nutrients should be balanced with the crop or other system that will use these nutrients. Manure storage and handling facilities should be properly designed to accommodate the numbers and types of pigs on the farm.

QUESTIONS AND ACTIVITIES

1. Define and discuss the terms manure, waste, lagoon, and nutrients.
2. Why might odors from hog farms cause problems with people? Is human perception important in understanding hog odor issues?
3. When you search the Web, do you find more sites for or against or neutral toward potential pollution problems associated with pig production units?
4. Describe a well-designed waste management system using a dry manure-handling system.
5. Describe a well-designed waste management system using a liquid manure-handling system.

LITERATURE CITED

Concentrated Animal Feeding Operations. 2001. Available at: *http://www.epa.gov/reg5oh2o/npdestek/npdcafa.htm*. Accessed May 22, 2001.

Mackie, R. I., P. G. Stroot, and V. H. Varel. 1998. Biochemical identification and biological origin of key odor components in livestock waste. J. Anim. Sci. 76:1331–1342.

Midwest Plan Service. 1987. Structures and environment handbook. MWPS-1. Ames, IA.

Melvin, S. W., F. J. Humenik, and R. K. White. 1979. Swine waste management alternatives. Pork Industry Handbook. Purdue University, West Lafayette, IN.

Schiffman, S. S. 1998. Livestock odors: implications for human health and well-being. J. Anim. Sci. 76:1343–1355.

INTERNET RESOURCES

A group of environmental activists who focus on the NC hog industry has a Web page at: *http://www.hogwatch.org/*

North Carolina State University provides research and education information about pig housing and waste management as well as other topics at: *http://www.bae.ncsu.edu/bae/* and *http://www.bae.ncsu.edu/programs/extension/proindex.html*

The University of Georgia provides a site that focuses on animal waste management at: *http://www.bae.uga.edu/outreach/aware/*

SECTION V

PIG PRODUCTION APPLICATIONS THAT MAKE BUSINESS SENSE

17

MANAGEMENT OF THE BREEDING HERD

INTRODUCTION

Several management factors maximize reproduction efficiency, measured as number of pigs produced per sow per yr (PPSY). This chapter provides examples from research and field data to illustrate management practices that lead to high herd performance in terms of increased PPSY. Some production units focus on related measures such as pigs per rate per yr.

AREAS OF OPPORTUNITY

The bottom line for the breeding herd is to maximize the number of pigs produced per sow per yr (PPSY). With the sow herd at a constant level, incremental increases in PPSY result in incrementally lower costs of production. Thus, higher profits are generally associated with greater numbers of PPSY.

Production units vary in PPSY. The former industry standard, the Pork Industry Handbook, used a PPSY of 16.0 in the economics section as an average of well-managed farms of the 1970s. By the 2000s, a farm that produced only 16 PPSY was a low-producing farm. A minimum acceptable level for PPSY now is 20.0. Many well-managed farms produce over 22 PPSY, while the best farms of the new millennium produce around 25 pigs per sow per yr.

Two main factors affect PPSY: litters per sow per yr and average litter size weaned. Litters per sow per yr are greater by lowering:

- Lactation length
- Nonproductive days
- Sow deaths
- Wean-to-estrus length

PPSY and litters per sow per yr are increased by improving:

- Estrus detection and mating
- Conception rates
- Parity distribution
- Lactation length

The number of pigs weaned/sow is a function first of the number of total pigs born. Total born includes those born alive (seek a high number), those stillborn (fully formed piglets at birth that did not breathe—seek a low number), and mummies (partially developed, partially decomposed piglets, usually black and shriveled—seek a low number).

After the number of pigs born alive is considered, the only other common measure that impacts number weaned is the number of piglet preweaning deaths (piglet mortality). A good piglet mortality value is less than 10%.

Figure 17–1 shows an example of a well-managed herd with 23 PPSY. This herd average PPSY is composed of 2.3 litters per sow per yr and 10.0 pigs weaned/sow. Of 49 best herds shown in Table 17–1, ranked by PPSY, only four herds weaned more than 10.0 pigs/sow. However, 41 herds listed in Table 17–1 had litters per sow per yr of 2.30 or more. Farms ranged in size from 235 to 3,530 sows. These data confirm that herds reach the top level of PPSY more often by having high litters per sow per yr than by increasing the number of piglets born alive or weaned/sow.

FIGURE 17–1
Productivity Tree Showing Items Important for High Productivity in the Sow Herd.

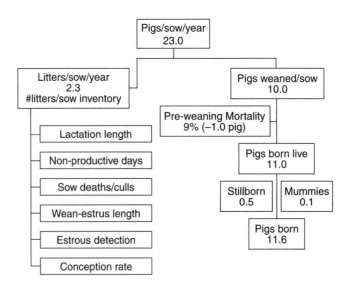

Table 17–1 lists the eight factors that contribute to placing a herd in the best or worst categories. The most significant factors that place a herd in the bottom 20% of herds are nonproductive days and weaning age. Nonproductive days are the days a sow is neither pregnant nor lactating. Normally, the nonproductive days are calculated by the average of the number of days from weaning to expression of estrus/mating multiplied by the numbers of litters per sow per yr. Using an example of 2.3 litters per sow per yr, 7 d from weaning to estrus, and an average of 80% conception/farrowing rate, the number of nonproductive days is 20.1 d (7×2.3 divided by 0.80), even under the best of circumstances. Still, the best herds have fewer than 60 nonproductive d. This figure suggests U.S. herds have about a three-fold higher number of nonproductive days than the biological minimum.

Better managers minimize the number of nonproductive days, especially by careful heat detection. A large part of the contribution to nonproductive days is failing to breed sows that could be bred. If estrus is expressed and the sow is not bred, another 21 d is added to the number of nonproductive days. Another significant contribution to nonproductive days is the gilt herd that enters the breeding area and awaits expression of estrus and mating. If the average gilt is bred on her second or later estrus, another 21 to 42 d are added to the nonproductive days measure.

Many newer units attempt to minimize the effect of the gilt herd on nonproductive days by use of a specialized building or area typically called a gilt breeding project. In this area, gilts are checked for estrus, bred on their second or later estrus, and kept until pregnancy check (d 30 to 60 after breeding). When they are confirmed pregnant, they are moved to the gestation barn as bred gilts. In this way, a single gilt breeding project can supply gilts to a number of networked units.

Weaning age is a second factor that is under the control of the production unit, but really not under the control of the workers. The production schedule will dictate the weaning age and the flow of sows through the facility. Most modern units wean pigs at less than 24 d. Common weaning ages in the 1990s and continuing into 2001 are 20 to 21 d, 18 to 19 d, 16 to 17 d, and 7 to 14 d. The earlier weaning age may be for a program of segregated early weaning (SEW) or medicated early weaning (MEW). Other ages, especially the 16 to 21 d weaning age, are common in modern pig units.

TABLE 17–1
Productivity Measures That, If Not Met, Result in the Herd Being in the Bottom 20% Rather than the Top 20% of the Rankings of Sow Herds. All Eight Productivity Measures Are Important Determinants of a Successful Sow Herd. Rank Refers to Relative Economic Importance (1 = largest economic impact).

Productivity Measure	Suggested Value	Relative Merit	Rank
Nonproductive days	< 60	61	1
Weaning age, days	< 24	55	2
Farrowing rate, %	> 80	12	3
Litter size, number born alive	> 10.0	11	4
Average sow parity	> 4.0	3	5
Prewean mortality, %	< 14.0	1.6	6
Matings/service	> 2.0	1.2	7
Sow mortality, %	< 8.0	1	8

Farrowing rate contributes significantly to PPSY. Herds with 80% or greater farrowing rate are more successful. A higher farrowing rate requires fewer sows in the unit to produce the same total number of pigs. A 1,000-sow unit, expecting to produce 22,000 pigs/yr (22 PPSY), would need about 200 more sows if its farrowing rate was reduced by 20%.

The number of sows needed to produce 22,000 pigs from a sow herd of 1,000 sows with a decreasing farrowing rate is shown in the following figure. A 100% farrowing rate is nearly impossible to reach on a commercial unit.

Farrowing Rate	Needed Sows
100%	1,000
90%	1,111
80%	1,250
70%	1,333
60%	1,667

Litter size contributes importantly to a high number of PPSY, but less so than does nonproductive days. The number of pigs born alive (and related factors of rates of stillborn and mummies) is still a very important determinant of a high number of PPSY.

Herds with a high PPSY tend to have a higher average parity, lower sow mortality, more than two matings/service, and a preweaning mortality of less than 14%. These factors, while important, should not be the focus of efforts toward increasing PPSY unless the farrowing rate and the number of nonproductive days are already well under control.

As a general rule, herds with low and marginal productivity problems should focus attention on the areas given below.

Herd Profile	Focus Improvements On
All herds	Lower wean age, increase average parity and matings/service
Low PPSY	Breeding techniques: Farrowing rate and nonproductive days

Medium PPSY	Breeding techniques:
	     Farrowing rate and nonproductive days
	     Litter size

High PPSY	Breeding techniques:
	     Farrowing rate and nonproductive days
	     Litter size
	 Preweaning mortality

SETTING PRODUCTION TARGETS—22 PPSY TODAY, 30 PPSY TOMORROW

Modern pig production units should wean at least 22 PPSY. Producers have the technology to do better than 22 PPSY, but this level remains a challenge on most farms.

The four main factors that contribute to PPSY are litters per sow per yr, total pigs born, stillborn rate, and preweaning mortality. Reaching 22 PPSY requires 11 pigs born, 2.3 litters per sow per yr, fewer than 0.60 stillbirths, and less than 10% preweaning mortality. If the litters per sow per yr are only 2.1 or total born is 10.0 or less, it is biologically impossible to wean more than 22 PPSY.

Well-managed farms that use quality genetics can produce at least 12 pigs born. With 12 pigs born and at least 11 pigs born alive, it should be possible to wean over 10 pigs/sow. With 12 pigs born and low preweaning mortality and low stillborn rate, the farm can wean over 25 PPSY. In fact, with less than 10% preweaning mortality and fewer than 1.0 stillbirths/sow, the farm will wean over 23 PPSY. The total number of pigs born is a function of activities in the breeding and gestation barns. A farrowing barn worker who has been delivered sows with 12 pigs/litter has the tools to wean a very high number of PPSY.

New research in breeding and genetics has identified breeds and at least one specific gene that codes for a high number of pigs born. With less than 10% preweaning mortality and fewer than 1.0 stillbirths, the farm that has 14 pigs born will wean over 27 PPSY. If a herd has 14 total pigs born and low stillbirths and preweaning mortality, the unit could potentially wean 30 PPSY!

SETTING BREEDING TARGETS

Every pig production unit should have goals, and one of the most important is to set the target number of sows to breed. Three main factors determine the breeding target:

- Number of farrowing crates/huts to be filled
- Farrowing rate
- Crate/hut occupancy rate

The formula to calculate the breeding target is:

$$\left(\frac{\text{no crates}}{\text{farrowing rate}}\right) \times \text{crate occupancy rate} = \text{breeding target}$$

In this example, the farrowing rate is a fraction (0.80, not 80%) and crate occupancy is a fraction (1.10, not 110%).

Farrowing barns and pastures have a certain number of crates or huts to fill. This fixed number reveals a lot about the size and production schedule of a farm. The greater the number of farrowing units, the greater the number of sows that need to be bred.

Farrowing rate refers to the number of sows giving birth compared to the number bred. If 100 sows are bred and 70 of those sows farrow, the farrowing rate is 70%. The situation is more complex, however, because a certain number of gilts are bred from a gilt pool and also because some sows fail to conceive and are re-bred in a different group. A higher farrowing rate is desired. The average farrowing rate is about 75% on commercial farms. The minimum target should be to have a farrowing rate over 80%.

Most large commercial farms have a crate occupancy rate well over 100%—that is, more sows farrow than the number of crates on the farm. Accountants like this concept because it means that the crates (an important asset) are fully utilized at all times.

As an example of a crate occupancy over 100%, assume the farrowing room has 30 crates. If 33 sows farrow, the crate occupancy rate is 110%. As a practical matter, when the last three sows are to farrow, the worst three litters (that is, those three with the least number of piglets) are weaned or crossfostered. Piglets are transferred to other sows and the sows return to the breeding area. The three new sows get a crate and farrow. Therefore, if the sows have fewer pigs born alive, the crate occupancy rate tends to increase.

For example, a farm is a 1,200-sow unit that has 54 farrowing crates that are used each week (actually two rooms of 27 crates each). How many sows should be bred for this unit? If the farrowing rate is 80% and if the crate occupancy rate is set at 110%, 74 sows should be bred each wk. The consequences of breeding 74 sows are in two directions. First, if the farrowing rate turns out to be 90%, the crate occupancy rate would have to be pushed over 115% to accommodate the added sows. Under the target (average) conditions, a producer expects to wean 10% of the sows early (five to six sows). If farrowing rate is 90%, the producer would have to wean 12 to 13 sows early. This could put a strain on the ability of the sows to nurse so many pigs, but they could probably handle it—this is about their biological limit. Estimating 10.8 pigs born alive and 10.0 piglets at 3 d on 54 sows, adding 120 pigs would add an average of two to three pigs/sow. This would leave each sow with 12 or 13 piglets.

In the other direction, the farrowing rate could be 65% (or even lower). With 74 sows bred and a 65% farrowing rate, the crate occupancy rate would be close to 100%. Considering the potential swings from very good to very poor farrowing rates, shooting for the 110% crate occupancy is a healthy target that does not over-tax the sows and fully utilizes the assets (the crates or huts).

SEASONAL EFFECTS

The pig is generally considered to be neither photo-sensitive nor a seasonal breeder, in contrast to some breeds of sheep and most birds. Some have said the sow expresses "normal" sexual behavior and reproductive function year-round. However, this is not entirely true. Sows do express a depression in return-to-estrus and conception rates in the late summer and early fall. The seasonal, reduced fertility is thought to be due to a combination of the effects of summertime heat stress, changing grain, and the changing photoperiods (as well as other seasonal cues).

> **SIGNS OF SEASONAL INFERTILITY IN SOWS:**
> - Decreased farrowing rate
> - Increased days from weaning to estrus

Young sows are especially influenced by summer infertility. In one study, parity one sows had an 18% decline in farrowing rate while older sows had only a 5% decline in farrowing rate during the late summer months (Love, 1981). The number of days from weaning until estrus increases in the late summer and fall as well.

To account for the less-than-perfect farrowing rates during the year, Levis calculated a seasonal coefficient to aid in setting breeding targets. A range of 11% to 35% more sows should be bred depending upon the time of the year. The most sows are bred in late summer to accommodate seasonal infertility.

SUMMARY

Reproductive efficiency is crucial to obtain profitable pork production. An accepted measure of reproduction efficiency is the number of pigs produced per sow per yr (PPSY). This chapter discussed the two main factors contributing to PPSY: litters per sow per yr and average number of pigs weaned/litter. Number of litters per sow per yr is increased by minimizing nonproductive days, lactation length, days from weaning to postweaning estrus, and sow deaths, and by improving estrus detection and conception rates. Realistic goals for attaining high PPSY records in a herd are: nonproductive days, <60; weaning age, <24 d; farrowing rate, >80%; litter size, >10.0 pigs born alive; preweaning mortality, <14%; mating/service, >2.0; and sow mortality, <8%.

> **AREAS TO ASSESS PROFICIENCY IN BREEDING AND GESTATION:**
> - Weaning to estrus length, days
> - Estrus detection
> - Breeding
> - Pregnancy check
> - Return to estrus
> - Moving and handling
> - Farrowing rate
> - Total number of pigs born
> - Total number of pigs born alive
> - Stillbirth rate
> - Rate of mummies

QUESTIONS AND ACTIVITIES

1. Considering two farms, could it be possible to produce less total PPSY on one farm, but actually make more profit than the other farm? How?
2. If a plant needed 16,000 pigs/d, and it processed pigs 250 days/yr, how many sows would be required if the farrowing rate was 90%, 80%, or 70%? Assume the other factors are at a set level (give assumptions).
3. If you have 100 farrowing crates in your facility/wk of production, how many matings should you make? What assumptions did you use to establish the number of sows mated?
4. For the previous question, how many sows would you breed in August to accommodate seasonal infertility?

LITERATURE CITED

Love, R. J. 1981. Seasonal infertility in pigs. Vet Rec. 109:407–409.

INTERNET RESOURCES

A swine business web site:
http://www.porknet.com

A breeding stock company with records:
http://www.pic.com

A list of software sites for livestock:
http://www.wisc.edu/anrsi/ANRSI_L2V0007.html

A pig-specific record-keeping program:
http://www.pigchampinc.com/

For current hog and pig inventories in the USA:
http://www.ers.usda.gov/briefing/hogs/

18

MANAGEMENT OF SOWS AND PIGLETS: BEFORE, DURING, AND AFTER FARROWING

INTRODUCTION

The producer's objective in the farrowing barn is to wean as many heavy pigs as possible. Farrowing is a key area that needs to be managed closely—failures and accomplishments are felt in the production cycle for at least 6 mo.

Good or poor breeding management interacts with good or poor farrowing unit management. Table 18–1 lists typical production records from four farms. A few outstanding farms can produce over 26 pigs per sow per yr (PPSY). At the other end, some farms have less than 14 PPSY. The most serious evidence of low PPSY in the United States is overall inventories. About 100 million market hogs are produced/yr in the United States with less than 7 million sows. This works out to an average productivity—all farms included—of about 15 PPSY (USDA, 2002). If hobby farmers and other less-than-serious producers are excluded, the average PPSY still is often reported at less than 20 PPSY in the United States. Actually, the range of PPSY on commercial farms whose owners are *really trying* is between 14 and 26 PPSY. This is an incredible 85% range in output/sow! This chapter focuses on the space and thermal needs of the farrowing sow and litter, management, and husbandry aspects of meeting the physical and behavioral needs of the sow and piglet, and the role and behavior of the personnel who care for the farrowing and lactating sow and her litter. The role of the stockpersons in contributing to the productivity and profitability of the enterprise is emphasized.

TABLE 18-1
Example Farms That Reflect a Range of Productivity From Poor to Good and the Effects on PPSY (pigs weaned per sow per yr). Note That Superior Performance Is Needed in Both Breeding and Farrowing to Result in a Large Number of PPSY.

MEASURE	POOR BREEDING		GOOD BREEDING	
	POOR FARROWING	GOOD FARROWING	POOR FARROWING	GOOD FARROWING
Litters/sow/yr	2.0	2.0	2.3	2.3
Total pigs born	10.0	10.0	12.0	12.0
Born dead/sow	1.2	0.4	1.2	0.4
Born live/sow	8.8	9.6	10.8	11.6
Mortality	20%	8%	20%	8%
Weaned/sow	7.0	8.8	8.6	10.7
PPSY	14.0	17.6	19.8	24.6

THE INTERDEPENDENCE OF THE BREEDING STAFF AND THE FARROWING STAFF

The breeding barn staff has a great influence on the farrowing rate and the number of total pigs born to sows in the farrowing area. The farrowing staff needs the following from the breeding staff:

1. Healthy, strong, well-conditioned sows
2. As many total pigs/sow as possible

It is then up to the farrowing area staff to take those sows and produce a large number of heavy pigs at weaning.

CRITICAL PRODUCTION VARIABLES

Not all production measures are under the control of the farrowing staff. The farrowing rate is determined in the breeding area. When each sow enters the farrowing facility, her reproductive tract contains a certain number of fetal pigs. Farrowing area staffers cannot influence the total number of pigs born to sows—they can only influence three critical production measures (CPMs):

- Number of piglets born dead
- % mortality or survival of pigs born alive
- Weaning weight/pig

With so many production measures and so many things that can go wrong and right, it is astounding that the farrowing staff actually has direct influence only on these three production variables.

Under normal circumstances, most of the fetal pigs are alive and well. A few die shortly before birth and must be delivered as dead piglets—stillbirths. The biological "normal" number of stillbirths is between 0.25 and 0.50 pigs born dead/sow. The average on most farms is between 0.5 and 1.0 pigs born dead per sow per litter. In the absence of disease, facilities and people's behavior contribute to elevated rates of stillbirths.

Preweaning mortality is 20% on average in the United States, but under 15% on well-managed farms. This translates into millions of piglets born alive each year that die before weaning. Preweaning piglet deaths may be the single most costly problem in the swine industry around the world. A recent report suggested that preweaning mortality adds $1.42 to the cost of production of every pig marketed in the United States (Ott et al., 1995).

The best farms have piglet mortality rates of 5% to 10%, whereas some farms have over 30% piglet deaths. One reason for the variation in field data is the presence of enteric and respiratory diseases. Most piglet deaths are due to the syndrome that includes starvation, chilling, and crushing. Attaining preweaning deaths consistently under 10% requires careful management of the physical and biological assets.

Piglet weaning weights are, to a large degree, under the control of the farrowing staff. Piglet weaning weights are controlled by the genetic potential of the piglets, the milk production of the sows, the feed intake of the sow, the litter size, and the physical facilities. Sows that produce more milk have piglets that eat more and, thus, grow faster. Sows nursing larger litters may not be able to consume enough feed to produce sufficient milk to satisfy the needs of each piglet. Thus, sow feed intake is a production variable that directly influences an important CPM. Maximizing sow feed intake is very important to obtain heavy weaning weights.

Two achievable targets are 20 and 22 PPSY. The level of 20 PPSY is an absolute minimum in today's economic climate, but well-managed farms of small or large scale ought to consistently reach 22 PPSY. When the breeding program and each of the three farrowing CPMs are well-managed, the farm can produce a target of over 22 PPSY. If either component falters, the 22 PPSY is nearly impossible.

Table 18–2 summarizes production figures from simulated herds that produce 10 total pigs born/sow (live plus dead). To reach 20 PPSY with only 10 pigs born, the farrowing staff must have under a 10% preweaning mortality and less than 1.0 stillbirths/farrowing. These are difficult numbers to attain consistently. With only 10 pigs delivered/sow, the 22 PPSY goal is impossible (at 2.5 litters per sow per yr).

TABLE 18–2
Reaching Farrowing Area Targets with 10 Total Pigs Born/Sow. Table Values Are Pigs per Sow per Yr (PPSY). Litters per Sow per Yr = 2.3. Note How Difficult it Is to Make Over 20 PPSY with Only 10 Pigs Born Live/Litter and That Producing 22 PPSY Is Nearly Impossible.

	% PREWEANING MORTALITY								
STILLBORN	4	6	8	10	12	14	16	18	20
0.4	21.2	20.8	20.3	19.9	19.4	19.0	18.5	18.1	17.7
0.5	21.0	20.5	20.1	19.7	19.2	18.8	18.4	17.9	17.5
0.6	20.8	20.3	19.9	19.5	19.0	18.6	18.2	17.7	17.3
0.7	20.5	20.1	19.7	19.3	18.8	18.4	18.0	17.5	17.1
0.8	20.3	19.9	19.5	19.0	18.6	18.2	17.8	17.4	16.9
0.9	20.1	19.7	19.3	18.8	18.4	18.0	17.6	17.2	16.7
1.0	19.9	19.5	19.0	18.6	18.2	17.8	17.4	17.0	16.6
1.1	19.7	19.2	18.8	18.4	18.0	17.6	17.2	16.8	16.4
1.2	19.4	19.0	18.6	18.2	17.8	17.4	17.0	16.6	16.2
1.3	19.2	18.8	18.4	18.0	17.6	17.2	16.8	16.4	16.0
1.4	19.0	18.6	18.2	17.8	17.4	17.0	16.6	16.2	15.8

Table 18–3 shows PPSY starting from 11.0 piglets/litter total pigs born. In this case, with fewer than 1.0 stillbirths and lower than 10% preweaning mortality, the farrowing unit can wean over 22 PPSY. To hit the mark of 20 PPSY, the preweaning mortality needs to be 18% and sows need to have less than or equal to 1.4 stillbirths/sow. If the farrowing staff is provided with properly bred sows, 20 and even 22 PPSY are fairly attainable targets.

MANAGEMENT OF PHYSICAL ASSETS

The farrowing area requires suitable, yet unique, environments for the sow and piglets. Sows need more space, food, and water, but a lower temperature, than do piglets. With such diverse physical requirements, the farrowing environment becomes an engineering challenge in which sows and piglets must survive, grow, and be healthy. The physical needs of the sow and piglets are divided into needs for space and temperature.

THE NEED FOR SPACE

Sows clearly need more space than do piglets. Space needs consist of static space, dynamic space, and social space.

Static Space Needs

Static or physical space is the area needed to contain the body of an animal. The static space is measured by recording the individual length, width, and height in a population of sows. Sows require a space about 22 in (59 cm) wide and 6.5 ft long (2.0 m) (about 12 ft^2/sow (1.1 m^2/sow)).

TABLE 18–3
Reaching Farrowing Area Targets with 11 Total Pigs Born/Litter. Table Values Are Pigs per Sow per Yr (PPSY). Litters per Sow per Yr = 2.3.

	% PREWEANING MORTALITY								
STILLBORN	4	6	8	10	12	14	16	18	20
0.4	23.4	22.9	22.4	21.9	21.5	21.0	20.5	20.0	19.5
0.5	23.2	22.7	22.2	21.7	21.3	20.8	20.3	19.8	19.3
0.6	23.0	22.5	22.0	21.5	21.0	20.6	20.1	19.6	19.1
0.7	22.7	22.3	21.8	21.3	20.8	20.4	19.9	19.4	19.0
0.8	22.5	22.1	21.6	21.1	20.6	20.2	19.7	19.2	18.8
0.9	22.3	21.8	21.4	20.9	20.4	20.0	19.5	19.0	18.6
1.0	22.1	21.6	21.2	20.7	20.2	19.8	19.3	18.9	18.4
1.1	21.9	21.4	20.9	20.5	20.0	19.6	19.1	18.7	18.2
1.2	21.6	21.2	20.7	20.3	19.8	19.4	18.9	18.5	18.0
1.3	21.4	21.0	20.5	20.1	19.6	19.2	18.7	18.3	17.8
1.4	21.2	20.8	20.3	19.9	19.4	19.0	18.5	18.1	17.7

Dynamic Space Needs

The dynamic space is the area that allows for "normal" postural changes. Dynamic space needs are met if a sow can stand up, eat, lie down, or sleep without bumping into the side and front walls. The dynamic space allowance for American-style sows is about 35 in (89 cm) wide and 7 ft (2.1 m) long (about 20 ft^2/sow (1.86 m^2/sow)).

Social and Behavioral Space Needs

Beyond normal space for single-animal behaviors, animals require more space for social interaction. When sows touch and interact with one another in a group pen, their need for social space is met. Interestingly, if a producer provides enough space to meet dynamic space needs/sow by holding sows in groups, the sows need no more space than their dynamic space requirements. A group of four sows with 20 ft^2/sow (1.8 m^2/sow) (80 ft^2 in total, or perhaps 10 × 8 ft; 7.4 m^2 or 3 × 2.4 m) provides the sows with plenty of space for apparently "normal" social interactions. However, they might be sufficiently crowded to cause excessive aggressive behavior, but this has not been studied extensively.

If a producer provides sows with space for static, dynamic, and social space needs, there will be plenty of room for the piglets to meet their static, dynamic, and social needs. Thus, when it comes to space for the sow and litter, accommodating the sow typically accommodates the piglets.

All of these engineering specifications are fine, but how do sows and litters perform in different environments? Of particular interest is how the three farrowing area critical production measures (CPMs) differ in environments that provide for differing amounts of space.

The data in Table 18–4 show that the physical environment has its greatest effect on piglet survival during and after farrowing (see % mortality column). The rate of stillbirths requires a greater number of sows in a given experiment to show an effect. Even still, sows kept outdoors typically have fewer pigs born dead than do sows housed indoors (Johnson et al., 2001; McGlone and Hicks, 2000). Perhaps the added room afforded outdoor sows or the provision of bedding reduces the stillbirth rate.

Preweaning mortality is highly responsive to the physical environment. Providing sows more quantity and *quality* of space can reduce preweaning deaths of piglets.

In the case of the turn-around crate described by McGlone and Blecha (1987), the sows had 58% more floor space—and their dynamic space needs were met. The full-height guardrails (bars from a standard farrowing crate flared outward) surrounding the sow gives the piglets protection from crushing—the greatest cause of piglet deaths in every environment examined.

Guardrails are not the total answer to reducing piglet crushing. The bars in the farrowing crate usually prevent a large degree of piglet crushing (McGlone and Morrow-Tesch, 1990). Given even more room (as in the 5 × 7-ft (1.52 × 2.13 m) pen in McGlone and Morrow-Tesch [1990] or the 5 × 7-ft hut in McGlone and Hicks [2000]), the sow proceeds to crush a large number of piglets whether or not the guardrails are present (guardrails were present in the level pen). These data support the notion that guardrails only save piglets from crushing in very small spaces like the 2 × 7-ft area they are normally restricted to in the standard farrowing crate. But larger pens, with certain quality features, support very low piglet crushing rates. The three

TABLE 18–4
Farrowing Environments That Influence CPMs.

				CRITICAL PRODUCTION MEASURES		
REF	TREATMENTS	TOTAL AREA, FT²	SOW AREA, FT²	BORN DEAD, #	MORTALITY, %	WEANING WT, LB
1	Crate, 5 × 7 ft	35	14	0.74	29.3	10.6
	Turn-around crate, 5 × 8.5 ft[a]	42	22	0.61	9.4[b]	10.3
2	Level crate, 5 × 7 ft	35	14	1.08	10.8	9.9
	Level pen, 7 × 7 ft[a]	49	35	0.64	27.1[b]	11.4
	Sloped crate	35	14	0.51	17.2[b]	11.2
	Sloped pen[a]	49	35	0.96	9.1	10.8
3	Crate, 5 × 7 ft	35	14	0.60	10.7	13.9
	Tether, 5 × 7 ft	35	14	0.60	12.5	15.4
4	Indoor crate, 5 × 7 ft	35	14	1.47	12.3	14.3
	Outdoor hut, 5 × 7 ft[a]	35	35	0.38[b]	27.5[b]	15.3
5	Indoor crate, 5 × 7 ft	35	14	1.00	19.7	15.3
	Outdoor hut, 5.5 × 9 ft[a]	49	49	0.69[b]	11.2[b]	14.4[b]

[a] The dynamic space needs of sows were met.
[b] Treatments differ significantly from control.
[1] McGlone and Blecha (1987).
[2] McGlone and Morrow-Tesch (1990).
[3] McGlone et al. (1994).
[4] McGlone and Hicks (2000).
[5] Johnson et al. (2001).

best examples of quality space improving piglet survival are the turn-around crate, the sloped pen, and the English-style arc hut (see Table 18–4 for data). The 8% floor slope in the sloped pen causes the sow to stand up and lie down much more carefully. In the English-style arc hut, the sow has straw bedding and she has access to a rectangular-shaped area that provides 40% more space than the lesser-effective hut designs. So for now, one can only conclude that piglets can be saved from crushing by added space plus either (1) full-height guardrails, (2) floor slope, or (3) a large bedded area of a certain shape (see Table 18–5).

THERMAL NEEDS

Thermal needs for the sow and piglets are quite different. They are so different, in fact, that they do not even overlap. The only way to meet the thermal needs of both the sow and piglets is to zone-cool or zone-heat the sow or piglet areas.

Table 18–6 shows the NPB Swine Care Handbook's (National Pork Board, 2002) recommended thermal environment for sows and piglets. The sow's preferred range in air temperature is from 60° to 80°F, while the piglet's preferred range at birth is from 90° to 100°F—two zones without overlap. The thermal environment would be very easy to manage if it were only necessary to provide one air temperature that met the needs of both sow and piglets.

TABLE 18–5
Minimal Space Needs for Sows and Litters in the Farrowing Area.

Item	Indoor in Stall	Indoor in Pen	Outdoor Hut
Sow area of crate or pen	9.2–14 ft^2 (2 × 7 ft)	22–49 ft^2 (5 × 8.5 ft)	49 ft^2 (5.5 × 9 ft)
Piglet area of crate or pen	10 ft^2 (1.5 × 7 ft)^a	10 ft^2 (1.5 × 7 ft)^a	Same as for sow^a

^{a}Guardrails were shown to reduce piglet deaths due to crushing indoors in crates or pens. Outdoors, where space allowances are much more generous for the sow, guardrails are not as effective.
Source: Adapted from the Swine Care Handbook, NPB (2002) and Johnson et al. (2001).

TABLE 18–6
Thermal Needs of the Sow and Piglets.

Age of Pig	Preferred Range, °F	Lower Intervention	Upper Intervention
Sows	60–80	50	90
Piglets at birth	90–100	90	None in practice
Piglets at 21–28 d	80–90	80	None in practice

Source: Adapted from the NPPC Swine Care Handbook.

TABLE 18–7
Effects of a Chill (40°F) on Illness of Weanling Pigs After Exposure to TGE Virus. Note That the Piglets Kept All the Time at 86°F Did Not Get Sick.

Days Before TGE		After TGE	% Sickness
14–18 d	0–14 d		
86°F	86°F	40°F	100
86°F	40°F	40°F	40
86°F	86°F	86°F	0

Source: Adapted from Shimizu et al. (1978).

The most common way to accommodate the diverse thermal needs of the sow and the piglets is to set the air temperature in the sow's preferred range and then zone-heat a creep area for the piglets. If the area for the piglets is not warm enough, they are much more likely to get chilled and crushed.

The stress of chilling causes increased risk of disease because piglets have such a poorly developed immune system. Every type of opportunistic and virulent microorganism will have a more serious effect if the piglets are chilled.

Table 18–7 illustrates the drastic effect of chilling (Shimizu et al. 1978). In this work, 2-mo-old pigs were kept at either 86°F (30°C) or 40°F (4.4°C) and exposed to Transmissible Gastro Enteritis (TGE) virus. Pigs kept at 86°F (30°C) after exposure to TGE were unaffected, whereas all pigs kept at 40°F (4.4°C) after exposure became sick. Similar outcomes for the most ordinary bacteria that nursing piglets are exposed to can be expected—a chill can be devastating.

Shimizu et al. (1978) demonstrated nicely that the rate of illness for viral-exposed pigs was reduced (in fact, it was zero) when they were kept warm. The 86°F (30°C) they used was warmer than the 75° to 80°F (24 to 27°C) that represents the lower critical temperature for these pigs. When pigs were exposed to a drop in air temperature to 40°F (4.4°C), all of the pigs became sick. If the piglets had a 14-d acclimation period before viral exposure, the rate of illness was reduced, but still present. While pigs can acclimate to a cold air temperature, the biggest concern with a cold chill is that resident microorganisms will become more of a problem when the piglet experiences the cold. The key is to plan ahead and prevent the cold experience. If illness begins, a still warmer temperature will probably help reduce the rate of illness.

The solution in the farrowing barn and the farrowing huts on pasture is to keep the room or hut warm. This is fine on "normal" days. The workers must also anticipate cool weather and prevent the sudden onset of a cold draft in the farrowing barn. The consequence of allowing even the smallest draft is greater enteric disease and slower weight gains. With longer exposure to a cool draft, preweaning mortality can be increased.

The nursing piglet is very tolerant of warm weather; the sow is not. Warm temperatures reduce sow feed intake (thus reducing milk production and piglet weight gains) and, in severe cases, heat can kill sows. Studies show that an air temperature of as little as 90°F (32°C) can be lethal to the sow (McGlone et al., 1988), particularly during farrowing (when tremendous heat is produced). Many methods of cooling sows have been studied over the years. The most effective cooling method is the water drip, but this technique only works well when sows and litters are on fully slotted floors (to prevent pooling of water near piglets).

Figure 18–1 shows that when sows were heat-stressed at or above 85°F (29°C) (not a very severe heat stress), feed intake was reduced by 47% compared with the intake of cooled sows. Concrete flooring provided some cooling, but the water drip provided even better results. The consequence in terms of a CPM is that heat-stressed sows that eat less feed produce less milk and wean smaller piglets. The same outcome is observed if the farrowing manager does not feed the lactating sows enough—weaning weights will be suppressed.

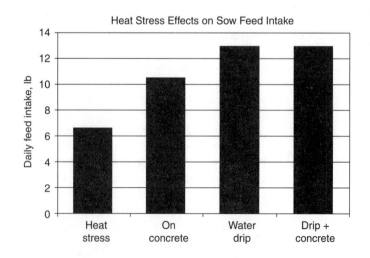

FIGURE 18–1

Effects of Cooling Methods on Sow Feed Intake. In This Study, All Sows Were Heat-Stressed with Air Temperatures at or Above 85°F. With No Cooling in Place and Sows on Plastic-Coated Expanded Metal, Sow Feed Intake Was Reduced by 48% Compared with Sows Provided a Water Drip. Weaning Weights Are Reduced in Proportion to Sow Feed Intake.

Source: Adapted from McGlone et al. (1988a).

It is significant, in the heat stress situation, that piglets thrive. Piglets are nearly impossible to heat-stress in production environments during warm weather because they are usually never too warm. Thus, piglet survival tends to increase in the summer. In one of the heat stress studies done at Texas Tech University in which preweaning deaths were carefully monitored, piglet mortality was 13.4% among control litters and only 7.1% among heat-stressed sows whose litters were not allowed to get cooler than 86°F (McGlone et al., 1988b).

The take-home message from this review of the thermal environment is simple: Do not chill piglets, and keep them even warmer when scours are present in the farrowing room. Keeping piglets warm can be accomplished by use of heat lamps, heat pads, and housing and bedding.

> **CRITICAL MANAGEMENT OBSERVATIONS OF PHYSICAL ASSETS IN THE FARROWING AREA**
>
> The good stockperson must observe and manage the physical assets. This means being acutely observant and planning ahead. The stockperson should look for following items during a walk-through of facilities:
>
> - Observe the fans and inlets.
> - Feel for air currents to be sure they are operating according to design.
> - Read the air temperature on a thermometer at pig level.
> - Confirm the animals experience of their thermal environment by observing:
>   - Sow respiratory rate and feed intake—sure signs of heat stress are increased respiratory rate and reduced feed intake.
>   - Piglet lying patterns—a sure sign of chilling is a pile of huddling piglets.
> - Check feeders and waterers for proper functioning.
> - Check flooring and penning materials for integrity.
> - Warm the environment when scours are found.

MANAGEMENT OF THREE BIOLOGICAL ASSETS: SOWS, PIGLETS, AND PEOPLE

Biological assets in the farrowing area include the sows and piglets. Producers sometimes forget that these assets are actually organisms with a nervous system that have behavioral and physical needs. Little is known about the behavioral needs of pigs. The workers in the farrowing area are another asset—they make or break the unit. The best livestock workers are able to manage the animals with superior skill, leading to a high degree of success—maximizing the CPMs.

What does it mean when the management of a given farm is referred to as "good"? It means that something the workers *do* results in success, particularly in weaning as many heavy pigs as possible. The physical environment can be marginal, but the "good manager" can still make it work well.

In this day and age, it is quite surprising that an agreement is not often reached on the qualities associated with "good management." Yet, each profession, from engineer to veterinarian to ethologist, often has a unique definition of "good" management.

Good management is the ability to integrate components of the physical and biological assets. The best integration is accomplished by the best managers. The best stockpeople have developed their skills, either through intuition or by learning through experience or training.

To be a good manager of biological assets, the stockperson must start by being a keen observer. Each time the room or field is entered, the stockperson should observe each sow and each piglet. With experience, the stockperson quickly notices departures from "normal." Some call this "management by exception." Management intervenes if a sow or piglet shows behavior out of the accepted range. Defining the proper range of "normal" requires an understanding of pig behavior.

SOW BEHAVIOR

Sows show differing behaviors during four phases in the farrowing area: (1) entry, (2) nest-building, (3) parturition, and (4) lactation (see Table 18–8).

The entry phase lasts a few d or even a few h in tightly regulated production systems. Sow behaviors during the entry phase are much like those exhibited during gestation. Sows will explore their farrowing environment after arrival, but their level of exploration is much less than that for younger, more active pigs. Still, this initial exploration should not be confused with the more active, often violent, nest-building behaviors. If sows are observed in nest-building behaviors in the gestation area, they should be moved immediately into the farrowing area.

When sows begin nest-building behavior, they are anywhere from 18 to 4 h from giving birth to the first piglet. Nest-building behavior can last from a few min up to 12 h. Individual sows vary in onset and duration of nest-building. When the nest-building behavior stops, on average, sows will farrow in about 4 h. At this point, milk should be easily expressed from a teat in a stream (as opposed to only a drop expressed earlier).

The sign of hefty milk flow is a sign that the hormone oxytocin is being released from the pituitary gland. Generally, sows will be having uterine contractions at this time because oxytocin stimulates both uterine contractions and milk ejection. Oxytocin is a neuroendocrine hormone that is very sensitive to the effects of stress. Stress during nest-building or parturition will literally shut off oxytocin release. Without oxytocin, milk ejection stops and uterine contractions stop. This is one factor that may result in stillbirths. During this sensitive period, handlers should minimize noise,

TABLE 18–8
Behavioral Stages Observed in the Farrowing Area. Times Given in the Second Row Are Relative to Parturition on Day Zero.

Entry	Nest-Building	Parturition	Lactation
−7 D to 0	−18 to −4 H	0	0 to X Days of Lactation
• Low feed intake	• No feed intake	• No feed intake	• Increasing feed intake
• Feed eaten in one meal in about 30 min	• Much activity: pawing, pushing, chewing	• Lying on side, occasional standing, drinking	• Cycling behavior, focused on nursing
• Relatively inactive	• Use of bedding, if available	• Quiet, calm, and in pain	

rough handling, and disturbances of any kind. Thus, human behavior and proper equipment operation have a direct effect on an important CPM (stillbirths).

Genotypes and environments can influence pig behavior. Most sows are brought up to full feed in the few d after parturition. In common production systems, lactating sows are fairly consistent in their behavior. They spend the majority of their time in the lying-down state (see Figure 18–2). When lying/inactive and nursing/lying states are combined, sows in farrowing crates spend an amazing 83% of their time in the lying-down position (making the presence of sores on their shoulders more understandable).

Sows stand up about 15% of their time. Just over one-half of their standing time is spent feeding. Drinking, sitting, and eliminating feces and urine represent a minor part of the sow's time budget activities—all very important activities.

Sows that are annoyed by their physical environment or by people stand up and lie down more often. A normal rate of standing up for lactating sows is from 6 to 12 times/d (McGlone and Morrow-Tesch, 1990). Sows make from 3 to 15 postural adjustments/h (Fraser et al., 1995). An excessive number of standing up and lying down bouts or postural adjustments increases the risk of crushing piglets.

Sows that are nursing normally will nurse their litters an average of every 45 to 50 min—about 30 nursing bouts each d. Fewer complete nursing bouts can be a sign of mastitis or other milking problems. Another sign of sow discomfort is for them to spend a large amount of time lying on their belly. In this position, piglets cannot nurse. If the sow's mammary glands are inflamed, belly-lying can be a coping mechanism. Sows normally spend only 3% of their lying-down time on their belly. The more common posture is to lie on their side (laterally).

> An excessive number of standing up and lying down events or postural changes can cause an excessive risk of crushing—6 to 12 times/d is the normal rate.

When looking for exceptional sow behaviors, the stockperson is keyed to changes in any of the major behaviors. If feed intake drops or if water consumption increases or decreases, the sow may be sick. If sick, she will not stand as often nor for as long a

FIGURE 18–2
Sow Behavior Indoors During Lactation (the fourth behavioral phase in the farrowing area). Sows in the Outdoor Systems Show Very Similar Time Budgets.

Source: Adapted from McGlone and Blecha (1987) and McGlone and Morrow-Tesch (1990).

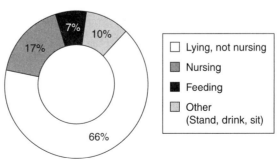

duration. If nursing bouts are too few or too many, the sow might not be nursing properly. Changes in nursing frequency or a decrease in milk production is most common in the case of mastitis (inflammation of the udder), agalactia (no milk production), or hypogalactia (low milk production). In these cases, sow and piglet behavior will clearly indicate to the stockperson that intervention is required. Alert stockpeople will notice these behavioral changes earlier and, thus, treat the problem in its early stages when there is a greater chance of success.

Normal interventions include giving antibiotics to the sow and perhaps starting a regimen of hourly oxytocin injections to stimulate milk production and let-down. In the case of hypogalactia or agalactia, the piglets, if greater than 4 or 5 wk old, should be provided creep feed. Younger piglets should be fed a liquid milk substitute by stomach tube.

PIGLET BEHAVIOR

Piglets also show very consistent behavior. They sleep more and spend more time lying down at birth than they do at a few wk of age. As they develop, they spend less time sleeping and more time exploring their environment (see Figure 18–3). Piglet and sow behaviors are synchronized when everything is normal and milk production is good.

From time to time, individual piglets miss a nursing, and from time to time, a given sow may have an "unsuccessful" nursing in which she delivers no milk. These failures are not of grave concern as an occasional event. However, if piglets or the entire litter miss more than one nursing bout, a problem is developing. The stockperson should be on the look-out for such failures from the earliest possible moment.

A common problem, due in large part to behavior, is that sows will crush a piglet. This is the leading cause of loss of life in the pig operation. Sows and piglets that are uncomfortable are more likely to crush or be crushed. Sows that stand and lie down often crush piglets more often. Piglets that are cold, hungry, or feel ill are more likely to lie against the sow, thereby increasing her chance of rolling over and crushing them.

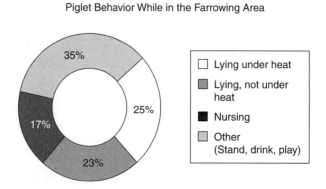

FIGURE 18–3
Behavior of Piglets During Late Lactation (about 21 d of age) Recorded in the Farrowing Crates. Data Would Be Similar for Piglets Nursing Sows Outdoors.

Piglets are crushed when sows move from:

- Standing to lying
- Sitting to lying
- Lying on their sternum to rolling on their side

Edwards et al. (1986) found that about half the crushed piglets in his study were caused by the second behavioral postural change—sows moving from sitting to lying. Furthermore, research at Texas Tech University (McGlone et al., 1991) shows that sows with greater sitting behavior are three times more likely to crush baby piglets than sows who rarely sit. It is speculated that environments that induce greater sitting behavior may increase the rate of crushing. Some sows are genetically more predisposed to sitting (McGlone et al., 1991).

> **CRITICAL MANAGEMENT OBSERVATIONS OF BIOLOGICAL ASSETS IN THE FARROWING AREA**
>
> The good stockperson must observe and manage the biological assets—sows and piglets. This means being acutely observant and planning ahead. The stockperson should look for the following items during a walk-through of facilities:
>
> - Observe each sow for normal behavior patterns.
> - Observe each piglet in each crate or pen.
> - Check to be sure the sow and piglets are interacting in a normal manner, particularly during nursing bouts.
> - Observe how other stockpeople interact with the sows and piglets.
> - Investigate the cause of any exception to "normal" behavior (check physical environment and person-pig interactions).

STOCKPERSON BEHAVIOR

People vary in their ability to care for sows and piglets. Most people can be trained to be better managers if they understand fundamental pig biology and some things about the physical environment. In large farms with multiple, identical production units (same feed, breeding stock, administration, etc.) there is still a vary large variation in unit-to-unit productivity.

Hemsworth and his colleagues in Australia (Hemsworth et. al. 1995) examined potential human factors that might cause variation in pig productivity:

- Attitude toward the job
- Attitude toward the pigs
- Technical skills

Hemsworth et al. (1995) suggested that a sequential relationship develops, starting with the worker's attitude toward the pigs. Whether positive or negative, human attitude translates into human behavior and related interactions with the pigs. Human behavior causes a relative amount of fear (or lack of fear) of humans in pigs.

Hemsworth et al. (1995) cited evidence of relationships between pig fear of humans and reproductive problems, particularly in the breeding/gestation area.

The logic in the farrowing area goes something like this: If the sows are afraid of people, they are more nervous and uncomfortable when people are around. When they are uncomfortable, they stand up and lie down more often and make more postural adjustments. These movements cause a greater incidence of piglet crushing.

Hemsworth et al. (1995) examined the times piglets were crushed during a 24-h d. They found that 7% of the sows crushed piglets when people were not working. Surprisingly, 16% of the sows crushed piglets while people were working. These data support the idea that people make sows nervous, which leads to more sow activity and therefore greater crushing of piglets.

One possible solution to prevent piglet deaths is to keep people away from the sows. This, of course, is not possible. Workers must be present to check equipment and care for the animals. It is important to make the sow and piglets experience with people as positive as possible. This is quite a challenge, especially in the farrowing area. When litters are processed, stockpersons remove the piglets, trim their teeth, dock their tails, and give injections. Piglets are squealing and upset and the sow gets upset. Making this procedure a positive experience is not easy, but most other interactions between people and pigs should be positive.

A NEW BREED OF DEDICATED STOCKPERSON

Fifty years ago, the job of livestock worker was a specialty that was recognized among livestock producers. Back then, a good "pigman" could be sure of a job managing a "large" farm of maybe even 100 sows. On large commercial farms, where work is more specialized and more time demanding, one farrowing worker may be present for a sow inventory of about 500 sows (or two people/1,000 or more sows).

The dedicated stockperson has the following features:

- A good work ethic
- Job satisfaction and a positive attitude
- A healthy, positive attitude toward pigs
- A basic understanding of pig biology
- The ability to develop skills in providing animal care
- The ability to develop interpersonal skills useful in people management
- The ability to plan complex tasks

SUMMARY

High survival rates and high weaning weights are key components of a successful farrowing unit. This chapter discussed the role and behavior of the stockperson in meeting the physical and behavioral needs of the sow and litter. Space and thermal needs for optimum sow and litter performance were described and recommended standards were provided. Piglet mortality in the preweaning period is reduced by minimizing stillbirths

CHAPTER 18 MANAGEMENT OF SOWS AND PIGLETS: BEFORE, DURING, AND AFTER FARROWING

and by providing an environment and housing that minimize piglet crushing, chilling, underfeeding, and disease.

QUESTIONS AND ACTIVITIES

1. On a given farm, if the litters per sow per yr was 2.3 and the number born alive was 11.0 pigs per sow, what would the PPSY be if preweaning mortality was 20% or 10%? Assume 1.0 stillborn/litter.

2. How many more weaned pigs would be produced on a 1,000-sow unit by lowering the preweaning mortality of a farm doing 20 PPSY with a 15% (or average) preweaning mortality of 7.5%?

3. Provide some example target values for number of pigs born total and born alive, stillbirth rates, and preweaning mortality for a herd that weans 24 PPSY.

4. Describe how the needs for a certain air temperature differ for the sow and the piglet. Which is more susceptible to cold stress?

5. How do piglet lying patterns change when they are warm, comfortable, and cold?

6. Describe the stages of sow behavior as she starts parturition through when she has a litter nursing.

7. How can the stockperson influence herd productivity?

LITERATURE CITED

Edwards, S. A., S. F. Malkin, and H. H. Spechter. 1986. An analysis of piglet mortality with behavioral observations. Anim. Prod. 42:470 (Abstr.).

Fraser, D., P. A. Phillios, B. K. Thompson, E. A. Pajor, D. M. Weary, and L. A. Braithwaite. 1995. Behavioural aspects of piglet survival and growth. In: M. A. Varley (ed.) The Neonatal Pig: Development and Survival. CAB International, Oxon, UK.

Hemsworth, P. H., G. J. Coleman, G. M. Cronin, and E. M. Spicer. 1995. Human care and the neonatal pig. In: M. A. Varley (ed.) The Neonatal Pig: Development and Survival. CAB International, Oxon, UK.

Johnson, A. K., Julie Morrow-Tesch, and J. J. McGlone. 2001. Behavior and Performance of Lactating Sows and Piglets Reared Indoors or Outdoors. J. Animal Science. 79:2588–2596.

McGlone, J. J. and F. Blecha. 1987. An examination of behavioral, immunological and productive traits in four management systems for sows and piglets. Appl. Anim. Behav. Sci. 18:269–286.

McGlone, J. J., W. F. Stansbury, and L. F. Tribble. 1988a. Management of lactating sows during heat stress: Effects of water drip, snout coolers, floor type and a high energy-density diet. J. Anim. Sci. 66:885–891.

McGlone, J. J., W. F. Stansbury, L. F. Tribble, and J. L. Morrow. 1988b. Photoperiod and heat stress influence on lactating sow performance and photoperiod effects on nursery pig performance. J. Anim. Sci. 66:1915–1919.

McGlone, J. J. and J. L. Morrow-Tesch. 1990. Productivity and behavior of sows on level and sloped farrowing pens and crates. J. Anim. Sci. 68:75–81.

McGlone, J. J., C. K. Akins, and R. D. Green. 1991. Genetic variation of sitting frequency and duration in pigs. Appl. Anim. Behav. 30:319–322.

McGlone, J. J., J. L. Salak-Johnson, R. I. Nicholson, and T. Hicks. 1994. Evaluation of crates and girth tethers for sows: reproductive performance, immunity, behavior and ergonomic measures. Applied Animal Behaviour Science. 39:297–311.

McGlone, J. J. and T. A. Hicks. 2000. Farrowing hut design and sow genotype (Camborough-15 vs. 25% Meishan) effects on outdoor sow and litter productivity. J. Animal Science 78:2832–2835.

Morrow-Tesch, J. and J. J. McGlone. 1990. Sources of maternal odors and the development of odor preferences in baby pigs. J. Anim. Sci. 68:3563–3571.

National Pork Board. 2002. Swine Care handbook. National Pork Board, Des Moines, Iowa. *http://www.porkboard.org/publications/pubIssues.asp?id=85*

Ott, S. L., A. Hillberg Seitzinge, and W. D. Hueston. 1995. Measuring the national economic benefits of reducing livestock mortality. Preventative Vet. Med. 24:203–211.

Shimizu, M. Y., Y. Shimizu, and Y. Kodama. 1978. Effect of ambient temperatures on induction of transmissible gastroenteritis in feeder pigs. Infect. Immun. 21:747–752.

USDA. 2002. US Hog Inventory, December 1–June 1, 2001. *http://www.ers.usda.gov/data/sdp/view.asp?f=livestock/94006/* Accessed April 11, 2002.

INTERNET RESOURCES

Feeding and caring for sows during lactation:
http://eru.usask.ca/saf_corp/livestok/pigs/feedlact.htm

For placing a value on farrowing workers and their skills:
http://www.nppc.org/PorkReport/SpecRepSPR00.htm

Sow and piglet management:
http://muextension.missouri.edu/xplor/agguides/ansci/g02500.htm

19
MANAGEMENT OF GROWING PIGS

INTRODUCTION

The stockperson cares for the growing pigs (the term "growing pig" in this chapter refers to both nursery [wean plus 4 to 12 wk] and growing-finishing pigs [8 to 12 wk post-nursery until market]). Workers on commercial farms have a significant challenge to care for individual pigs. More often, workers may view pens and barns full of pigs rather than viewing individual pigs. While a small operation may have only a few growing pigs to care for, modern commercial farms have one employee per 1,000 pigs per hr of work. In traditional pork production, a worker grinds and mixes the feed, cares for sows and nursing pigs, repairs fences and pens, manages the manure, and takes care of any other problems that may arise. In modern corporate pork production, the workers who are assigned care of the growing pigs provide animal care as their primary job responsibility. If they need feed, they call the feed mill; if they need plumbing or electrical repairs, they perform triage and then call the maintenance staff. Manure is the responsibility of others when it leaves the barn.

MANAGEMENT OF THE MODERN GROWER/FINISHER UNIT

Modern technology and efficient pen and barn design have made such advances that one person can care for up to 8,000 pigs in a day. If events such as a disease outbreak or a major equipment failure occur, more than 8 hr/8,000 pigs is required.

To spend 8 hr/d caring for 8,000 pigs or 1 hr/1,000 pigs and to provide good animal care requires an alert, well-trained individual. Assuming that 1/2 hr is spent treating individual pigs or equipment problems, the worker must look at 33 pigs/min, or about 2 sec/pig. The keen observer will find exceptions to the normal in such a quick scan of the pigs and equipment. Noticing and dealing with these exceptions in a timely manner are the keys to success in the growing pig barns. This management style is appropriately called management by exception.

NORMAL PIG BEHAVIOR

Understanding normal pig behavior requires some appreciation of how the pig evolved. The pig's ancestor was a forest-dwelling creature that also foraged in grass if necessary. These pigs preferred to be in social groups and avoided, rather than confronted by other species.

The life of the pig's ancestor was spent searching for food. The search included feeding off the ground and on low plants, and rooting in the ground in search of roots and insects. If these pigs happened on a bit of carrion, they were happy to eat it in its entirety. They were slightly neophobic (fear of eating new things), but they consumed any food that did not make them sick. If they were to come across a field of grain plants, they ate until the grain was consumed. They did not travel unless it was necessary.

To manage the exceptions, producers must be intimately familiar with what is normal. Normal pig behavior is well-known to experienced stockpersons, who often cannot explain what normal behavior is, but they can recognize it and move their eyes to other areas in their search for abnormal exceptions. Understanding normal pig behavior is in the best interest of the growing pig caregiver.

MAINTENANCE BEHAVIORS

Figure 19–1 shows data on the maintenance behaviors (feeding, drinking, lying) of indoor pigs. They spend a large amount of their time (80% or more) lying down. About half the lying-down time is spent sleeping. Over time, as the pigs consume larger, but less frequent, meals they spend more and more time lying down and less time eating. Ad libitum-fed pigs spend 20% of their time active when in the nursery, but only 12% active 4 mo later in finishing. The increase in time spent lying comes almost entirely from a reduction in time spent feeding.

The pattern of behavior changes with time of day. Pigs show clear cycles in behavior with some active and some inactive times. The hours of activity depend on the feeding system, the thermal environment, and, to some degree, the genetic line. Genetics, however, is more likely to change the overall level of behavior and not the daily rhythm. Some lines, such as the Meishan, are less active than European breeds (see Figure 19–2). Even when the Meishan influence is diluted to 25% through crossbreeding, the crossbred pig is less active than more traditional crossbred pigs.

FEEDING BEHAVIORS

Pigs feed in discreet meals throughout the day (see Figures 19–3 and 19–4). A pattern of nursing is established in about 24 hr. Nursing occurs about once/hr (each 45 to 50 min)

FIGURE 19–1
Behavior of Nursery and Growing-Finishing Pigs Fed Ad Libitum and Housed on Totally Slotted-floor Pens. Each Pen had Access to Four Toys (two chains and two hoses) Suspended from the Ceiling.

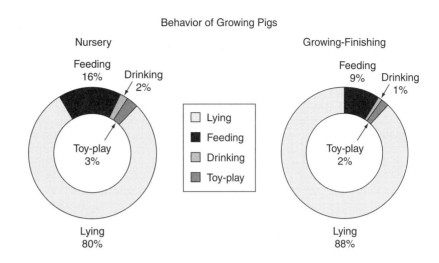

FIGURE 19–2
Normal Lying (top) and Feeding (bottom) Behaviors for Two Genetic Lines of Pigs. The C-15 Is the Camborough-15, a Common Crossbred Pig in the United States. The exp-94 Is an Experimental Crossbred Line Containing 25% Meishan. Note the Daily Rhythm in Behavior with Less Lying around the Early Morning (6 to 9 a.m.) and the Afternoon (2 to 10 p.m.). The Low Level of Lying Is Associated with the Feeding Bouts. Note Also That the Meishan Is Less Active Overall and Spends More Time Lying and Less Time Feeding During Many, but Not All, Hours.

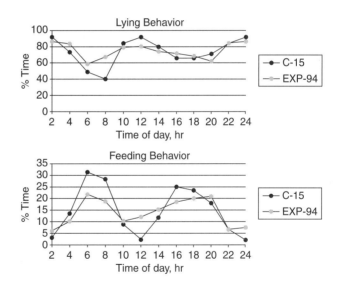

in a synchronized manner at least until 4 wk of age. Piglets can eat more milk in a given meal than the sow's udder can provide. After weaning, domestic pigs, unless provided with creep feed before weaning, are unwilling to eat the dry feed. Later, with ad libitum access to feed, pigs will eat larger meals less often during the day. They would not have enough space and time to eat once/hr with the amount of feeder space usually provided. Even with a wide-open trough, they would synchronize their feeding to some degree after weaning. Growing pigs naturally tend to eat in groups.

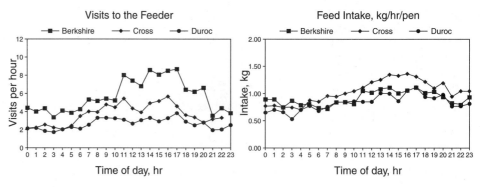

FIGURE 19–3
Feeding Behaviors of Pigs of Three Breed Types Over 24 hr/d. Pigs Were Housed 39 Pigs/Pen. Note the Greater Number of Feeding Bouts or Visits to the Feeder in the Afternoon for the Berkshires.

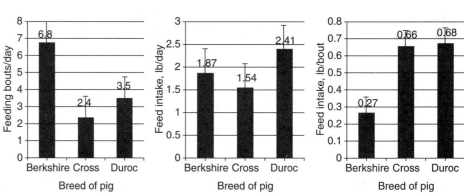

FIGURE 19–4
Daily Feeding Behavior and Feed Intake for Berkshires, Terminal Crossbred Pigs, and Durocs. Note That Although the Bershires Had More Meals/d, Ate Less/Meal, and This Resulted in about the Same Weight of Feed Eaten/d as the Other Genotypes of Pig.

The feeding behavior of pigs is facilitated by their pen mates. This increase in feeding associated with being in a social environment is called social facilitation. Social facilitation of feeding should take place among group-housed growing pigs as long as the feeding space is neither too limited nor in excess. Some individual pigs who are socially submissive will have socially induced lower feed intake, but the majority of pigs will be stimulated to eat more in the presence of other pigs.

In the late 19th century (1888), Harris described social facilitation among growing pigs destined for the fair. He described how he was able to get his prize pig to eat even more feed than its normal ad libitum intake. The prize pig was given enough feed to eat until it was full and stopped eating. A second, skinny and hungry pig was let in the pen and ran over and ate feverishly. The excited feeding of the hungry pig caused the prize pig to eat a bit more. Harris was using social facilitation to increase the prize pig's feed intake.

Pigs of a given genetic line fed a given diet will have a repeatable pattern of feed intake. Feeding patterns are described by the numbers of meals or feeding bouts/d, the feed intake/meal, and the total intake/d.

Feeding patterns may change with age. As pigs get older, they eat fewer, but larger, meals. If given enough room to eat simultaneously, pigs will develop a circadian pattern of feed intake—eating more meals during the day than at night. As feeder space becomes limited, more pigs must eat at night and the group pattern becomes flat (no peak of feeding in the day hours).

Growing pigs eat 1 to 8 meals/d. Each meal may be from 0.1 to 2 lb (0.05 to 0.9 kg) and will require 2 to 15 min/feeding bout for the usual grain-based feeds. With some less-dense feed, feeding may require several hr/d.

The fact that pigs of a given genotype eat more or less meals/d does not mean that the overall feed intake will be increased or decreased. In one investigation at Texas Tech University, Berkshires ate more meals/d than did pigs of Duroc or a commercial crossbred line (see Figure 19–4). Although the Berkshires ate more meals/d, they ate less feed/meal. The overall feed intake (lb/d) was similar for all three genotypes.

Managers must understand the normal feeding pattern of the pigs in their charge. When "normal" is understood, exceptions to normal can be identified more easily.

Feed intake declines during warm weather and is noticeably depressed above temperatures of 85°F (29°C) for pigs 100 lb (45 kg) and above. Younger pigs can grow and behave normally at warmer temperatures.

Feed intake tends to increase when the air temperature is cooler. Pigs use some of the feed to burn as fuel to keep their bodies warm. The increased feed intake in cool temperatures is associated with a progressively lower feed efficiency. In this case, from an economic point of view, the cost of feed as a fuel should be considered carefully.

DRINKING BEHAVIORS

Pigs require water to sustain life. They will drink from streams, cups, and bowls, bite nipples, and drink from almost any available source. Pigs do not unequivocally prefer clean water. For example, outdoor pigs are often observed drinking from a lived-in wallow when clean water is available nearby. However, clean water should be provided so they can remain healthy. Some have suggested that pigs prefer drinking clean water over dirty water. Preferences are complicated to interpret and other factors may interfere with a given choice (e.g., difficult, uncomfortable, or remote accessibility). Pigs may then seem to "prefer" to drink dirty water. Pigs often drink when they eat. Being normally fed dry feed, they alternate between eating and drinking. If water intake is limited, feed intake will become limited.

DUNGING BEHAVIORS

Pigs do not defecate and urinate in a random manner. They seem to have reasons, or at least preferences, for their dunging behavior. Pigs defecate when they are frightened or in a new environment, but in their home pen, pigs defecate and urinate in a relaxed and calm manner, away from most pigs and away from the feed. They may defecate against a wall or in a corner because the chances of disturbance are minimal. Other pigs investigate their pen mate's defecation and urination behaviors using their senses of smell, taste, and touch.

When given enough room, pigs establish clear areas of the pen for different uses. A well-designed pen has areas for feeding, drinking, resting, and dunging. Pigs ordinarily will not dung where they eat, but rather do so in a wet or cold area away from their feed. They may, however, dung near their water supply. During thermoneutral and cool temperatures, pigs urinate and defecate in the wet and cool areas of the pen.

During these times, they rest in the dry, warm parts of the pen. In warm weather, pigs may lie in the wet places in the pen and become dirty as the result of resting and wallowing in manure. Pigs do not avoid manure and will use feces and urine as they would use mud to cool their skin during warm weather.

When pigs live on totally slatted flooring, they have a difficult time establishing clear dunging areas, although they still try. On a totally slatted-floored pen, pigs may defecate against a wall to attempt to establish a clear area for dunging.

Changing the defecation pattern is difficult. The job of directing them to eliminate in the desired place is easier, if producers plan ahead, than it is to try to change their elimination behavior. When pigs are first introduced into a partially slotted pen or into a pen in which a producer wishes to shape the pig elimination pattern, water should be applied where defecation is desired (e.g., over the slatted area). Directing cool air currents over the desired area will help make the area cool. Feed should be sprinkled on the floor where elimination is discouraged. These actions will discourage the pigs from defecating in the area with feed—they will rest in these places.

CHEWING/ROOTING AND PLAYING

Pigs spend 10% to 50% of their waking, active times chewing and rooting with objects and animals in their environment. Pigs explore their environment through their senses and they will taste any new object in the pen. While it is easier for them to chew and root on soil, they will chew and root on any material available to them, including concrete flooring and stainless steel feeders—any material in the pen will become the focus of their attention.

A moderate amount of chewing, rooting, and playing with equipment is normal pig behavior. An excess of chewing may be the result of unintended selection or some dietary deficiency. Mineral deficiency has been blamed for excess chewing, but this has not been clearly demonstrated. When chewing and exploration become either too intense or too reserved, the stockperson should seek causes in the diet or the environment.

SICKNESS BEHAVIORS

When pigs become sick, their behavior changes. During the early hours after exposure to a pathogenic microorganism, a cascade of metabolites is produced that includes some immune cell cytokines. These cytokines travel to the brain to cause sickness behavior.

The behavior of sick pigs varies with the type of illness, but some general trends can be described. Sick pigs eat less and may drink less. They spend more time lying and their appearance is often different. They may have a hunched-up or hunched-over look. Their hair may be raised and appear rough. Sick pigs often lie down and their apparent resting behavior can be mistaken for healthy resting when they are actually ill. The stockperson with a keen eye is able to detect the very early signs of sickness and will begin interventions (such as antibiotic treatment or adjustment of the thermal environment) during the early stages of sickness.

DAILY PIG MANAGEMENT

Each operational unit on a swine farm should develop standard operating procedures (SOPs) for the daily care of the animals. The SOP for growing pigs should include at least a careful, daily observation of each pig. A second, quicker walkthrough later in the day is desirable.

On some farms, stockpersons look at the pigs from the aisle while on others, they enter each pen each day for a closer observation of individual animals. On many well-managed farms, the caregivers enter each pen each day and ensure that each pig can stand, walk, and move with ease. The SOPs should state exactly how much human contact is to be provided each day.

WHAT IS STOCKMANSHIP?

Peter English and his colleagues (1992) have given considerable study and thought to the description of stockmanship, especially with pigs. They suggest good stockmanship involves a combination of:

- A sound knowledge of the pigs and their requirements
- A basic attachment and patience for the pig
- The ability and willingness to communicate and develop a good relationship with the pigs (empathy)
- The ability to recognize individual animals and to remember their particular eccentricities
- A keen sensitivity for recognizing slight departures from normal behavior of individual pigs (perceptual skills)
- An ability to organize working time well
- A keen appreciation of priorities and a ready willingness to attend to individual animals even if it disrupts the schedule

Successful animal workers provide sound stockmanship or animal care. They know their pigs and their pigs know them. They know when to follow the routine and when intervention is needed. They know how to handle pigs in a routine manner and how to handle the occasional difficult pig. The good caregiver has a work-day flow that is logical and efficient, but is flexible enough to be interrupted for reasons of providing sound animal care.

People are not born with skills of good stockmanship, but some people are better suited to provide animal care than others. With proper training, people with the predisposition to care for pigs can develop into outstanding pig caregivers. These stockpeople are an important asset to the successful farm.

Today, many large farms have multiple units of 1,200 or 2,400 sows. These units have identical pig genetics, nutrition, and housing. The difference between the best and worst units is in the people. How they care for the pigs and how they interact with each other (people skills) determine their success. Some of the success is a function of how well they were trained—people do not often become outstanding caregivers without some training.

POSITIVE HUMAN-PIG INTERACTIONS

How human caregivers interact with their pigs can have significant effects on pig behavior and productivity. Pigs on each farm will have a pattern of behavior that is shaped by the animal workers. Some of the behaviors exhibited by pigs include:

- Fear, due to being abused (even mildly)
- Fear, due to no human exposure
- Neutral toward people
- Comfortable around people
- Aggressive toward people

Pigs that are dealt with in a crude or mean manner, even in passing, will develop a fear of people (see Figure 19–5). Pigs have good memories and they will generalize their fear of one person to other handlers. Pigs that have an intense fear of humans may have been shocked, prodded, kicked, slapped, or beaten. These fearful pigs will avoid people and some individual pigs may actually scream when people are near or when people touch them. Some of this extreme pig behavior may have a genetic component, so one should not automatically assume that screaming and fearful pigs were abused. Most likely, these behaviors are caused by a combination of a predisposed genotype and hostile treatment.

Pigs that are not handled often by people show a different type of fear of humans. These pigs have a very large flight zone (the closest a person can get to a pig before it flees). When a person approaches these pigs, they run away in fear of the unknown. Pigs on some commercial farms show this extreme flight zone. These pigs are easy to handle in well-designed facilities, but they can be very difficult to handle in poorly designed facilities where an individual pig can get turned around and, thus, disrupt the flow of animals in the desired direction.

Pigs that are neutral toward people are very uncommon. These pigs will have an intermediate flight zone and express a middle level of fear toward people.

Pigs that are comfortable around people have had many experiences with people and the majority of experiences were positive in nature. Positive experiences include people touching, stroking, petting, and talking to the pigs. The pigs may actually move

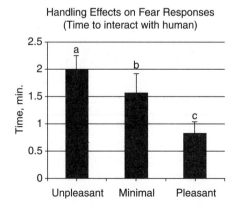

FIGURE 19–5
Fear Responses of Pigs to Unpleasant, Pleasant, and Minimal Human Interactions.

Source: Adapted from Hemsworth et al. (1986).

toward people and, therefore, they may have a negative flight zone, which may cause an increased time required to move pigs through chutes and races.

Some pigs that are very comfortable around people are problematic. These pigs seem to think people are something to chew on, play with, or eat. They do not so much bite, as chew, on people. They start with chewing on shoes, boots, or pant legs and graduate to chewing on legs or arms. These pigs increase their aggressive chewing the longer people are in the pen. Working with equipment in the pen is very difficult with aggressive pigs. Some genetic lines may show this increased aggressive behavior more than other lines.

The Fear Test

Australian researchers developed a simple test to determine the relative level of fear among pigs (Coleman et al., 2000). In the fear test, a person abruptly places a hand in the direction of the head of an awake pig. The pig's normal reaction is to step back. The person then times how long it takes for the pig to return to the still hand and to touch the hand (or alternatively, to get within 6 in (15 cm) of the hand). The greater the pig's fear of people, the longer it will take for the pig to return to the hand. Pigs that have had limited positive experiences average about 1 min. Pigs that take longer than 2 min have a significant fear of people. Some genetic lines and some environments may require an adjustment of normal and fearful times.

How much human interaction is best for growing pigs? This question has not been answered fully, but the simple answer is that minimal human contact is needed. When human interaction is required, handlers should make the interactions as positive as possible.

> When human interaction is required, make the interactions as positive as possible.

Researchers have examined how different levels of human interaction impact pig behavior and productivity. In one study, Hemsworth et al. (1987) showed that pigs handled in a positive and pleasant manner had improved ADG and F:G ratios compared with pigs handled in an unpleasant manner (see Figure 19–6). Inconsistent handling was nearly as bad for pig performance as unpleasant handling.

One interesting insight into the Hemsworth work was that pigs handled in a minimal way had fairly good ADG and F:G ratios. This finding supports the argument that if some unpleasant experiences are necessary (injections, ear tagging, etc.), it may be better to handle pigs infrequently or in a minimal manner to minimize the stress reaction when pigs see people approaching. When faced with the need for pleasant handling in a consistent manner, many farmers choose to provide only minimal human-pig interactions.

Scientists at Texas Tech University have studied the effects of positive human interactions and use of toys (suspended chains and hoses) in growing pigs. Hill et al. (1997) recently completed an investigation of the efficacy of toys and/or positive human interaction. They compared the following treatments:

- Negative (Neg) control (nearly zero human contact)
- Control (Con) (once/d quick walkthrough the pen; once/wk a longer time was spent adjusting equipment)

FIGURE 19–6

Effects of Level of Human Handling on Pigs ADG and F:G Ratio. Pigs Handled in an Unpleasant Manner Had Lower ADG and Worse F:G Ratios than Pigs Handled Either in a Positive Manner or a Minimal Manner.

Source: Adapted from Hemsworth et al. (1987).

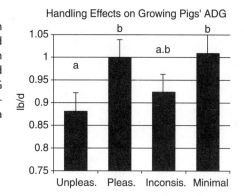
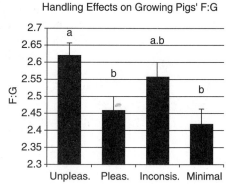

FIGURE 19–7

Effects of Different Levels of Human Interaction and Provision of Toys on Pig Time to Contact People in Their Home Pen. Pigs in the Isolated (neg) Treatment Took Much Longer to Contact People than Pigs in the Other Treatment Groups.

Source: Hill et al. (1998).

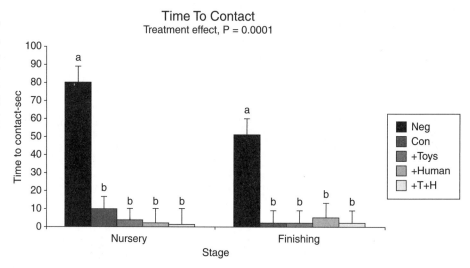

- Toys (same as control, plus two hanging chains and two hanging hoses)
- Human (positive human interactions, 2 min per pen per day, 5 d/wk)
- Toys + human (same as each of the above)

In this work, there was a large effect of handling on pig behavior during standardized tests. For example, the data for pigs that had little or no human contact showed clearly that they avoided humans (see Figure 19–7). They did not avoid people because they were abused, but rather because they were unfamiliar with people, especially at close range.

Hill et al. (1997) used two genotypes in their work. The Camborough-15 (C-15) is a common, crossbred maternal genotype. The Exp-94 contains 25% Meishan. The C-15 showed no performance boost or decline with different levels of enrichment (see Figure 19–8). The Exp-94 line did respond to enrichment in certain ways. The 25% Meishan genetic line increased its ADG and improved its F:G ratio with provision of toys. However, the body fat percentage increased with more enrichment for both

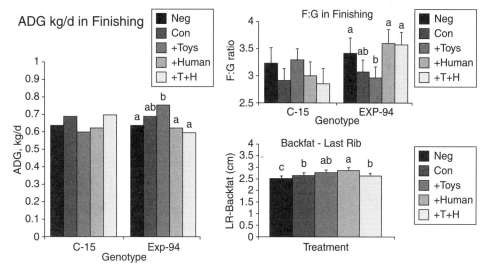

FIGURE 19–8
Effects of Different Levels of Enrichment from Toys or Humans on Pig Productivity (ADG, F:G ratio, and backfat thickness). The C-15 Genotype Did Not Change ADG or F:G Ratios with Enhanced Enrichments. The Exp-94 Genotype Had Improved ADG and F:G Ratios When Given Toys. Pigs Handled the Least (Neg) Had the Least Amount of Fat (last rib backfat thickness).

Source: Hill et al. (1998).

genotypes. In fact, provision of toys increased backfat by 12%. Packers would pay a premium for carcasses with lower backfat thickness and, therefore, producers would be discouraged from providing toys. Also, the "Neg" treatment, wherein human caregivers did not enter pens, would be the lowest labor input. Furthermore, these pigs were easy to handle in that they moved away from people in handling tests.

Pork producers must weigh animal welfare concerns against economic concerns (pig performance, body composition, and labor costs). The balance between pig welfare and economics is not yet clear, although there is general agreement that achievement of animal well-being is often associated with economic costs.

BEHAVIORAL PROBLEMS

TAIL BITING AND EAR CHEWING

The majority of growing pigs in pens and buildings have few behavioral problems. Most farms have an occasional behavioral problem that expresses itself as an acute outbreak varying in degree and with unknown cause. Other farms have chronic behavioral problems.

Chronic behavioral problems include, but are not limited to, tail biting, ear chewing, navel sucking, belly nosing (persistent inguinal thrusts), massage, excessive human-directed chewing/biting, hyperactivity, and hypersensitive syndrome.

The most pervasive form of behavioral problem is the tail-biting and ear-chewing syndrome. Van Putten (1969) suggested that tail biting occurs when pigs take the tail of another in their mouth sideways, while cannibalism involves pigs chasing a victim around the pen and chewing as they chase. Tail biting leads to cannibalism. Cannibalism occurs when an animal of one species eats part or all of the body of an animal of the same species. When pigs show excessive tail or ear chewing, they are clearly exhibiting cannibalism. Opinions differ as to when idle chewing turns to cannibalism. However, pigs with bloody stumps for tails or ears indicate that a tail biting-ear chewing episode is clearly cannibalistic. After healing from an extreme outbreak, these pigs will have no tails and no ears.

Pigs groom, rub, nibble, and chew on one another as forms of social and maintenance behaviors. Most oral-body interactions are around the head and the anal-genital regions. The head grooming focuses on the ears and eyes. The posterior interactions focus on the tail and anus (in prepubertal pigs). The natural chewing, rooting, and grooming sometimes progress to a more persistent chewing. When the skin is broken and blood is spilled, the chewing pig receives a reward and his/her chewing may increase.

As the syndrome develops, it moves from occasional chewing to persistent chewing. The chewed-upon pig seems to accept most of the chewing, with only an occasional scream of discomfort. Meanwhile, other pigs may join the chewing of the unfortunate victim. At this point, if the victim is not removed, it may bleed to death or it may incur such a massive wound that irreversible infection leads to death or the need for euthanasia. Sometimes the tail infection reaches the spinal cord or the victim suffers direct spinal cord injury. If the chewed-upon pig becomes paralyzed, the others will continue chewing on it and they will certainly kill it. In this case, intervention is needed.

Genetic lines vary in the basal rate of tail biting and ear chewing. In one field investigation at Texas Tech University (McGlone et al., 1992), there was a tail-biting incidence of 5.4% among Camborough gilts and 10.2% incidence among Camborough-15 gilts. Air-flow patterns did not influence the incidence of tail biting, confirming that poor air quality is not a primary cause. Lop-eared pigs (Landrace and Welch breeds) had a higher incidence of cannibalism than did erect-eared pigs (Yorkshire and Large White) (Penny and Hill, 1974).

When genetic lines predisposed cannibalism are uncomfortable, they will begin tail biting. When tail biting occurs, the pigs' tails or tail stumps (if docked) will be down, rather than in the usual up position. Stockpeople will notice the uncomfortable look among their pigs; this look includes a tail down or tucked between their legs.

Older reports from the United Kingdom showed that the incidence of tail biting was 10% to 12% among undocked pigs (Penny and Hill, 1974). Tail docking clearly lowers the incidence of tail biting. While over 40% of the pigs with intact tails may have tail-biting scars or wounds, only 2% of the docked pigs would have tail scars or wounds. In a recent UK survey, tail-docked pigs had a 21% incidence of tail biting. That the incidence of tail biting seems to be higher in the United Kingdom today than in the United States is speculative, but this may be the case because UK producers use more intact males, they meal feed (rather than ad libitum), and they use more wet feeding systems.

The incidence of tail biting is about twice as common among castrated males as females (Colyer, 1970; Chambers, 1995). The suggestion is that barrows are more hungry and when they have to wait to eat, they are more likely to chew on tails.

Concrete slats clearly increase the incidence of tail biting compared to pigs on bedding. Field studies show that the incidence of tail biting is several-fold higher when pigs are on concrete slats than when they are in straw-bedded pens. However, floor surface and bedding are confounded factors in this syndrome. Most solid-floored pens use bedding and, among finishing pigs, most pigs on slatted floors are on concrete slats. Thus, to be on concrete slats often means to be without bedding. The concrete slats increase the incidence of tail injury (inciting an apparent thirst for blood) and the lack of bedding means a preferred chewing object is not available.

Theorists report (Fritchen and Hogg, 1983; Colyer, 1970; Helms, 1961) causes of the tail-biting and ear-chewing syndromes are often multi-factorial, drawing from one or more categories of:

- Genetic predisposition
- Human management
- Pathogens
- Temperature, humidity, air pressure changes
- Dietary inadequacy
- Wet feeding
- Injury
- Facility design or operation
- Boredom

The entire area of pig behavioral problems suffers from a lack of solid scientific information and controlled studies. One series of studies (Fraser, 1987) showed that a complete deletion of minerals or salts from the diet increases chewing of model tails. However, few cases of cannibalism in the field can be linked to severe mineral deficiencies.

The few studies in the literature on cannibalism often contradict one another. Therefore, determining the cause, prevention, and treatment of behavioral problems becomes a matter of professional judgment. Of the putative causative factors, the following factors have been shown to *not* cause tail biting in controlled studies:

- Crowding (McGlone and Nicholson, 1992)
- Limited feeder space (McGlone and Nicholson, 1992)
- High atmospheric ammonia or CO_2 (Ewbank, 1973)
- High nutrient density (Ewbank, 1973)
- A lack of toys (giving hanging chains or hoses did not reduce tail biting; Hill et al., 1998)
- Lack of dietary magnesium (Krider et al., 1975)

Factors that predispose pigs to tail biting include:

- Viral or bacterial infections (ex., TGE, PRRS, hemolytic *Streptococcus*)
- Dirty pigs with feces covering their bodies for various reasons
- Concrete slotted floors/lack of bedding
- Meal or limit feeding and wet feeding
- A genetic predisposition

Some potential remedies to an outbreak of tail biting or ear chewing include:

- Provision of a small amount of chewing material, such as straw or earth in a handful on the floor or in a hanging basket
- Moving the biter pigs out of the pen and isolating the badly chewed pigs
- Addition of MgO (0.1% to 1%) to the diet

OTHER BEHAVIORAL PROBLEMS

Other behavioral problems that are less of an industry-wide concern than tail biting and ear chewing include pacing, rocking, rubbing, pawing, sham chewing, bar or fence biting, excessive drinking (polydipsia), self-mutilation, urine or prepuce sucking, belly nosing, anal massage, mounting, unresponsiveness, and its opposite, hysteria (Fraser and Broom, 1990). The suckling-related behaviors (urine sucking and belly nosing) are often symptoms observed in early-weaned pigs. These behaviors often pass as the pigs mature. Other abnormal behaviors are shaped by features of the environment; for example, frequent drinking develops if the water pressure is low. After the problem is resolved, polydipsia may have been created or shaped. If the equipment has sharp edges, pigs may seem to self-mutilate to obtain resources (especially feed and water). Removing the sharp edges should solve the problem.

On a given farm, some of these behaviors may be a significant problem. However, their incidence on most farms is insufficient to call them a behavioral problem.

MIXING, SIZING, AND SORTING

When pigs that have no experience with each other are first grouped (mixed), they fight to establish a dominance order. Pigs in group sizes up to about 20 individuals have a clear dominance hierarchy. Dominant pigs occupy preferred resting, feeding, and drinking places. Some pigs are more dominant over certain commodities within the pen (such as a given resting place). With about a 75% accuracy, pigs dominant for one resource are dominant for other resources.

Pigs will fight for a period ranging from hours to days after they are regrouped. Pigs in small, group-sized pens will establish a dominance order in 4 to 8 hr. Intermediate group sizes (8 to 30 pigs/pen) will establish a dominance order in about 48 hr. Very large groups may never fully establish a stable dominance order—they may fight a very small amount of time during most waking hours, or they will establish subterritories in the pen in which they maintain stable dominance orders.

The dominance order is linear in small group sizes with a clear, socially dominant, intermediate, and submissive pig. In larger group sizes, many pigs occupy the middle of the social order and participate many social triangles. In any group size above three pigs/pen, the dominance order is nonlinear.

When pigs of uniform body weight are grouped, some individuals become dominant. Even a small difference in body weight is likely to lead to the slightly heavier pig becoming dominant. The dominant pigs clearly gain more weight faster than do subdominant pigs. Figure 19–9 presents data from Hicks et al. (1997) that document

FIGURE 19-9

Effects of Pig Social Status on Body Weight Change After Weaning. Pigs of Each Social Status Had a Statistically Similar Body Weight at Weaning; However, the Dominant Pigs Were Heavier. Each Data Point Represents 44 Pigs.

Source: Adapted from Hicks et al. (1997).

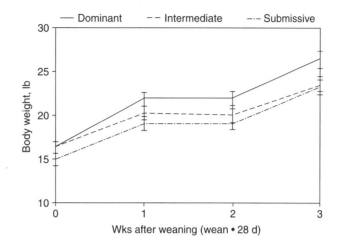

TABLE 19-1
The Costs of Mixing Growing Pigs.

WEIGHT AT MIXING	DURATION OF GAIN DEPRESSION	ADDED D TO MARKET	OTHER PROBLEMS
5–40 lb (0.07–18.1 kg)	Not measurable	0	Wounds
40–80 lb (18.1–36.3 kg)	7 d	0	Wounds
80–150 lb (36.3–68.0 kg)	28 d	0	Wounds and injury
Over 150 lb (over 68 kg)	28 d+	7+	Wounds and injury

Source: Adapted from Grandin et al. (1989).

the effect of dominance status on post-weaning weight gain. This relationship between weight (or weight gain) and social status is difficult to quantify in medium to large group sizes because the dominance hierarchy is not totally linear. However, in any group size, the social status of pigs has a powerful influence on pig growth.

The effects of mixing and sorting pigs on pig performance are well documented. During the initial stages when pigs are fighting, their feed intake and weight gains are reduced compared to those of pigs of the same weight and age who were not mixed. In the period immediately after weaning, the pigs are not eating much feed anyway and, in fact, the effect of weaning on weight gain is much greater than is the effect of mixing and fighting.

The consequence of mixing increases as pigs get larger. Larger pigs have a greater slowdown in weight gain due to mixing than do smaller pigs. The "costs" of mixing are summarized in Table 19–1. Pigs should not be regrouped after 125 to 150 lb body weight—this will cause a significant growth check. At all times, mixing pigs causes injuries and wounds that can become infected. Mixing pigs beyond weaning should be avoided when possible.

Methods of reducing fighting among growing pigs have been studied for the past few decades. Two solutions to fighting have been published. The first involves use of hiding areas for pigs—areas where pigs can place their head and ears. With their heads hidden, other pigs will not attack them (McGlone and Curtis, 1985).

The odor of a boar (5-α-androsten-16-en-3-one) in very low concentrations will reduce fighting among growing pigs (McGlone et al., 1986). However, this compound is not approved for use with commercial pigs and the economic return to develop this market is not great enough to cause development of the product. Other remedies such as masking odors are not effective in reducing fighting.

In some countries, tranquilizers are given to pigs to reduce fighting or to reduce the stress of transport. Mind-altering drugs provide only a temporary solution to the problem of fighting. When the pigs recover, they may resume fighting and, in the end, the same total amount of aggressive behavior may occur.

TAIL ENDERS

Due in large part to the social hierarchy, pigs will establish a certain amount of variation in body weight. Even in pigs that are tightly grouped by size, the social order will cause a spreading of weights in the group. For this reason, sorting pigs by size is recommended only to group the very large and very small pigs together. Mixing pigs is only recommended at weaning. After that, further mixing does not have an economic advantage.

Regardless of the amount of mixing of pigs, those of a common age will vary in weight at the same number of days of growth. At 160 d of age, pigs may weigh more than 70 lb. The normal way to send pigs to market is to send them over a 3- or 5-wk period. Then the last group of pigs, called the tail enders, will be variable. These pigs should either be sold for a reduced price for slaughter or they should be moved to another site to add weight. The entire building's biosecurity should not be compromised to allow these few pigs to grow. The tail enders must be dealt with in a manner that utilizes expensive building space and does not compromise biosecurity (finishing buildings should be managed all-in-all-out, if possible).

HANDLING PIGS

Some general principles should be understood when handling pigs during weighing, sorting, or moving:

- Pigs have good memories (they remember good and bad experiences).
- Pigs will follow other pigs.
- Pigs will explore as they go (they will explore unique lighting, smells, surfaces, sounds, and other animals).
- Restraint is very stressful to pigs of all ages.
- Pigs respond to handling or fear by vocalizing and attempting to escape.
- Response: vocalize and escape.
- The reason for restraint has to outweigh the stress reaction.
- There are strong correlates between human attitude and pig productivity.
- A positive attitude toward pigs is preferred.
- Producers can be too positive and this may hurt the pigs and the bottom line.

- Training improves worker attitude and herd performance.
- Touching is good—make sure each experience is positive.
- The stockperson's behavior translates into the pigs' reproductive performance and growth performance.
- Some genotypes are more or less fearful of humans.

When designing loading and moving areas, producers need to consider that pigs have features of the environment that make them comfortable and other features that make them uncomfortable. Pigs move easier if the floor is a solid color with no drains, discolored areas, or other variations. A uniformly lit walkway is also needed for smooth animal flow. Data from a recent study of floor color effects on pig movement are given in Table 19–2. A solid, white flooring caused the easiest, quickest movement onto and over the floor pattern. Any variation in floor pattern slowed the pigs. These data add confidence in the suggestion that floors should be solid-colored and uniform in appearance.

The texture and feel of the flooring to the pigs' feet should be comfortable—a compromise between smooth and nonslip. A slightly abrasive floor is best for sure footing, which allows pigs to walk easier and with more confidence, and reduces the chance of injury.

In addition to the color, lighting, and texture needs, the area into which pigs are moved should be uniform in air currents. Pigs will balk (stop moving or even turn) if they receive a cold air draft while they are moving forward. This represents a special challenge when pigs are going out of a building and into a truck. The doorway may have bright light from the sun and a breeze from the open door. This combination will cause pigs to balk.

DEVICES TO HELP MOVE PIGS

The caregiver has several choices for devices to help move pigs. The handlers' natural (but not the best) method is to clap their hands, whistle, shout, and wave their arms. This technique can do in a pinch, but should not be the usual method used to move pigs.

The traditional sorting board (hurdle) is a solid piece of wood, plastic, or aluminum that has a handle (see Figure 19–10). The board is placed in front of the worker

TABLE 19–2
Time-Naïve Finishing Pigs Required to Step on the Floors of Different Patterns and the Total Time to Walk, Unassisted, Over 4 ft of Floor Length.

FLOOR STYLE	TIME TO ENTER, S	TOTAL TIME TO MOVE 4 FT, S
Solid white	3.8	16.7
Solid black	3.5	28.3
Longitudinal lines	10.9	52.4
Horizontal lines	22.3	58.1
Chevron forward (>)	10.6	42.2
Chevron back (<)	16.6	48.5

Source: Adapted from data from Texas Tech University (Song and McGlone, 1997, unpublished).

FIGURE 19–10
A Typical Sorting Board with Handles. The Length and Width will Vary with the Size of Aisles in the Facility. Some Sorting Boards Have Curved Sides and Some Are Flexible in Length. The Most Important Feature Is That the Board Be Solid and Not Allow the Pig to Escape the Desired Flow Pattern.

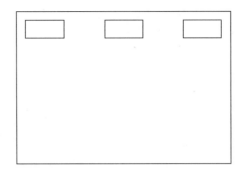

FIGURE 19–11
Paddles Are a Preferred Tool for Moving Pigs.

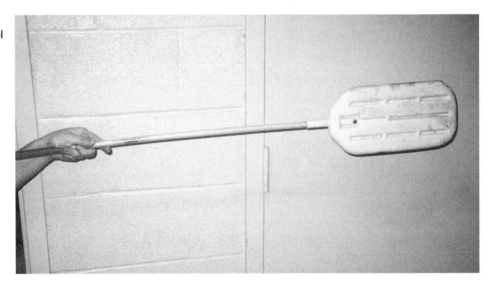

and if a pig turns, the board is placed squarely on the ground, giving the pig the impression that a solid wall is behind it. Pigs will more likely move forward if they are convinced a solid wall is behind them.

Other devices that can be used to move pigs are slappers, paddles, and electric prods (see Figure 19–11). Electric prods are not recommended for pigs of any size. However, an electric prod may be needed when pigs must be moved quickly for their own safety. Farms and processing plants should be designed so electric prods are not needed.

TRANSPORTING PIGS

Pigs need to be transported at ages varying from early weanlings to adults. Care must be taken when transporting pigs to avoid injuries and illness. Shipping is one of the most powerful stressors to a pig and, at times, shipping cannot be avoided. Everything must be done to make the shipping experience as nonstressful as possible.

FIGURE 19–12
Chart Showing When Pigs Are at Risk of Death or Discomfort Due to Excessive Thermal Load (heat stress).

Source: Adapted from the Midwest Plan Service (MWPS-1) and the Livestock Conservation Institute.

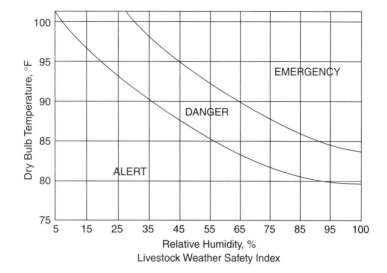

Trucks should be bedded when pigs are transported. Bedding should be sand or, in the summer, a wetable material. Straw or sawdust is a good bedding in the winter. Transporting pigs in cold weather is usually not a problem, except for very young pigs that must be kept warm. Transporting pigs during very warm weather is a hazard (see Figure 19–12). Pigs should not be transported in the emergency phase of the Livestock Weather Safety Index. Pigs can be moved at night and in the morning hours to avoid very warm temperatures.

SUMMARY

This chapter focuses on the care and management of pigs during the period of growth from weaning at a few weeks of age to slaughter at 5 or 6 mo of age. Pigs exhibit normal behaviors related to maintenance (eating, drinking, lying), excretory activities, chewing/rooting/ playing, and sickness. Daily pig management, in terms of providing appropriate care and husbandry to satisfy the pigs' normal behavioral patterns and animal well-being, was discussed. Producers must be able to recognize abnormal behavior patterns in pigs and address the causes of such behaviors. The chapter reviewed the general principles of handling pigs and appropriate design and layout of pens, floors, and transport facilities and pig movement devices.

QUESTIONS AND ACTIVITIES

1. Growing pigs react differently if they have been handled extensively in a positive, negative, or neutral manner. A fear test is performed to assess pig "fear" of humans. This is done most easily to sows in gestation crates. The observer places his or her

hand in the crate while the sow is standing, but not eating. The observer starts a stop watch. The sow steps back. The observer measures the time it takes for the sow to return within a short distance ($\sim$ 1 in) of his or her hand. Sows with a large amount of fear of people will take longer to return. Using available pigs at your farm, perform the fear test on 10 sows. Calculate the mean and standard deviation of the response time.

2. Find a price grid from a pork packer near you. Weigh a group of late finishing pigs (on two occasions, if possible, 7 d apart). Estimate when the average pig will reach the market weight. How many pigs will fall in the price range that pays the best? Develop a spreadsheet that includes a discount or premium based on the variation of the group of pigs. How many batches of pigs should be marketed (e.g., in a single batch, or over 1 to 4 wk) to optimize dollar income?

3. The flow of pigs through a chute can be very pleasant or it can be very stressful for the pigs and the workers. What is the rate of pig movement in pigs/min if you wish to achieve 1,000, 500, 250, or 100 pigs/hr? What can be done to improve pig flow assuming the physical facility is well-designed?

4. In a study by Sarignac et al. (1997), litters were managed either as individual litters or in a communal manner (piglets could interact) during lactation. After weaning, litters were mixed and the following average number of fights were observed:

Isolated-Indoors	Social-Indoors	Isolated-Outdoors	Social-Outdoors
5.0	0.0	0.0	0.0

Explain these results. What "normal" agonistic behaviors should weanling pigs exhibit?

LITERATURE CITED

Chambers, C., L. Powell, E. Wilson, and L. E. Green. 1995. A postal survey of tail biting in pigs in Southwest England. Vet. Rec. 136:147–148.

Coleman, G. J., Hay, M., Hemsworth, P. H., and Cox, M. 2000. Modifying stockperson attitudes and behaviour towards pigs at a large commercial farm. Appl. Anim. Behav. Sci. 66:11–20.

Colyer, R. J. 1970. Tail biting in pigs. Agriculture 77:215–218.

English, P., G. Burgess, R. Segundo, and J. Dunne. 1992. Stockmanship: Improving the care of the pig and other livestock. Farming Press, Ipswich, UK.

Ewbank, R. 1973. Abnormal behaviour and pig nutrition. An unsuccessful attempt to induce tail biting by feeding a high energy, low fibre vegetable protein ration. Br. Vet. J. 129:366–369.

Fraser, A. F. and D. M. Broom. 1990. Farm animal behaviour and welfare. Bailliere Tindall, London.

Fraser, D. 1987. Mineral-deficient diets and the pig's attraction to blood: Implications for tail-biting. Can. J. Anim. Sci. 67:909–918.

Fritschen, R. and A. Hogg. 1983, January. Preventing tail biting in swine (anti-comfort syndrome). NebGuide.

Grandin, T., K. Ernst, D. Ernst, and J. McGlone. 1989. Handling Hogs. Pork Industry Handbook. PIH-116.

Harris, 1888. On the Pig. Guilford: Globe Pequot Press. Guilford, CT.

Helms, H. T. 1961, May 15. Tail-biting in swine: The ecologic approach. Modern Vet. Practice: 56–60.

Hemsworth, P. H., J. L. Barnett, and C. Hansen. 1987. The influence of inconsistent handling by humans on the behaviour, growth and corticosteroids of young pigs. Appl. Anim. Behav. Sci. 17:245–252.

Hicks, T. A., J. J. McGlone, C. S. Whisnant, H. G. Kattesh, and R. L. Norman. 1998. Behavioral, endocrine, immune and performance measures for pigs exposed to acute stress. J. Anim. Sci. 76:474–483.

Hill, J. D., J. J. McGlone, S. D. Fullwood, and M. F. Miller. 1998. Environmental enrichment influences on pig behavior, performance and meat quality. Appl. Anim. Behav. Sci. 57:51–68.

Krider, J. L., J. L. Albright, M. P. Plumlee, J. H. Conrad, C. L. Sinclair, L. Underwood, R. G. Jones, and B. G. Harrington. 1975. Magnesium supplementation, space and docking effects on swine performance and behavior. J. Anim. Sci. 40:1027–1033.

McGlone, J. J. and S. E. Curtis. 1985. Behavior and performance of pigs in pens equipped with hide areas. J. Anim. Sci. 60:20–24.

McGlone, J. J., W. F. Stansbury, and L. F. Tribble. 1986. Aerosolized 5-α-androst-16-en-3-one reduced agonistic behavior and temporarily improved performance of growing pigs. J. Anim. Sci. 63:679–684.

McGlone, J. J. and R. I. Nicholson. 1992. Effects of limited floor and feeder space on pig performance and tail biting. Texas Tech University Animal Science Research Report T-5-317.

McGlone, J. J., W. Vermette, and G. Larson. 1992. Field investigations of tail biting episodes on a pork production unit. Texas Tech University Animal Science Research Report T-5-317.

Penny, R. H. C. and F. W. G. Hill. 1974, March 2. Observations of some conditions in pigs at the abattoir with particular reference to tail biting. Vet. Rec. 94:174–180.

Sarignac, C., J. P. Signoret, and J. J. McGlone. 1997. Relation mere-jeune, comportement et performances en fonction du systeme de logement et de l'environnement social [Sow and piglet performance and behavior in either intensive outdoor or indoor units with litters managed as individuals or as small social groups]. Journees Rech. Porcine en France. 29:123–128.

Van Putten, G. 1969. An investigation into tail-biting among fattening pigs. Br. Vet. J. 125:511–517.

INTERNET RESOURCES

Growing pig handling sites:
http://www.gov.mb.ca/agriculture/livestock/pork/swine/bab10s06.html
http://www.grandin.com/
http://www.pighandling.com/
http://www.gov.mb.ca/agriculture/livesock/pork/swine/bab02s23.html

20
MANAGEMENT OF PIG HEALTH

INTRODUCTION

Profitable pork production in the commercial pig unit requires the development and maintenance of a herd health program based on the prevention and control of infectious and noninfectious swine diseases. Team effort is required between the producer and technical experts, including veterinarians, agricultural engineers, and swine husbandry specialists. The past emphasis on "putting out fires" has been replaced by the current emphasis on disease prevention. Veterinarians increasingly serve their clients primarily by working with them to develop herd health programs centered on sanitation, restricted human and animal traffic onto the farm premises, tailored vaccination and immunization programs, and routine monitoring of overall herd health and production. Early diagnosis, isolation, and treatment of sick animals is critical to the success of individualized herd health programs. The producer, therefore, is a major player in the management of pig health.

This chapter discusses the most common and important infectious diseases, noninfectious toxicants, and metabolic and nutritional disorders that affect swine. Some of the infectious diseases are of special significance because the pathogenic organisms can be spread to humans. Such diseases are termed zoonotic diseases or zoonoses. The chapter provides an introduction to the nature and clinical signs of swine diseases and disorders. No attempt is made to detail the etiology, diagnosis, or treatment of specific infectious diseases. For a more comprehensive discussion, see the benchmark text, Diseases of Swine (edited by Straw et al., 1999).

APPROACHES TO ENSURING PIG HEALTH

Pork producers on a given farm have unique health concerns. The pathogens present, and their virulence, are influenced by the microenvironment, the level of sanitation, the stockmanship, and the flow of people and animals through the facility. When farms are started and facilities are new, the only microorganisms on the farm that are pathogenic to the pigs are those that are brought in on the pigs themselves. After that, pathogenic organisms can be brought in by newly introduced pigs, people, equipment, feed, pets, and wildlife.

To keep pathogenic organisms under control, an effective program of biosecurity, sanitation, and vaccination should be in place (Biehl et al., 1997). The particular organisms that the farm vaccinates against will depend on the pathogens present. It is unwise to vaccinate against microorganisms not found on the farm, unless the farm's pigs are moved to a site that has those microorganisms.

BIOSECURITY

To prevent introduction of new pathogens, an effective biosecurity program should be in place. Biosecurity refers to the barriers established to separate the pigs from the outside world. Some barriers are real and some are not. If two neighbors have pigs that are only 100 m (300 ft) apart, it is virtually impossible in the long run to keep pathogenic organisms from passing between the farms unless extensive and costly measures are in place.

The first rule of biosecurity is to put as much distance as possible between two groups of pigs. Authors argue about how much distance is enough to prevent spread of disease, but the greater the distance, the greater the barrier. Biosecurity protection is proportional to the square root of the distance to the nearest pig of a different origin.

The second rule of biosecurity is to isolate, test, and acclimate incoming pigs. Incoming pigs should be isolated at a great distance (1 mi or more would be better than shorter distances) from the resident pigs. While the new pigs are isolated, they should not have the same caretakers as the main herd to prevent early potential pathogen spread. Isolated pigs should be tested about 30 d (or 15 to 45 d) after arrival to ensure they do not have new pathogens. The exact testing to be conducted depends on what microorganisms the farm wishes to exclude, local regulatory requirements (e.g., testing for pseudorabies or brucellosis), and upon the advice of the attending veterinarian. It is a good idea to put a young pig from the resident herd in with the new animals to acclimate the new pigs to the microorganisms on the farm.

The third rule of biosecurity is to control traffic flow of people, feed, and equipment through the facility and near the pigs. People flow through the facility is often controlled by a shower-in practice that requires visitors and workers to remove their street clothes, shower, and wear clothes that are kept on the farm. This practice requires on-farm clothes-washing facilities. Facilities without walkthrough showers should at least require that visitors wear boots or footwear that remains on the farm.

Visitors should avoid contact with pigs for a period of 48 hr or more before their visit. Supplies that enter the farm should be sanitized. Trucks that enter the farm should follow a route not used by the farm workers and, if possible, the wheels of the trucks should be sanitized.

SANITATION AND PIG FLOW

Pigs naturally shed microorganisms; some are pathogens and some are not. In some cases, animal stress leads to increased shedding of microorganisms and, thus, a normally healthy herd can begin to shed pathogenic microorganisms. To minimize the effective dose of bacteria, viruses, and parasites, an effective sanitation program should be in place.

Many newer farms use an all-in-all-out feature that helps to control pathogen spread. If a batch of pigs remains together in the same air space, and if the pigs are healthy to start with, they will probably remain healthy as long as the biosecurity program is not violated (there are rare exceptions). Typically, growing-finishing, nurseries, and farrowing barns are operated on an all-in-all-out basis. Sow breeding and gestation units are commonly in a mixed-batch, large air space. Thus, if the sows due to farrow next week are infected today by older sows, they will be sick and their new piglets may be in peril. Thus, all-in-all-out systems start at birth, often with mixing of the breeding herd. Because the breeding herd is often not managed as an all-in-all-out system, the breeding herd can incubate pathogens.

Effective sanitation (Becker et al., 1990) is performed between batches of pigs in an all-in-all-out system. Marginal sanitation controls can be used while the pigs are present, but clearly these measures will not be effective if the pigs are in the facility and able to re-infect the site immediately. Marginal, but necessary, sanitation measures include sweeping the aisle; timely removal of dead pigs and manure; cleaning; dusting cobwebs; and general housekeeping.

Effective sanitation begins with removal of all organic matter. This is best accomplished with a detergent, warm-water pressure sprayer. Pressure sprayers can be 200 to 1,000 psi, and while greater pressure removes organic matter more easily, it can chip paint and damage equipment. Steam-cleaning is even more effective than using just warm water. Sanitizing kills a large portion (99%) of the microorganisms. Sterilizing kills 100% of the organisms. Sterilization is difficult, but possible, if pressure washing and disinfecting are followed with a gas sterilizing agent. Sanitizing is performed by power washing, drying, disinfecting, and allowing the surfaces to dry.

Among outdoor units, the sun is a powerful sanitizing agent. Farrowing huts should be turned over between uses (flipped) to allow the sun to sanitize the inside of the hut. Pastures should be rotated to allow the sun to impact soil-borne microorganisms. The sun is not very effective against parasites that are in bedding or below the surface of the soil.

Sanitation with agents other than steam heat or sunlight can cause development of resistant microorganisms, which may reduce effectiveness of the agents and lead to food safety concerns. For these reasons, the sanitizing agents should be rotated periodically (at least every 6 m). Table 20–1 provides a list of sanitizing agents, adapted from the Pork Industry Handbook (Meyerholz and Gaskin, 1981). Consult vendors for up-to-date information—new agents are continually developed.

TABLE 20–1
Selected, Common Disinfectants and Their Use in Commercial Swine Operations.

	CHLORTEXIDINE	**FORMALDEHYDE AND OTHER ALDEHYDES**	**CHLORINE HYPOCHLORITES CHLORAMINES**	**IODOPHORS**	**SODIUM HYDROXIDE**	**QUATERNARY AMMONIUM COMPOUNDS**	**CRESOLS PHENOLS**
Spectrum of Activity							
Gram + bacteria	S.A. not pyogenic cocci	Yes	Yes	Yes	Yes	Yes	Yes
Gram − bacteria	S.A., not pseudomonads	Yes	Yes	Yes	Yes	S.A.	Yes
Tuberculosis bacill	S.A.	Yes	S.A.	S.A.	S.A.	No	S.A.
Bacterial spores	S.A. at 1% concentration	Yes	S.A.	S.A.	Yes (5–10% solution)	No	No
Fungi	S.A.	Yes	Yes	Yes	Yes	S.A.	S.A.
Viruses	S.A. not parvovirus	Yes	S.A.	S.A.	Yes	S.A.	S.A.
Special Properties							
Resistance to organic debris	Good	Good	Very poor	Poor to fair	Good	Fair	Excellent
Effect of hard water	None	None	None[2]	None[2]	None	[3]	None
Detrimental effect of heat	No	[4]	[5]	[5]	No	No	No
Residual activity	Yes	[6]	[7]	Yes	Yes	No	Yes
Most effective pH range	Alkaline	Not affected by pH	Acid	Acid	Alkaline	Alkaline	Acid
Compatibility with anionic surfactants (soaps)	Yes	Yes	Yes	Yes	Yes	No	Yes
Compatibility with nonionic surfactants	Yes	Yes	Yes	Yes	Yes	Yes	No
Disadvantages	Reduced activity against certain organisms	Irritating fumes[8]	Inactivation by organic debris	Inactivation by organic debris	Caustic	Incompatible w/soaps— limited spectrum	[9]
Commonly Used Concentrations							
Disinfecting solution	1%	2–8%	Hypochlorites 3–5%[10,11]	50–75 ppm	2–10%	400–800 ppm	Variable
Sanitizing solutions	0.5%	1–2%	Hypochlorites 2–3%[11]	12–25 ppm		200 ppm	
Appropriate Uses E - Equipment CE - Clean Equipment P - Premises F - Foot baths	E,P,F	E,P,F	CE	CE	P	CE	E,P,F
Common Brands and Names[12]	Nolvasan[3]	Cidex® DC & R® Formaldegen® Formatin	Chloramine-T® Chlorox® Halazone®	Betadine® Iofec® Isodyne® Losan® Tamed Iodine® Weladol®	Lye	Germex® Hi-Lethol® San-O-Fec® Warden® Zephiran®	Cresl-400® Environ® Laro® Lysol® Orthophen- ylphenol Sodium or- thophenyl- phenate

[1] S.A. - Some Activity.
[2] Unless hard water is alkaline.
[3] Reduces speed of kill.
[4] Formaldehyde gas works best at 80–140°F.
[5] Use at less than 110°F, active principal driven off by heat.
[6] No, except slow-release formulas.
[7] Hypochlorites: No, chloramines: Yes.
[8] Glutaraldehyde is less irritating and is superior to formaldehyde as a germicide.
[9] Strong odor with coal and wood tar distillates.
[10] 3.3% Chlorox inactivates parvovirus on clean surfaces.
[11] Chloramines variable.
[12] Products listed are intended as examples, not endorsement; many suitable products are not listed.

Source: Adapted from Meyerholz and Gaskin (1981).

HERD HEALTH PROGRAM

THE OVERALL HERD HEALTH PROGRAM

Each commercial pork production unit should have an overall herd health program (Biehl et al., 1997). A herd health program is typically designed by an attending veterinarian with training and experience with pigs, who reviews the health status of the herd and recommends methods to treat current diseases and methods to prevent introduction of new diseases. The attending veterinarian may be in residence on large farms or may visit as infrequently as once/yr. Even if on-site visits by a veterinarian are infrequent, producers should conduct their disease surveillance efforts and biosecurity program in conjunction with their veterinarian. A valid and open producer-veterinarian relationship will benefit the pigs.

SLAUGHTER CHECKS AND OTHER MEANS OF DISEASE SURVEILLANCE

A method of disease surveillance should be in place (Smith et al., 1990). The veterinarian and the producer should discuss the particular diseases to be monitored. Over-testing is costly, but under-testing can be deadly. The common methods of disease surveillance include:

- Necropsy of ill (euthanized) or dead pigs on the farm and (or) at a diagnostic laboratory
- Blood collection for serology or organism isolation
- Fecal collection for internal parasite evaluation
- Urine collections for metabolic or infectious problems
- Skin scrapings for determination of external parasites

A common method of disease surveillance is to perform a routine slaughter check (Meyer et al., 1990). Some countries perform slaughter checks routinely. Pointon et al. (1999) recommend quarterly slaughter checks to account for seasonal effects. While "normal" pigs are being processed at a plant, the following features can be evaluated:

- Lung lesions, pneumonia, and pleurisy
- Liver lesions due to ascarids (roundworms)
- Snout turbinate damage (atrophic rhinitis)
- Illeitis
- Mange or lice on the skin
- Kidney problems such as nephritis (perhaps leptospirosis)

The sample size needed to estimate the population of parasites depends on the incidence of the disease in the herd. If the disease incidence is low, more animals must be sampled. From 10% to 45% of the animals must be sampled at slaughter to get a representative sample. The person doing the sampling can use two scales: present or absent and degree of severity. For most conditions in a healthy herd, the owner simply wants to know if the problem is present or absent. Owners who are attempting to clean a herd of a problem should use a severity score, which better tracks how well the problem is being resolved. If the producer has little hope of resolving an endemic disease, a severity score is important so that control measures can be assessed. For these herds, the severity of disease problems at slaughter are more a reflection of management practices put in place to control the disease.

EUTHANASIA

Euthanasia refers to humane methods of causing death (true or good death). Pigs that are severely ill or in significant or irresolvable pain should be euthanized. Timely euthanasia prevents the spread of disease because pigs that are ill from an infectious disease are likely to spread the disease organisms at a rapid pace.

Several methods of euthanasia are available, but the key result, regardless of the method, is a quick and painless 'death'. The American Veterinary Medical Association (AVMA) endorses a number of acceptable methods of euthanasia of pigs. Some methods that are acceptable to the AVMA are not practical on the farm (i.e., injection with lethal controlled drugs). For each method of euthanasia, the stockpeople should be trained until they reach an acceptable skill level. The acceptable and practical methods of euthanasia on the farm include:

- Any method of humane slaughter used at a processing plant; stunning, and exsanguination
- Penetrating captive bolt (while highly lethal, this method can cause personal injury)
- Lethal gunshot
- Blunt trauma (this method was allowed by the National Pork Board and the American Association of Swine Veterinarians in their recommended methods of euthanasia on the farm)
- CO_2 asphyxiation

VACCINATIONS

Vaccinations, when available, should be used to control pathogenic microorganisms. Vaccinations can be given by injection (intramuscular [i.m.], subcutaneous [s.c.], or other means), orally, intranasally, or to the sow, which passes the antibodies to the piglets through the milk. Once the target disease is understood, the route of administration may be obvious. However, many vaccinations are given systemically for enteric diseases because the stomach and small intestine may lower the effectiveness of vaccines.

Some vaccinations are less effective than others. Vaccines can lose their effectiveness if stored or handled improperly. Producers should follow the guidelines on the product's label or seek advice from their veterinarian. In addition, the particular strain of microorganism on a given farm may be different enough from the vaccine strain to offer less-than-optimum protection. If pigs are stressed around the time of vaccination, they may be less able to build protective antibodies. If pigs are crowded in a non-sanitary pen and stressed, the microorganism challenge may overwhelm the pig's immune system.

ANTIMICROBIALS

Antimicrobial products are used in pigs to improve health and performance. The antimicrobials used most often are the antibiotics. Other antimicrobial agents include high levels of dietary copper or zinc and various "natural" immunomodulators including vitamins, some minerals, yeasts, bacterial cultures, and certain plant materials.

Antibiotics are given for two purposes. Subtherapeutic levels of antibiotics are often placed in the feed as a method of preventing disease and (or) stimulating pig productivity (weight gain and feed efficiency). Therapeutic levels of antibiotics are given to treat a particular disease. Regardless of the use of the antibiotic, withdrawal times must be followed. Each antibiotic has an FDA-approved number of days that the pigs must not be given the drug before slaughter (see Table 20–2). Some antibiotics have a zero-day withdrawal time, which means feeding may continue right up to slaughter.

Feeding antibiotics to pigs, particularly subtherapeutic antibiotics, is a source of controversy. The administration of antibiotics to animals and humans leads to an increase in the numbers of antibiotic-resistant strains of microorganisms. When resistant strains of a pathogen develop, antibiotics are not effective in treating human and animal diseases that formerly were treatable with a given antibiotic. Some scientists and physicians claim that the development of resistant strains of pathogens as a result of long-term feeding of antibiotics to animals creates a public health problem. The well-documented problem of resistant strains of pathogens in hospitals has led to a practice of frequently changing therapeutically administered antibiotics. It is still unclear whether the use of subtherapeutic levels of antibiotics in swine production contributes to the resistance problem in humans.

Clearly, the best option is to avoid the use of antimicrobials when the pig herd is extremely healthy. Healthy pigs do not need therapeutic antimicrobials. Subtherapeutic antibiotics are less effective at stimulating performance in very healthy pigs housed in a sanitary environment. If vaccines are available to control a particular disease, vaccination is the preferred method to prevent the disease. Still, in some cases, withholding antimicrobials is not in the best interest of sick pigs.

GIVING INJECTIONS

Handlers should perform needle injections carefully and only after receiving training. Table 20–3 shows needle sizes (length and gauge) commonly used for intramuscular and subcutaneous injections. Producers should select the smallest needle that can get the job done, read the instructions that come with the medications, and follow the recommended practices.

Injections should be administered (a) where the least discomfort will result, (b) in the neck or shoulder, but not in the ham, and (c) in the route recommended (under the skin, s.c., or in the muscle, i.m.) (see Figure 20–1).

TABLE 20-2
Withdrawal Periods for Antibiotics Used in Commercial Pork Production in the United States. Consult the FDA for Updates to This Information.

Drugs for Labeled Use	Days
Parenteral unit dosage form drugs	
Ceftiofur	0
Erythromycin	2
Gentamicin (neonatal pigs)	40
Lincomycin	2
Oxytetracycline (varies by label)	18–26
Procaine penicillin G	7
Tylosin	14
Oral unit dosage form drugs, water soluble	
Ampicillin trihydrate	1
Bacitracin methylene disalicylate	0
Chlortetracycline	5
Chlortetracycline and sulfamethazine	15
Levamisole	3
Lincomycin	6
Neomycin	20
Spectinomycin, oral pump suspension	21
Spectinomycin, water soluble	5
Sulfachlorpyridazine	4
Tetracycline hydrochloride	4
Thiabendazole	20
Tiamulin (see labeling)	3–7
Tylosin	2
In-feed dosage forms (maximum, as times may vary with dose)	
Apramycin	28
Carbadox	70
Chlortetracycline	0
Chlortetracycline, procaine penicillin, and sulfamethazine	15
Chlortetracycline, procaine penicillin, and sulfathiazole	7
Hygromycin B	15
Levamisole	3
Lincomycin	0–6
Neomycin and oxytetracycline	10
Penicillin and streptomycin	0
Tetracycline	5
Tiamulin	0–2
Tilmicosin	7
Tylosin	0
Tylosin and sulfamethazine	15
Virginiamycin	0

Source: Adapted from Henry and Apley (1999).

TABLE 20-3
Needle Sizes for Intramuscular and Subcutaneous Injections.

Intramuscular Injection		
	Gauge	Length
Baby pigs	18 or 20	1/2 in. or 5/8 in.
Nursery	16 or 18	5/8 in. or 3/4 in.
Finisher	16	1 in.
Breeding stock	14 or 16	1 in. or 1 1/2 in.

Subcutaneous Injection		
	Gauge	Length
Nursery pigs	16 or 18	1/2 in.
Finisher	16	3/4 in.
Breeding stock	14 or 16	1 in.

20x1/2" 18x5/8" 16x3/4" 16x1" 16x1 1/2" 14x1" 14x1 1/2"
(actual size)

Source: Adapted from NPPC.

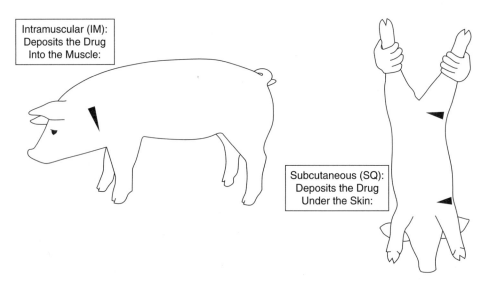

FIGURE 20-1
Injection Sites for Pigs.

Source: Adapted from NPPC (1999).

INFECTIOUS AND NONINFECTIOUS DISEASES

Swine diseases can be classified in many ways, but for the purposes of this chapter, these diseases are grouped into three categories:

- Highly infectious diseases, not presently found in the United States, that would cause significant national and international concern if they were introduced; these diseases include African swine fever, hog cholera, and foot-and-mouth disease

- Highly infectious diseases under federal oversight with an eye toward eradication; these diseases include brucellosis and pseudorabies
- All other diseases found in the United States, including respiratory, enteric, reproductive, metabolic, and noninfectious diseases; these diseases are found on many farms and, in many cases, pork producers accept their presence and live with the conditions in their pigs

Infectious diseases can be classed according to the type of pathogenic agent (bacteria, virus, mycoplasm, and mycotic organisms). Table 20–4 lists the major infectious diseases of swine according to the three categories and the agents responsible for each. Table 20–5 lists other rare infectious diseases of pigs (not discussed here; detailed information on each is available in Straw et al., 1999). Some infectious diseases of swine are also zoonoses—diseases that can infect people as well as pigs. Because of the potential for zoonoses, workers should wash their hands regularly and practice other means of personal hygiene to minimize exposure to potential zoonotic agents.

HIGHLY INFECTIOUS DISEASES NOT FOUND IN THE UNITED STATES

African Swine Fever

African swine fever (ASF) is caused by a DNA virus that is the sole member of a separate viral family. ASF is not found in the United States, Europe, or Asia. ASF virus replicates primarily in the pig's macrophages (but also in other cell types in other organs and tissues). First described in 1921, ASF also infects other nondomestic pigs such as the warthog. Ticks are involved in spread of the virus.

ASF-infected pigs show clinical signs such as internal organ and skin hemorrhage, a very high fever, and loss of appetite. The condition can be confused with hog cholera or erysipelas.

Hog Cholera or Classical Swine Fever

Hog cholera, also called classical swine fever, was eradicated in the United States in 1976, following a 14-yr eradication program. Parts of Europe, most notably The Netherlands (1997), have had recent outbreaks of classical swine fever.

Classical swine fever is a highly contagious RNA virus, sharing the viral genus *Pestivirus* with bovine viral diarrhea. Symptoms include a very high fever of 41° to 42°C (106°F) and respiratory and enteric symptoms. The pig's eyes may be crusted shut.

Foot-and-Mouth Disease

Foot-and-mouth disease (FMD) is caused by an RNA virus. FMD was last reported in the United States in 1929. The same picornavirus that causes FMD in pigs also infects ruminants like cattle and buffalo. Pigs shed FMD virus at a greater rate than do cattle.

TABLE 20–4
Major Infectious Diseases of Swine.

Disease	Agent
Bacterial	
Brucellosis	*Brucella suis, B. abortus, B. melitensis*
Bordetellosis (Atrophic rhinitis)	*Bordetella bronchiseptica*
Colibacillosis	*Escherichia coli*
Edema disease (ED) (a form of enterotoxemic colibacillosis)	*Escherichia coli* (hemolytic)
Erysipelas	
Haemophilus infections	*Erysipelothrix rhusiopathiae*
Polyserositis and arthritis (Glasser's disease)	
Pleuropneumonia	*Haemophilus parasuis*
Leptospirosis	*Haemophilus pleuropneumonia, H. parahaemalyticus*
Pasteurellosis (hemorrhagic septicemia)	*Leptospira pomona* (and other serovars in some countries)
Salmonellosis	*Pasteurella multocida*
Streptococcal	*Salmonella choleraesuis* and *S. ryphimurium*
Cervical lymphadenitis (jowl abscess)	
Septicemia, arthritis	*Streptococcus beta hemolytic* (Group *E. Streptococcus*)
Arthritis, septicemia, meningitis	Group C (*S. equisimilis*) and Group L (*Streptococcus*)
Swine dysentery	*Streptococcus suis* (Group D)
	Treponema hyodysenteriae
Viral	
African swine fever	*Pestes africana suum*
Congenital tremors	Infectious congenital tremors virus
Foot and mouth disease	Foot and mouth disease virus
Hog cholera (swine fever)	*Togaviridae Pestivirus suis* or hog cholera virus
Porcine parvovirus infection	*Parvovirus* species
Porcine rotavirus infection	*Rotavirus* species
Pseudorabies (Aujeszky's disease)	Pseudorabies virus, a herpes virus
Swine influenza	Type A *influenza suis*
Swinepox	Swinepox virus
Transmissible gastroenteritis (TGE)	TGE virus
Mycoplasmal	
Arthritis	*Mycoplasma hyosynoviae*
Polyserositis and arthritis	*Mycoplasma hyorhinis* (see also *Haemophilus parasuis*)

Disease	Agent
Mycoplasma pneumonia	*Mycoplasma hyopneumoniae*
Mycotic	
Aspergillosis	*Aspergillus flavus*
Moniliasis	*Candida albicans, Oidium albicans, Monilia albicans*
Ergot	*Claviceps purpurea*
Mold toxicity	*Fusarium (Gibberella) roseum, F. tricinctum, Gibberella zeae*
Ochratoxicosis	*Aspergillus ochraceas* and *Penicillium* species
Miscellaneous	
Atrophic rhinitis	*Bordetella bronchiseptica, Mycoplasma hyorhinis,* noninfectious irritants, nutritional factors
Enterotoxemia (edema disease, gut edema, intestinal edema, gastric edema)	Toxin from *E. coli*
MMA-complex (metritis, mastitis, agalactia)	*Streptococcus* species, *Staphylococcus* species, *Actinomyces bovis, Actinobacillus lignieresi, Corynebacterium pyogenes, Mycobacterium tuberculosis, Spherophorus necrophorus,* possible endocrine factors

	Parasite	Residence of Adult
Internal Parasites		
Ascariasis	*Ascaris suum* (ascarid)	Small intestine
Trichinosis	*Trichinella spiralis*	Small intestine
Whipworm infection	*Trichuris suis*	Cecum and large intestine
Nodular worm infection	*Oesophagostomum* species (*Strongylid nematodes*)	Cecum and large intestine
External Parasites		
Sarcoptic mange	*Sarcoptes scabiei* var. *suis*	
Lice	*Haematopinus suis*	

TABLE 20-5
Other Infectious Diseases of Swine.

Disease	Agent	Type
Anthrax	*Bacillus anthracis*	Bacterial
Botulism	*Clostridium botulinum (Bacillus botulinus)*	Bacterial
Corynebacterial infections	*Corynebacterium* species	Bacterial
Listerosis	*Listeria monocytogenes*	Bacterial
Malignant edema	*Clostridium septicum*	Bacterial
Necrotic rhinitis	*Spherophorus necrophorus*	Bacterial
Tetanus	*Clostridium tetani*	Bacterial
Tuberculosis	*Mycobacterium tuberculosis avium, bovis, hominus*	Bacterial
Actinomycosis	*Actinomyces bovis*	Mycotic
Encephalomyocarditis	*Cardiovirus* species	Viral
Eperythrozoonosis	*Eperythrozoon suis*	Rickettsial
Hemagglutinating encephalomyelitis	Hemagglutinating encephalomyelitis virus	Viral
Porcine adenoviruses	*Adenovirus* species	Viral
Porcine enteroviruses	Teschen virus, Virus T80, Virus F7, Virus Californianos	Viral
Porcine cytomegalovirus infection (inclusion body rhinitis)	*Cytomegalovirus* species	Viral
Rabies	*Rhabovirus* species	Viral
Reovirus infection	Reovirus	Viral
Vesicular exanthema	Vesicular exanthema virus	Viral
Vesicular stomatitis	Vesicular stomatitis virus	Viral
Coccidiosis	*Eimeria* and *Isopora* species	Protozoan
Toxoplasmosis	*Toxoplasma gondii*	Protozoan

An outbreak of FMD in Taiwan in the late 1990s and in the United Kingdom and some countries in continental Europe in 2001 devastated the local pig industry and essentially halted pork exports. Hogs in parts of Asia remain infected.

FMD starts with vesicular lesions in the mouth and between the toes. A related, yet clearly different, disease is vesicular stomatitis (VS, one of several vesicular diseases). VS resembles FMD in the mouth lesions, but the morbidity and mortality do not occur in VS. Other symptoms of FMD include fever, chomping and salivation, and lesions of the feet, toes, and skin. Mortality in animals with FMD can be low (<5%) or higher (50% to 100%) depending on the strain of virus.

INFECTIOUS DISEASES THAT ARE BEING ERADICATED FROM THE UNITED STATES

Brucellosis

Brucellosis is a bacterial disease of swine that also infects cattle, sheep, wildlife, and humans (Leman, 1979). The *Brucella* genus contains six species, one of which is *Brucella suis*. The species of *Brucella* that affects pigs is not thought to be infectious to people.

Most U.S. states have a low or zero incidence of *Brucella* infection. When a herd is discovered to have brucellosis, it is quarantined and an attempt is made to eradicate the disease from that herd. One major problem with eradication is that some wild or feral pigs in the southern United States are infected with *Brucella.* These reservoirs are difficult to eradicate. However, intense efforts at eradication of brucellosis from U.S. swine herds have been highly successful. Brucellosis-free status has been attained in many states and the prospects seem excellent for complete eradication of brucellosis in the entire U.S. swine population.

Symptoms of brucellosis in pigs include abortion, infertility, lameness, and variable signs of fever. Symptoms may not persist and death is not a major symptom. Bacterial shedding can be significant for 30 d but may persist for over 2 yr.

Pseudorabies, or Aujeszky's Disease

Pseudorabies (PRV) is caused by a DNA virus. Pigs are considered to be the natural host (Maré et al., 1991). PRV can infect cattle and other ruminants, causing a "mad itch." Pseudorabies has been present in the United States for nearly a century, but its importance was not fully appreciated until the 1980s. PRV is caused by a member of the herpes virus group and, unlike most herpes viruses whose host range is limited to a few species, PRV affects most mammals and many birds. Pigs are the major reservoir for PRV because most other animals, including cattle, sheep, dogs, cats, and mice, are killed by the infection. Mortality in swine infected with PRV is highest in baby pigs, but is usually nil in mature swine. The incubation period is about 30 hr. The first signs of the disease are sneezing, coughing, and a slight fever, followed by excessive salivation, muscle spasms, convulsions, and paddling, and finally death after 4 to 9 d. Pigs less severely affected may recover after a few days of fever. Pseudorabies virus crosses the placenta. If infection occurs prior to 30 d gestation, embryos are resorbed; infection after 30 d gestation may result in death of some or all fetuses or the birth of stillborn or weak-infected piglets. Many diagnostic procedures have been developed and vaccines are available.

Most U.S. states have a low or zero incidence of PRV infection. When a herd is discovered to have PRV, it is quarantined and an attempt is made to eradicate the disease from that herd. Wild or feral pigs in the southern United States are infected with PRV. Effective vaccines are available for PRV, including a genetically modified virus whose inoculation can be separated from natural infection. A program of eradication of PRV in the United States is underway and, if successful (only 312 U.S. herds remained quarantined for PRV in November 1999), the effort will be a significant success story and will represent a major contribution to the future of commercial pork production.

Viral strains of PRV differ in symptoms. The most common symptoms include respiratory, enteric, and nervous system symptoms (especially ataxia) and reproductive problems (infertility, increased stillbirths).

INFECTIOUS DISEASES THAT SOME U.S. PORK PRODUCERS "LIVE WITH"

Respiratory Diseases

The primary mode of transmission of infectious respiratory diseases of swine is nose-to-nose contact. Pigs expose one another through social contact. In addition, aerosol transmission is possible for most respiratory pathogens. When respiratory pathogens are present, the air quality of the facility becomes especially important. Reducing dust and noxious gases can reduce respiratory symptoms.

Respiratory pathogens travel on particles of dust or water. The particles of smaller size, particularly those smaller than 10, 5, and 2.5 microns, will embed deep in the lung. Dust particles of greater than 10 microns are likely to be caught in the turbinates. Dust particles, with attached bacteria, that are less than 2.5 microns are considered a health risk. Lung alveoli are excellent culture media for bacteria. Every effort should be made to reduce respiratory pathogen concentrations and dust in the air to reduce respiratory symptoms.

PRRS

Porcine respiratory and reproductive syndrome (PRRS) symptoms were first reported in the United States in 1987. The syndrome may have been present much earlier but was called other names. PRRS virus was isolated and reported in 1991 and, in that year, the name was agreed upon by the international community. PRRS is a troubling disease because it is now widespread in the United States and Europe, even though the virus has unique characteristics on each continent.

PRRS virus causes reproductive problems such as abortion and infertility. A lower farrowing rate and a slightly lower number of pigs born alive are the major symptoms in the sow herd. Among nursery and growing-finishing pigs, the major symptom is a respiratory syndrome. Increased morbidity and mortality associated with other respiratory pathogens will be observed if the PRRS virus is present.

Because the genetic variability of the PRRS virus is so great and because it is known to mutate rapidly, vaccines are less effective than for other pathogens. Also, because the PRRS virus infects the pig's macrophages, the virus is effective at lowering the pig's ability to mount an immune response.

PRRS can be controlled by strict biosecurity. To maintain a PRRS-negative herd, incoming animals should be held in isolation and tested before entry into the herd. No therapeutic agents are available. Modified live-virus and killed vaccines against PRRS are available and effective in controlling the respiratory, but not the reproductive, component of the disease. The effectiveness depends on the strain of PRRS virus involved. PRRS is widespread in the United States and interacts with *Mycoplasma hyopneumoniae* (discussed under Mycoplasmal Diseases) in affecting the clinical signs and lung lesions produced. Research at Iowa State University by Thacker et al. (cited by Carlton and Miller, 1999) showed that pigs inoculated with *M. hyopneumoniae* 21 d before PRRS virus inoculation had no observable *Mycoplasma* pneumonia, but did have PRRS-induced lung lesions that persisted for 4 wk after the PRRS virus inoculation. In contrast, PRRS-related lesions were resolved within 4 wk in pigs inoculated with only PRRS virus. This suggests that the chronic inflammatory response produced during the course of *Mycoplasma* pneumonia aggravated the adverse effect of PRRS virus on the lung lesions. Therefore, controlling *M. hyopneumoniae* infection may be critical in reducing the impact of PRRS-induced pneumonia. The respiratory syndrome, involved when PRRS virus and *M. hyopneumoniae* are present together, has resulted in the coining of the porcine respiratory disease complex (PRDC). Because vaccines against PRRS are less than totally effective, prevention of exposure is preferred. Cleaning up a herd without depopulation has been performed, but is very difficult.

Atrophic Rhinitis

Atrophic rhinitis (AR) is a disease of the snout of the pig with complications of pneumonia and other respiratory ailments. The bones of the snout contain four (two on each side of the nasal septum) spiral-shaped cartilages that serves to warm, humidify, and filter incoming air. In particular, respiratory pathogens are caught in the convoluted, moist turbinates and prevented from entering the lungs.

A pig with severe AR has a snout that is either twisted or snubbed. As the turbinates are damaged by bacteria, and the snout continues to grow, its growth will favor the less-affected side of the snout. If turbinate damage is very severe, both sides will be stunted resulting in a snubbed look.

AR is caused by more than one microorganism. Two leading candidates as the cause of AR are toxigenic *Bordetella bronchiseptica* and toxigenic *Pasteurella multocida.* Each organism alone causes variations in the AR syndrome, with both microorganisms, the full disease syndrome is present. When *B. bronchiseptica* infects the pig's nasal cavity, the organisms attach to the epithelial surface, proliferate, and destroy the cells in the region and diffuse into the nasal turbinates, which become distorted due to altered bone metabolism. The first clinical signs are sneezing, followed in about 3 wk by turbinate atrophy (see Figure 20–2) and nasal bleeding (turbinate atrophy is not detectable in the live animal until facial distortion begins weeks later). Positive diagnosis is made by post mortem examination of a cross section of the nasal cavity. There is no cure for the nasal turbinate destruction and facial distortion associated with severe

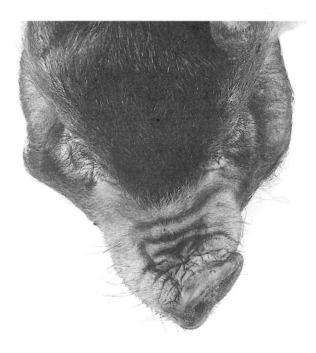

FIGURE 20–2
A Pig with a Distorted Snout Caused by Nutritional Secondary Hyperparathyroidism (NSHP) That Was Induced by a Dietary Ca-P Imbalance.

AR. Noninfectious irritants, such as dilute acetic acid infused into the nasal cavity, can produce turbinate atrophy in young pigs free of *B. bronchiseptica* (Logomarsino et al., 1974). High levels of ammonia in poorly ventilated swine buildings may aggravate turbinate atrophy in *B. bronchiseptica*-infected pigs (Curtis, 1981). Cats, rodents, and other animals are carriers of *B. bronchiseptica*. Vaccination against *B. bronchiseptica* controls bordetellosis and AR in swine herds.

Because AR involves changes in bone metabolism, the possibility of a nutritional component in the syndrome has been examined. Distorted and abnormal turbinate bone and twisted snouts indicative of AR have been produced in growing pigs by dietary calcium levels below the requirement and phosphorus levels much higher than the requirement (Brown et al., 1966). Bone lesions are characterized by excessive bone resorption and replacement of bone tissue by fibrous tissue (fibrous osteodystrophy). Dietary calcium-phosphorus imbalance produces nutritional secondary hyperparathyroidism, resulting in excessive calcium removal from bone (Brown et al., 1966). When toxigenic *Pasteurella multocida* is present, the disease is considered progressive with significant morbidity and performance problems.

The primary symptoms of AR are respiratory tract problems, such as runny noses, sneezing, watery eyes, and bleeding nose and increased incidence of pneumonia. The primary diagnostic tool is the slaughter check, when the snouts are cut in cross section between the molars and the turbinates examined and scored for damage.

AR increases mortality somewhat due to respiratory tract infections, but more commonly, an entire herd has chronic respiratory morbidity. The loss due to reduced pig performance can be very costly.

Secondary respiratory diseases are common for pigs with AR. Mycoplasmal hyopneumonia is common among AR-affected pigs, but the two diseases can be found independent of each other.

Environmental problems in pig facilities exacerbate the symptoms and performance problems associated with AR. Management or facility practices that increase the severity of AR include a large number of pigs in a single air space, crowding, poor ventilation, high levels of dust and ammonia, continuous pig flow, poor sanitation, and indoor housing in general.

Vaccines are available to treat AR, but are considered only partially effective. Some antibiotics are effective, again in part. Even with the best medication a herd with AR will typically have measurable lung lesions and turbinate damage upon slaughter check.

Mycoplasmal Pneumonia

Formerly known as enzootic pneumonia, mycoplasmal pneumonia is caused by *Mycoplasma hyopneumonae* (Hogg et al., 2001). Mycoplasmal bacteria do not have a cell wall, as do many other bacteria. Because many antibiotics inhibit bacterial growth by interference with growth of the bacterial cell wall, many antibiotics, particularly in the early years of therapy, were not effective in control of mycoplasmal pneumonia. Also, due to its unresponsiveness to antibiotics, the causative agent was thought earlier to be a virus; researchers now know the disease is caused by a *Mycoplasma*.

Mycoplasmal pneumonia is a chronic respiratory disease that shows high morbidity and relatively low mortality. The most common clinical sign is a nonproductive cough. Upon slaughter check or necropsy, characteristic lung lesions are observed. The typical lung lesions are at the tips of the lobes where small dust particles lodge and inoculate the lungs with the microorganism.

Some antibiotics are effective at reducing the symptoms. Vaccines are available. Complete control is not possible because herds that both vaccinate and use antibiotics still have morbidity. Medicated, very early weaning programs have been used to eliminate the causative microorganism in some herds.

The primary concern with mycoplasmal pneumonia is the performance set-back associated with chronic infection. In addition, when other respiratory pathogens are present, such as PRRS or AR, the herd has significant respiratory symptoms.

Other mycoplasmal microorganisms cause different problems. The second most reported mycoplasmal problem is associated with arthritis. The arthritis is responsive to antibiotics and cortisone. Infected pigs develop immunity to later infection.

Actinobacillus Pleuropneumonia

Formerly known as *Hemophilus* pleuropneumonia, *Actinobacillus* pleuropneumonia is widespread and of particular concern in intensive pig units. The causative agent is the bacterium *Actinobacillus pleuropneumoniae.*

The onset of the disease is very rapid. Typically, certain pigs become ill very quickly, with a high fever and loss of appetite. There may be some diarrhea or vomiting. Pigs develop respiratory distress rapidly, with breathing through the mouth and may have a foamy, bloody discharge through the respiratory tract. The onset of the disease can be just hours after infection. Stressing the infected pigs will bring on a storm of morbidity and mortality. In the chronic form, there may be no fever and only an occasional death. The disease symptoms are much worse when other respiratory pathogens are present.

Upon necropsy, large areas of the lungs are affected, showing large, dark-red areas of damage and associated pleurisy.

Certain antibiotics are very effective; however, antibiotic-resistant strains have been reported. Vaccines are available. Most farmers prefer to keep the disease out by strict isolation of incoming animals.

Swine Influenza

Swine influenza (SI) is caused by a type-A influenza virus. The causative agent is an RNA virus in the Orthomyxoviridae family. The most prevalent strain of swine influenza in the United States is H1N1. A new strain, H3N2, previously found only in Europe, has recently been identified in the United States (Miller, 1999). There is evidence of cross transfer between pigs and people (as well as among other species).

SI has a rapid onset and a rapid recovery. Pigs will spike a fever and have respiratory symptoms. Pigs will have labored breathing and will not be willing to move. The mortality is usually low, but the morbidity can be 100% of the pigs in a given air space. Symptoms in an infected herd may be absent 7 d after its onset.

SI has no effective treatment. The symptoms can be treated by providing a warm, draft-free, low-dust environment. Vaccines are available. Affected pigs develop active immunity to subsequent exposure to a particular strain of the virus.

Pneumonic Pasteurellosis

Pneumonic pasteurellosis is caused by *Pasteurella multocida,* a-gram negative bacterium widespread in the industry. *Pasteurella* is often associated with chronic respiratory disease in affected herds, especially in the porcine respiratory disease complex (PRDC).

A variety of antibiotics can be effective; however, many antibiotic-resistant strains are found. Antibiotic therapy is often reported to be ineffective at control of PRDC when *Pasteurella* is involved. Early weaning can control the disease. Some vaccines are available, with limited effectiveness.

ENTERIC DISEASES

The primary mode of transmission of infectious enteric diseases of swine is nose-to-nose and nose-to-feces contact. Pigs expose one another through social contacts as well. The primary symptom is diarrhea, which varies in color and consistency with the causative agent. Sanitation is especially important when enteric diseases are present. When enteric disease occurs, pigs become dehydrated, and dehydrated pigs are easily chilled. Replacement hydration and a warm environment are important.

TGE

Transmissible gastroenteritis (TGE) is a viral enteric disease of swine (Haelterman and Bohl, 1993). TGE virus is a coronavirus containing RNA. Surface antigens are similar on a related virus called porcine respiratory corona virus (PRCV) and laboratory assays may cross react.

When a herd is first infected, the onset of TGE often begins with vomiting, followed by severe diarrhea. A very high mortality (up to 100%) is observed among young pigs (nursing and recently weaned). If sows have protective antibodies in their milk, the symptoms will be worse among weaned pigs. While enteric symptoms are found among growing-finishing pigs, the mortality among older pigs is lower. After pigs are infected, surviving pigs have damaged villi in their small intestine and may have a severely reduced growth rate and feed efficiency. The incubation period is short (24 to 48 hr) and antibiotic treatment is ineffective, so the prevention of large losses due to an outbreak of TGE depends on diligent management and sanitation. The organism is carried mechanically on boots, and vehicle wheels contaminated with excreta from infected animals; strict isolation of infected from susceptible animals is essential. The economic impact of an infected herd is significant.

Survivors develop long-term immunity. Vaccination of the sow against TGE provides immunity of the newborn pig against the disease. In retrospect, it is of interest that perhaps one of the first oral vaccines of practical use in swine disease control was discovered empirically by pork producers who found that a degree of immunity against TGE in newborn pigs was attained by feeding the ground carcasses of pigs that died from TGE to pregnant sows.

Effective treatments for TGE are not available. The symptoms should be treated by attempting to keep the pigs warm, dry, and hydrated. Prevention of entry of the TGE virus is important. Vaccines are available. Some herds will re-feed dead or euthanized pigs to the sow herd to build immunity.

Collibacillosis

Escherichia coli is a natural inhabitant of the gastrointestinal tract. However, some strains of gram-negative *E. coli* are pathogenic. Other names for various forms of *E. coli* infections are collibacillosis (Kohler et al., 1994) and gut edema (Bergeland and Kurtz, 1992). Gut edema is believed to be a manifestation of an anaphylactoid response to the absorption of *E. coli* toxins absorbed from the intestinal tract. The syndrome often occurs following a sudden change in diet or management. Severely affected pigs may exhibit circulatory distress as indicated by dyspnea (labored breathing) and cyanosis (reddish patches on body surface and/or deep-red or bluish ears due to poor oxygen supply to tissues). *E. coli* infection can lead to a deadly septicemia. Other forms of *E. coli* cause mastitis or urinary tract infections.

E. coli most typically causes significant enteric problems for nursing piglets and weaned pigs. In weaned pigs, the condition is called milk scours or white scours. Morbidity is the primary economic concern, but short-term mortality can be observed. In gut edema, a severe edema is evident, including edematous eye lids. Post-weaning diarrhea is common when the maternal milk antibodies are no longer available to protect the gastrointestinal tract of pigs.

Antibiotics are effective in many cases. Vaccines are available. The environment contributes to the severity of the disease—a cool, drafty environment will cause greater problems for the affected pigs.

Swine Dysentery

Swine dysentery, also known as bloody scours or vibrionic dysentery, is caused by the bacterium *Serpulina hyodysenteriae* (Harris et al., 1993). This gram negative bacterium causes a bloody diarrhea. Seven serotypes of the bacterium have been reported.

The incubation period (time from exposure to disease onset) is about 2 wk. Death losses may be 25% or 30% in growing pigs. Often, no fever is present in affected animals. The first indication of the presence of swine dysentery in a herd may be the sudden death of one or more animals with no previous symptoms. Infected animals that survive may be re-infected within a few weeks or months, suggesting that little or no immunity is acquired by infected animals. The primary symptom of bloody diarrhea is accompanied by signs of fever, loss of appetite, mucous discharge into the feces, arched back, and kicking indicative of gut pain.

Several antibiotics are available to treat the condition. Pigs are often treated through the water or the feed, but water medication is preferred. Sanitation and provision of a warm environment will prevent the disease from becoming severe after antibiotic therapy has started. Incoming animals should be free of this disease to maintain dysentery-free status of the herd.

Ileitis, or Proliferative Enteropathies

Ileitis is related to other conditions such as proliferative enteropathies, proliferative ileitis, and porcine intestinal adenomatosis (Lomax et al., 2001). The term proliferative enteropathies refers to the collection of diseases in this category—acute and chronic conditions with differing clinical signs and different causes. The primary microorganism is *Lawsonia intracellularis,* a bacterium that grows inside intestinal cells. Lesions found in ileitis include thickened mucosa in the distal small intestine and proximal large intestine.

The main symptom is loss of appetite and slowed growth with only occasional diarrhea in growing-finishing pigs. Some mortality may be observed, but the major problem is loss of performance due to loss of appetite and a wasting syndrome. Some antibiotics are reported recently to be effective against ileitis. Vaccines are not yet available.

Salmonellosis

Salmonellosis, known simply as *Salmonella,* is caused by one of several species of gram-negative bacteria, including *Salmonella cholerasuis.* Other species, including *S. typhimurium,* are zoonotic.

Salmonella is found most commonly in intensively kept, weaned pigs. In the form leading to septicemia, pigs develop a fever, loss of appetite, and some breathing difficulty. Stressful situations cause a rapid onset of the disease. In the enteric form, weaned pigs have a watery diarrhea that may later contain blood. Mortality is usually low, but morbidity is high.

While several antibiotics are effective, antibiotic-resistant strains are reported to be widespread. Sanitation and good management practices are important to control the disease. Most pigs are thought to be infected with one or more strains of *Salmonella.* Pathogenic *Salmonella* have been isolated from numerous species of birds and animals in addition to pigs and humans.

Coccidiosis and Toxoplasmosis

Coccidiosis is caused by the parasite *Isospora suis* and other parasitic species (Hall et al., 2001). *Isospora* must pass through the pig or other animals to complete its life cycle. The feces of infected pigs are contagious to other pigs.

Pigs with coccidiosis have a yellow-to-grayish diarrhea that turns very watery. Nursing and weaned pigs can be affected. Morbidity can be very high, but mortality is usually low.

Dehydration is the primary concern and supportive therapy should be provided. A warm, dry, clean environment is important. Some coccidiostats are available. Sanitation and prevention are important in the control of the condition.

Toxoplasmosis, a related parasitic disease, requires cats as an intermediate host. Toxo (as it is called) is caused by the protozoan *Toxoplasma gondii.* Most infections are subclinical. Toxo is a serious zoonotic disease. There is no reported effective treatment and vaccines are not available. Prevention of entry of the disease is important.

Clostridium perfringens and Related Tetanus

Clostridium is a genus of gram-positive bacteria that includes *Clostridium perfringens, C. tetani, C. botulinum,* and others. The most common form is *C. perfringens* Type C.

C. perfringens can take several forms. The most common form causes a rapid death associated with an acute fever and hemorrhagic diarrhea. The body temperature may drop prior to death. In one form, piglets die within 3 d of life. Older pigs may have a nonhemorrhagic diarrhea with yellow-gray, mucous feces. While death is unlikely in older pigs, the performance setbacks can be significant. Once clinical signs are observed, the intestinal damage is already severe. One form of the disease infects the skin and can cause gangrenous lesions.

Some antibiotics are effective. Vaccines are available. Antibodies in the sow's milk are important and piglets should be vaccinated at 3 wk of age to minimize postweaning effects.

Rotaviral Diarrhea

Rotaviral diarrhea is caused by the RNA-containing rotavirus (Saif et al., 1987). There are many serotypes and the virus is widespread in nature.

Rotavirus causes significant diarrhea in young piglets, especially those less than 1 wk of age. The diarrhea is watery-yellow to white. Mortality can be very high, but morbidity is always high. The rotavirus causes lesions on the small intestine epithelium. Rotavirus diarrhea appears to be aggravated by the presence of *E. coli* and other pathogens, and following stressful changes in the environment (crowding, environmental temperature, change to solid feed).

There is no effective treatment for rotavirus infection. Antibiotics can be given for secondary bacterial infections. A warm environment is very important and fluid therapy is recommended. Most adult swine have serum antibodies against rotavirus and provide passive immunity to their offspring via colostrum. Management practices should be directed at sanitation, generating passive maternal immunity, and reducing the viral load, if possible. Vaccines are available.

REPRODUCTIVE DISEASES

Reproductive diseases of swine cause two main problems: (1) reduced farrowing rate and lower numbers of pigs born alive and (2) abortions and failure to cycle (Leman, 1979). Diseases that cause abortion may do so in an obvious manner wherein the aborted fetuses are observed, or in a less-obvious manner wherein the abortion is early enough in gestation that the aborted embryos or fetuses are reabsorbed. Diseases that increase the numbers of mummies (partially decomposed) or stillborn pigs are obvious to the animal caregiver. Seasonal variations in fertility should be considered when one attempts to understand variation in fertility.

Leptospirosis

Leptospirosis is caused by at least eight species of spirochete bacteria of the genus *Leptospira.* Leptospirosis (lepto) is widespread in the world, found in all the major pig-producing countries. Leptospirosis infects the urinary and reproductive tracts of pigs, causing both kidney and reproductive failures.

The majority of signs of lepto are subclinical. In its acute form, symptoms include septicemia, loss of appetite, and fever. The primary signs of chronic leptospirosis are infertility, late abortions, stillbirths, and unthrifty piglets. The reproductive failures can have significant economic cost. Lepto is a potential zoonotic agent.

The incidence of leptospirosis is high in deer and other wildlife as well as in farm animals. Transmission occurs by coitus, by the ingestion of contaminated feed and water, by eating infected rodents or their tissues, or by entrance of the organism through skin abrasions or conjunctiva. While antibiotics can be effective, most producers control lepto through vaccination.

Parvo Virus

Parvo virus (parvo) is a DNA-containing virus that can cause significant reproductive problems. Parvo is widespread in the United States and Europe. The virus can replicate in pigs of any age, but the main clinical sign is reproductive failure in pregnant sows.

The primary clinical sign of parvo is an increase in the number of mummified and stillborn piglets. Primary control is through vaccination. There are no effective treatments. Prevention of entry into the herd is critical.

EXTERNAL AND INTERNAL PARASITES

Many parasites affect swine, but most do not survive in humans.

External Parasites

Sarcoptic mange mite (agent: *Sarcoptes scabiei*). The sarcoptic mange mite burrows under the outer layers of the epidermis and causes irritation and thickening of the skin (McKean et al., 1999). (The subspecies of S. scrapiei that affects pigs apparently does not affect humans and vice versa.) In severe cases, the entire body surface is covered with thick, scurfy skin and alopecia is common. Appetite and weight gain may be reduced. Severe lesions resemble those of parakeratrosis, the zinc-deficiency disease of swine, but the microscopic examination of skin scrapings of affected pigs allows identification of the mange mite. The drug ivermectin, administered orally, parenterally, or topically, affords complete elimination of the mange mite, along with other external parasites and many internal parasites. Application of solutions of currently approved insecticides using liquid spray under pressure to penetrate the outer layers of epidermis is also somewhat effective in control of mange. Legal requirements for application and withdrawal before slaughter must be followed.

Demodectic mange mite (agent: *Demodex hylliides* or *D. folliculorum*). This mange mite is less common than *S. scabiei,* but occurs frequently in humans, dogs, sheep, and other mammals. It is easier to control because it does not penetrate the skin layers so deeply. Control is the same as for sarcoptic mange.

Louse. One species of lice, *Haematopinus suis,* occurs in pigs. Continual skin punctures made by the louse to suck blood result in irritation and in rubbing by the pig against any available object for relief. Control is by topical application of dry powder or spray containing any currently approved insecticide. Ivermectin is also effective in controlling lice. Legal requirements for application and withdrawal before slaughter must be followed.

Other External Parasites. Several other external parasites are pests of swine, including fleas, flies, and ticks. Their control is the same as that for lice.

Internal Parasites

Several internal parasites affect pigs (Biehl et al., 1993), but the most important economically are *Ascaris suis* and *Trichinella spiralis.* Other parasites include the whipworm (*Trichuris suis*), nodular worm (*Strongulodes ransomi*), lung worm (*Metastrongylus* species), liver fluke (*Fasciloa hepatica*), kidney worm (*Stephanuses denatus*), and *Taenia solium,* whose larvae invade the cardiac muscle, diaphragm, tongue, and other muscles. Currently approved oral medications are effective in controlling each of these parasites. Ivermectin, administered orally or parenterally, is effective against most.

Ascarids (*agent: *Ascaris suis* or *A. lumbricoides). The ascarid (Stephenson et al., 1980) is ubiquitous in swine populations and is economically important because of its negative effect on growth and feed utilization and tissue damage caused by its migration through the lungs and liver during the larval stage of its life cycle. The cycle is diagrammed in Figures 20–3a and 20–3b.

The adult lives in the small intestine where it mates. The fertilized eggs are excreted in the feces and go through a period of maturation on the ground; the duration of maturation depends on environmental temperature and humidity. The larvae, if ingested by a pig several weeks later, are swallowed. After dissolution of the protective covering by the digestive processes of the host, the larvae are absorbed into the blood from the intestinal tract. The developing larvae lodge in the liver where scar tissue is formed at the local site of intrusion. Heavy infestation results in severe liver damage seen at slaughter as white fibrous tissue scars on the surface of and within the liver. Severe scarring results in condemnation of the liver for human consumption. For comparison, note the smooth, dark surface of a normal liver shown in Figure 20–4.

The adverse effects of ascarid infestation are manifested both by competition between the adult ascarid and the host for nutrients and by the liver damage caused by the migration of the ascarid larvae. Small intestine weight is increased in direct proportion to the ascarid load. Nutritional research shows that the intestinal tract represents a disproportionate share of the energy requirement of the pig. Therefore, a heavy burden of ascarids undoubtedly decreases the efficiency of pig growth by increasing the energy required to maintain the hypertrophied small intestine. From the liver, migrating larvae are pumped by the heart to the lungs where they are expelled into the bronchioli and coughed up into the mouth. They are swallowed and enter the small intestine where they mature and repeat the cycle. The coughing associated with the migration through the lungs may be confused with signs of infectious respiratory disease. The migration and

FIGURE 20–3
(a) Life Cycle of the Ascarid; (b) Journey of the Ascarid through the Pig.

Source: From USDA (1963).

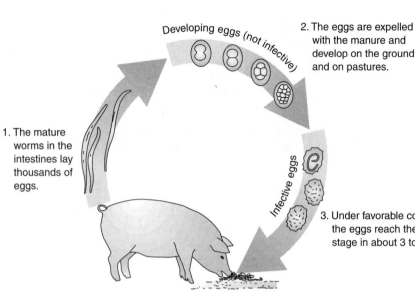

(a)

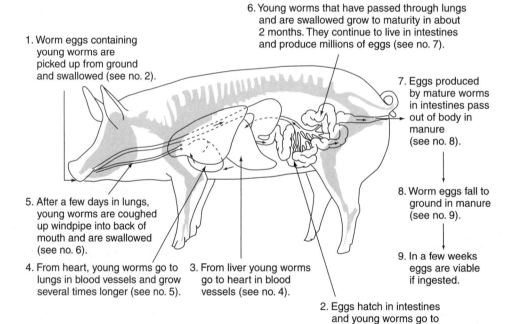

(b)

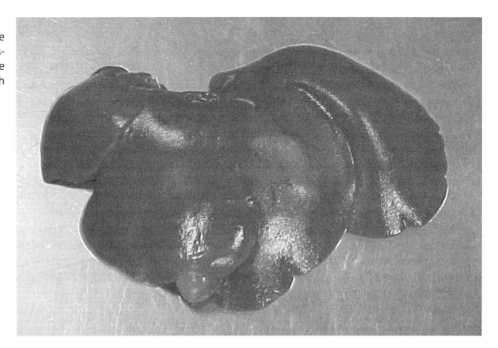

FIGURE 20–4
Normal Pig Liver. Note the Smooth, Dark Surface. A Diseased Liver will Have White Milk Spots Associated with Ascarid Infection.

associated coughing usually persist only a few days. Ascarid larvae pass the placental barrier and infect the fetus in utero. Effective control of ascarids is accomplished by a choice of several treatments, including piperazine, and currently approved feed additives. Most of these antihelminhtics are effective against other internal parasites as well. Ivermectin administered parenterally or orally is also effective against ascarids and other internal parasites.

Trichinosis (**agent:** *Trichinella spiralis*). This parasite is highly infectious to humans and, for this reason, remains a detriment to pork consumption, even though the incidence of trichinosis in humans in the United States is nearly zero. However, the incidence worldwide remains significant (this issue is discussed in more detail in Chapters 3 and 9). The life cycle is diagrammed in Figure 20–5 and is described briefly here. Adults live and mate in the intestinal tract of the host (pigs, humans, rodents, and other animals) and fertilized ova are deposited in the intestinal mucosal lining. The larvae pass into the lymphatic system and blood and become encysted in the diaphragm and skeletal muscle of the host. The time required from original ingestion of infected meat to encysting of the larvae in skeletal muscle of the host is 1 to 4 wk. The encysted larvae can remain viable in the skeletal muscle of the host for several yr. The ingestion of infected pork by humans results in severe nausea, anorexia, and digestive upset followed by muscle pain as the larvae migrate to the skeletal muscle. Effective treatment is available.

Rapid laboratory testing allows detection of the presence of *Trichinella* in pork, so the source of infection often can be identified. Trichina organisms are destroyed by heat (77°C (170°F) for 30 min) or freezing (20 d at 5°F or −15°C). Irradiation of pork

FIGURE 20–5
Life Cycle of the Trichina.
Source: Adapted from USDA (1963).

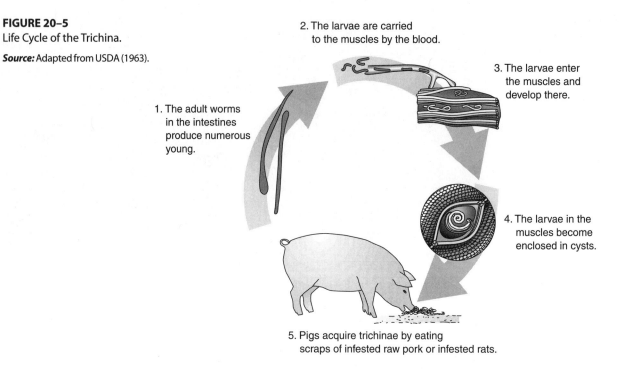

in U.S. slaughter plants has been approved recently. Irradiation destroys trichina and, therefore, represents yet another effective means of eliminating trichina from pork.

JOINT AND ARTHRITIC CONDITIONS

Erysipelas

The causative agent for erysipelas is the bacterium *Erysipelothrix rhusiopathiae,* a gram-positive bacillus that is widespread in the swine industry. Like brucellosis and leptospirosis, erysipelas is a zoonotic disease. In its acute form, the clinical signs include anorexia, apathy, and high fever that may persist for several days. The acute signs can be controlled by antibiotic treatment. The acute form may be followed by chronic erysipelas with inflammation, swelling, and stiffness of the joints. Occasionally, large, diamond-shaped red blotches may appear, accounting for the term "diamond skin disease." Chronic erysipelas can be treated and cured with antibiotics. Routine immunization against erysipelas is an effective control measure.

Other Joint Problems

Pigs can have swollen joints due to infection caused by a number of agents (Ross and Bailey, 1991) including streptococci, *Mycoplasma hyosynoviae* (Ross, 1981), or *Actinobacillus suis* as well as by physical injury or nutritional or genetic problems leading to osteochondrosis.

OTHER POTENTIAL PIG HEALTH PROBLEMS

PSS/PSE

Porcine stress syndrome (PSS) is a condition in live pigs in which the pigs are highly susceptible to stress. The term malignant hyperthermia syndrome (MHS) has also been applied to PSS pigs, because of the elevated body temperature during stress. Affected pigs become red-skinned, have a spiraling hyperthermia (called malignant hyperthermia), and shake, lie down, and often die. If the PSS pig is stressed immediately before slaughter, even mildly, it will have light-colored (pale), soft and watery (exudative) meat (PSE) (see Chapter 9). Carrier pigs that have one copy of the mutated gene show some symptoms and pigs with two copies of the recessive gene have a high incidence of the condition (see Chapter 6). Selective breeding can easily eliminate the disease.

MMA or PPDS

When lactating sows either stop milking or significantly reduce milk production (Wagner et al., 1993), the cause may be what was formerly known as the syndrome mastitis-metritis-agalactia (MMA). MMA is known now as postpartum dysgalactia syndrome (PPDS). PPDS is known for a hypogalactia (low milk production) more often than complete cessation of milk production (agalactia).

PPDS is often, but not always, associated with metritis (inflammation of the uterus). Metritis is common particularly in sows that had a difficult or assisted farrowing. The fever and loss of appetite leads to constipation and a general malaise. The enclosure of sows in farrowing crates that restrict movement does little to help the constipation (exercise reduces the likelihood of constipation).

When sows get mastitis, their piglets may starve to death. The follow-up to PPDS is unthrifty piglets and difficulty in rebreeding sows. Piglets on PPDS sows can be saved by feeding them either supplemental feed while with the sow or by early weaning.

Swinepox (agent: swinepox virus and vaccinia virus)

Swinepox causes skin eruptions that persist over the entire body for 1 to 2 wk and then disappear. Swinepox virus and vaccinia virus are similar. Vaccinia virus is a laboratory strain that was used for immunization against smallpox in humans until the eradication of smallpox eliminated the demand for the vaccine. Pigs with swinepox reduce feed intake for a few days, then usually recover without complications. Death is rare. A high level of permanent immunity follows the disease.

Porcine Respiratory Disease Complex (PRDC) (agents: PRRS and *Mycoplasma hyopneumoniae*)

This common syndrome involves an interaction of PRRS virus and *Mycoplasma pneumoniae* to produce clinical signs distinct from those observed when either of these pathogens is present separately. The clinical signs of PRDC are therefore variable, depending on poorly understood relationships between PRRS and *Mycoplasma* hyopneumonia and their respective vaccines. The control of PRDC appears to depend on the selection of appropriate timing of vaccination against

PRRS and *Mycoplasma* hyopneumonia, the strains of PRRS virus used in preparing the modified-live-virus vaccine, the level of maternal antibodies, the timing of seroconversion to PRRS, and other unknown variables. The complex and poorly understood interactions between pathogens (in this case, viruses and mycoplasmas) in pig disease syndromes can be expected to receive continued attention in herd health management.

Mycotoxins in the Feed

Feedstuffs commonly fed to pigs can contain a number of natural plant contaminants (Carlson and Lloyd, 1981) or poisonous plants (Kingsbury, 1964). The most common are the mycotoxins and among those, aflatoxin is the most common mycotoxin (NRC, 1998). Certain regions of the world are contaminated with mycotoxins probably because of local weather conditions. Many of these molds and fungi have hormonal properties, especially estrogenic properties. Mold-infected feed should not be fed to breeding stock. Some low level of mold is tolerable in the diet of growing pigs. Feed intake can be negatively affected if the levels are too high.

Aspergillosis (agent: *Aspergillus flavus*)

This fungus is common in grain, peanuts, corn, and other crops harvested under wet conditions and stored at a high moisture content and may be present in fish meal if the moisture content is high (>18%). The toxicity of *A. flavus* is due to the absorption from the intestinal tract of mycotoxins (aflatoxins) produced by the fungus. Clinical signs of *A. flavus* toxicity are anorexia, diarrhea, poor growth, pulmonary distress, and death in severe toxicity. Aflatoxins have been associated with liver cancer in humans. Aspergillosis can be prevented by avoiding the use of feed contaminated with *A. flavus*.

Mold Toxicity (agents: *Fusarium (Gibberella) roseum, Fusarium tricinctum, Gibberella zeae*)

Harvesting and storage of corn and cereal grains at high moisture content (above 18%) is conducive to the growth of molds. Mold-contaminated corn and cereal grains may cause poor growth and reproductive failure and lead to major economic losses when fed to swine.

Ochratoxicosis (agents: *Aspergillus ochraceas* and *Penicillium* species)

Concentration of as little as 200 parts per billion (ppb) of ochratoxin A in swine feeds have been shown to produce kidney, liver, and muscle lesions. Other signs of ochratoxicosis due to contamination of pig feed with *A. ochraceas* or *Penicillium* species include diarrhea, dehydration, and liver necrosis. Reduced weight gain and feed utilization may not appear at 200 ppb ochratoxin contamination, but have been reported when the diet contains ochratoxin at levels above 1,400 ppb. Tests are available to measure the level of ochratoxins and aflatoxins in feedstuffs. The cost of obtaining this information on feedstuffs in questionable harvesting and storage conditions is usually a good investment.

Ergot (agent: *Claviceps purpurea*)

This fungus invades rye and other cereal grains and its toxic alkaloids cause reproductive failure in sows and gangrene of extremities (feet, ears, snout, tail). Withdrawal of ergot-infected grain from the diet reverses the adverse effects.

Prolapse, Ulcers, and Hernias

Pigs are susceptible to a variety of physical problems, including prolapse, ulcers, and hernias. These conditions can have a genetic component and, certainly, some environments make the conditions worse. Stockmanship is important in finding the conditions early and attempting to minimize the affected animals.

Pigs can have a prolapse (protruding) rectum or vagina. Among growing pigs and sows, the prolapsed rectum is a problem. Prolapse of the vagina can be observed especially after farrowing. The prolapse is a particular concern for pigs that are group-housed because pen mates will attack the prolapse, resulting in damage to the tissue and a potential for bleeding to death. Prolapses can be surgically repaired, but this practice is time-consuming and not always permanently effective. Prolapsed animals should be isolated until they are at least partially healed. Once prolapsed animals are repaired, they should be sent to market as long as drug withdrawal protocols have been followed.

Ulcers are common among some pigs. Pigs can develop a stomach ulcer very quickly (in a matter of minutes or hours). The ulcers can result in pigs bleeding to death in the acute form or becoming anemic if the condition is chronic. The incidence of gastric (stomach) ulcers in pigs remains relatively high despite its recognition as a health problem in pigs for several decades. The secretion of gastric juice with high acidity and high pepsin concentration increases the susceptibility to ulceration. The process of ulcer development can occur within minutes during a 2 to 3-d period of stress. Consumption of heated corn is associated with a higher incidence of ulcers than is consumption of unheated corn. Oats tend to protect the pig from developing stomach ulcers. An alcohol-soluble extract from oats has been shown to inhibit gastric acid secretion and to protect against ulcers in the growing pig. Stresses associated with marketing (fasting, long duration of shipment and crowding) increase the incidence. Finely ground feeds (particle size <750 microns) tend to improve growth rate and feed utilization, but are also associated with increased incidence of esophageal ulcers. Feeding a finely ground diet following a 24-hr fast may increase the severity of the lesions induced by the fast (Lawrence et al., 1998). Changing the particle size of the diet to >750 microns average diameter and feeding the diet for as little as 3 to 7 d has been shown to reduce ulcer severity in the esophageal region and to result in complete healing of the lesion within 2 to 4 wk (Lawrence et al., 1998). Transport of pigs for 4 hr results in a 40% to 50% reduction in feed intake over the ensuring 3 d (McGlone et al., 1993), indicating that even a relatively short duration of transport has the potential to precipitate ulcer formation. Lactating sows fed finely ground diets are also subject to pars esophageal ulcers. The physical restraint of the farrowing crate does not, by itself, appear to produce stomach ulcers as long as a high level of feed intake is maintained.

Hernias can be of two forms: umbilical or scrotal. Some genetic lines may be predisposed to develop hernias. Hernias form when a sphincter weakens, loosens, and allows the intestines to fall through. The hernia grows when the influx of intestinal mass exceeds the outflow. Some hernias can be so large that they drag on the ground, posing a risk of being punctured. Hernias can be repaired surgically, but this is time-consuming and not totally effective. Affected pigs may need to be euthanized.

BEHAVIORAL PROBLEMS

Pigs can experience a number of behavioral problems, but the most common problem—tail biting—is resolved by docking the tail. The last two-thirds of the tail is removed near birth to prevent tail biting. If the tails are left on, and pigs are housed indoors on slats, the incidence of tail biting can be very high. Ear chewing is a related anomaly (see the discussion on behavioral problems in Chapter 18).

Early weaned pigs show some behavioral problems, including ear, navel, and prepuce suckling. The urine-drinking piglet—probably with a strong drive to suckle—can consume large quantities of urine, which is unhealthy.

When sows are hungry and particularly when they are waiting in line to eat, vulva biting can develop. The vulva is a blood-rich organ. If it is punctured, sows can lose a large amount of blood and experience discomfort.

Sows housed individually in indoor units show two distinct behaviors: (1) During gestation, limit-fed sows lick, chew, and bite the bars or other pen materials. It is not known why sows do this behavior, but some scientists have speculated that this is a sign of stress. Such claims have led to the banning of the crate for gestating sows. (2) Sows show nest-building behaviors in pens or paddocks if they have nesting materials. Interestingly, even if they do not have nesting materials, they still show nest-building behaviors. Pre-farrowing sows show what is called phantom nest-building behaviors. It is not known if this phantom behavior is a sign of a stressful situation for the sow.

Pigs that are penned in one environment for a long period (usually months) have a difficult time moving through novel environments. Thus, when some pigs are moved through chutes and onto and off of trucks, there is variation in the ease with which pigs are handled. Suggestions were made that some of the more lean, heavily muscled pigs are more difficult to handle due to inadvertent genetic selection for behaviors that make handling pigs difficult. Regardless of the origin of this behavior, difficult handling of pigs is a serious behavioral problem for the producer and the processor.

Each behavioral problem is a function of both genetics and environment. The selection program and the specific features of the environment (including stockmanship) should be examined closely when behavioral problems arise.

SUMMARY

This chapter provided an overview of pig health for use by nonveterinarians involved in pork production. Standard approaches to ensuring pig health in commercial pork production units include biosecurity, sanitation, control of pig flow, and overall planned herd health programs in collaboration with a herd veterinarian. Infectious disease-control programs include vaccination, subtherapeutic and therapeutic antimicrobeal administration, and biosecurity measures. Major infectious viral, bacterial, mycoplasmal, and fungal diseases and internal and external parasites were described, as well as their causative agents, signs of infection, and control. Pigs can experience noninfective physiological diseases such as stomach ulcers and behavioral problems.

QUESTIONS AND ACTIVITIES

1. Create a table of common enteric diseases of swine with the name of the disease in the left column. In the other columns, list the causative microbe and the major and distinguishing symptoms and the treatment or prevention tools available.
2. Create a table of common respiratory diseases of swine with the name of the disease in the left column. In the other columns, list the causative microbe and the major and distinguishing symptoms and the treatment or prevention tools available.
3. Select one disease and provide an in-depth review of the causative microbe, the clinical signs, treatment, and prevention.
4. Describe what might be the cause when the following symptoms are present (there is more than one answer to some symptoms):
    a. Bloody feces
    b. Vomiting among sows
    c. A transient respiratory problem
    d. Abortion early in pregnancy
    e. Increased numbers of mummified piglets
    f. Swollen joints

LITERATURE CITED

Becker, H. N., G. W. Meyerholz, and J. M. Gaskin. 1990. Selection and Use of Disinfectants in Disease Prevention. Pork Industry Handbook No. PIH-80. North Carolina Agricultural Extension Service, North Carolina State University, Raleigh.

Bergeland, M. and H. Kurtz. 1992. Edema Disease (Gut Edema, *E. coli* Enterotoxemia). Pork Industry Handbook No. PIH-40. North Carolina Agricultural Extension Service, North Carolina State University, Raleigh.

Biehl, L. G., R. F. Behlow, and E. Batte. 1993. Internal Parasites. Pork Industry Handbook No. PIH-44. North Carolina Agricultural Extension Service, North Carolina State University, Raleigh.

Biehl, L. G., B. Lawhorn, and L. Wernimont. 1997. Guidelines for the Development of a Swine Herd Health Calendar, Pork Industry Handbook No. PIH-68. North Carolina Agricultural Extension Service, North Carolina State University, Raleigh.

Brown, W. R., L. Krook, and W. G. Pond. 1966. Cornell Vet. 56 (Suppl. 1): 1–108.

Carlson, T. L. and W. E. Lloyd. 1981. Toxic chemicals, plants, metals and mycotoxins. In: A. D. Leman et al. (eds.) Diseases of Swine. 5th ed. pp. 603–616. Iowa State University Press, Ames.

Carlton, J. and M. Miller. 1999, March. Does mycoplasma link PRRS to PRDC? Pork '99: 42.

Curtis, S. E. 1981. Environmental Management in Animal Agriculture. Pp. 24–30. Animal Environment Services, Mahomet, IL.

Haelterman, E. O. and E. H. Bohl. 1993. Transmissible Gastroenteritis (TGE). Pork Industry Handbook No. PIH-47. North Carolina Agricultural Extension Service, North Carolina State University, Raleigh.

Hall, R. E., R. C. Meyer, and A. C. Todd. 2001. Swine Coccidiosis. Pork Industry Handbook No. PIH-81. North Carolina Agricultural Extension Service, North Carolina State University, Raleigh.

Harris, D. L., R. D. Glock, L. Joens, and I. T. Harris. 1993. Swine Dysentery (Bloody Scours, Vibrionic Dysentery, Black Scours). Pork Industry Handbook No. PIH-56. North Carolina Agricultural Extension Service, North Carolina State University, Raleigh.

Henry, S. C. and M. Apley. 1999. Therapeutics. In: B. Straw et al. (eds.) Diseases of Swine. 8th ed. pp. 1155–1162. Iowa State University Press, Ames.

Hogg, A., W. P. Switzer, and D. O. Farrington. 2001. Mycoplasmal Pneumonia and Other Mycoplasmal Diseases of Swine. Pork Industry Handbook No. PIH-29. North Carolina Agricultural Extension Service, North Carolina State University, Raleigh.

Kingsbury, J. M. 1964. Poisonous plants of the United States and Canada. Prentice-Hall, Englewood Cliffs, NJ.

Kohler, E. M., O. A. R. D. C. Wooster, and H. Moon. 1994. Enteric Collibacillosis of Newborn Pigs. Pork Industry Handbook No. PIH-30. North Carolina Agricultural Extension Service, North Carolina State University, Raleigh.

Lawrence, B. V., D. B. Anderson, O. Adeola, and T. R. Cline. 1998. Changes in pars esophageal tissue appearance of the porcine stomach in response to transportation, feed deprivation, and diet composition. J. Anim. Sci. 76:788–795.

Leman, A. D. 1979. Infectious Swine Reproductive Diseases. Pork Industry Handbook No. PIH-59. North Carolina Agricultural Extension Service, North Carolina State University, Raleigh.

Logomarsino, J. V., W. G. Pond, L. Krook, and D. Kirtland. Effect of Dietary Calcium-Phosphorus and Nasal Irritation on Turbinate Morphology and Performance in Pigs. 1974. 39:544–549.

Lomax, L., R. Glock, H. Kurtz, and L. Thacker. 2001. Porcine Proliferative Enteritis. Pork Industry Handbook No. PIH-93. North Carolina Agricultural Extension Service, North Carolina State University, Raleigh.

Maré, C. J., D. P. Gustafson, and L. W. Schnurrenberger. (1991). Pseudorabies (Aujeszky's Disease, Mad Itch). Pork Industry Handbook No. PIH-38. North Carolina Agricultural Extension Service, North Carolina State University, Raleigh.

McGlone, J. J., J. L. Salak, E. A. Lumkin, R. I. Nicholson, M. Gibson, and R. L. Norman. 1993. Shipping stress and social status effects on pig performance, plasma cortisol, natural killer cell activity, and leukocyte numbers. J. Anim Sci. 71:888–896.

McKean, J., D. Holscher and S. Quisenberry. 1992. External Parasite Control on Swine. Pork Industry Handbook No. PIH-40. North Carolina Agricultural Extension Service, North Carolina State University, Raleigh.

Meyer, K., L. Biehl, and D. Reeves. 1990. Slaughter Checks—An Aid to Better Herd Health. Pork Industry Handbook No. PIH-93. North Carolina Agricultural Extension Service, North Carolina State University, Raleigh.

Meyerholz, G. W. and J. M. Gaskin. (1981a). Environmental Sanitation and Management in Disease Prevention. Pork Industry Handbook No. PIH-79. North Carolina Agricultural Extension Service, North Carolina State University, Raleigh.

Miller, M. 1999, December. A new flu virus uncovered. Pork'99: 46.

National Pork Board. 2002. Swine Care Handbook. NPB, Des Moines, Iowa.

NRC. 1998. Nutrient Requirements of Swine. 10th ed. National Academy Press, Washington, D.C.

Pointon, A. M., P. R. Davies, and P. B. Bahson. 1999. Disease surveillance at slaughter. In: B. Straw et al. (eds.) Diseases of Swine. Pp. 1111–1132. Iowa State Univ. Press, Ames, IA.

Pond, W. G. and J. H. Maner. 1984. Swine Production and Nutrition. AVI Publishing Co., Westport, CT.

Ross, R. F. 1981. Mycoplasmal diseases. In: A. D. Leman et al. (eds.) Diseases of Swine. 5th ed. pp. 535–549. Iowa State University Press, Ames.

Ross, R. and J. Bailey. 1991. Swine Arthritis. Pork Industry Handbook No. PIH-36. North Carolina Agricultural Extension Service, North Carolina State University, Raleigh.

Saif, L. J., O. A. R. D. C. Wooster, J. G. Lecce, and A. Torres. 1987. Rotaviral Diarrhea in Pigs. Pork Industry Handbook No. PIH-61. North Carolina Agricultural Extension Service, North Carolina State University, Raleigh.

Smith, W. J., D. J. Taylor, and R. H. C. Penny. 1990. Color Atlas of Diseases and Disorders of the Pig. Iowa State University Press, Ames.

Stephenson. L. S., W. G. Pond, M. C. Nesheim, L. Krook, and D. W. T. Compton. 1980. *Ascaris suis:* Nutrient absorption, growth and intestinal pathology in young pigs experimentally infested with 15-day old larvae. Exp. Parasitol. 49:15–25.

Straw, B. E., S. D'Allaire, W. L. Mengeling, and D. J. Taylor. 1999. Diseases of Swine. 8th ed. Iowa State University Press, Ames.

Wagner, W. C., R. G. Elmore, R. F. Ross, and B. B. Smith. 1993. Lacatation Failure in the Sow. Pork Industry Handbook No. PIH-37. North Carolina Agricultural Extension Service, North Carolina State University, Raleigh.

INTERNET RESOURCES

General pig health sites:
http://www.aasp.org/
http://www.porkboard.org/Home/default.asp
http://www.pighealth.com/

Major Pig Pharmaceutical sites:
http://www.elanco.com/
http://www.alpharma.com/ahd/index.html
http://www.4animalhealth.com/
http://www.bi-nobl.com/
http://www.schering-plough.com/prod/prod07_animal.html
http://www.roche-vitamins.com/
http://www.intervetusa.com/
http://www.monsanto.com/Monsanto/default.htm

SECTION VI

APPENDICES

A

HUMAN RESOURCES

INTRODUCTION

Of all the resources in a pig production unit, the human resources are the most valuable. People will either make a unit fly or they will send it down the tubes. Fully automated pig production units with no human workers are unlikely to ever be achieved and are certainly not desirable. Efforts are underway to automate the more mundane tasks on a pig unit, but these efforts are very far from being implemented.

The difference between a successful pig production unit and an unsuccessful one is in the labor. Problems arise from the following categories:

1. Finding motivated workers
2. Training workers to have *consistently* high productivity
3. Retaining skilled workers

Two types of employees are sought by commercial pig units: those with experience and those without experience. Those new employees who have pig industry experience must be retrained to follow the procedures required by the new employer. Sometimes it is easier to train naive workers than to retrain people who have learned skills that are inappropriate to the new unit.

Workers can be further divided by level of education. New recruits may not have finished high school or may be high school graduates, 2-year college graduates, or 4-year graduates. They may have graduate or professional degrees. Workers with various degrees may or may not have pig experience.

Workers can be found in local communities, in the region, from nearby and distant universities, and even from overseas (Europe, in particular). Two common philosophies in pig production units are: (1) to hire the best people and move them to the region, especially for middle and upper management, or (2) to hire local people and train them starting with lower-level positions and move them up to middle and upper management positions. At this time, each philosophy has variable success rates.

Training workers is a critical component of both herd productivity and worker retention. When workers are trained properly, they feel the company is looking out for their continued development and giving them opportunities to be productive. Workers who feel the company cares about their needs in the workplace (and at home) are more likely to have a positive attitude about their job. Workers tend to be in either an improving or a declining productive phase. To keep workers improving or maintaining a high level of productivity requires both initial training and continued education and retraining.

Retention of workers has become a major problem, especially in large-scale pork production. Annual turnover rates over 50% are common. The first step in retaining workers is to keep them motivated. Once workers lose their motivation for the job, productivity slips and their attitude spirals down.

The following pointers may help keep motivation high for pig industry workers:

- Employees like the management to know them on a personal basis. The management should know the worker's name, family members, and a few interests of the employee.
- Although managers know parts of the worker's personal life, the employees need to have a private life that company doesn't see. Personal problems should be left for the employee to bring up in conversation. The management can then take an interest in the problem without prying.
- Off times, recreation, and family times are important for today's workers. Family time should be quality time. If workers stay at the farm for many hours and are tired or even exhausted when they get home, they cannot get enough *quality* family time. This leads to worker burnout.
- Families should know where the workers spend their time. Tyler suggests holding an open house so family members can see where their relative works.
- Managers should ask prospective employees about hobbies and special interests during the interview. People who have regular hobbies often enjoy talking about them.
- Employees should be trained in the history and development of the company. This builds loyalty and mutual respect.
- Employees and employers need to look at the relationships from the other side of the fence. They should explore ways to make it a better work experience and a more profitable enterprise.

HOW MANY WORKERS ARE NEEDED?

The restructuring of American pig production has resulted in an improvement in labor efficiency—fewer people are required to run a pig production unit. This trend results in the supervision of more and more animals per worker.

On a modern farrow-to-finish unit, the number of full-time workers is in the range of one worker per 200 to 300 sows. For very small units (1 to 200 sows), more than one person might be needed, at least for some task and to cover all 365 d/yr. The basic unit of production in modern pig units is the 1,200-sow unit or multiples of this size. A 1,200-sow unit (breeding and farrowing only) will wean and market about 1,000 pigs/wk with five on-farm employees. A 2,400-sow unit should wean and market about 1,000 males and 1,000 females/wk using 9 or 10 on-farm employees. A 4,800-sow complex might use 18 to 20 on-farm employees. The larger the farm, the more specialized (and presumably proficient) labor can become.

This appendix lists the jobs needed in a large-scaled pig unit. In smaller units, all of these job responsibilities may be held by one person. By looking at the large number of jobs and the specialized skills required, one can see that few individual people would be qualified for each and every job on the unit.

Following the list of job titles and responsibilities is a study of areas in the broad category of human resources that people should be familiar with if they are to be effective workers and leaders.

JOBS IN A PIG UNIT

A list of job titles and responsibilities for a modern farm spans over 20 job descriptions (see organizational charts, Figures A–1 to A–5). These jobs could be held by one

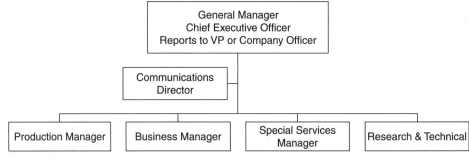

FIGURE A–1
Organizational Chart Level 1.

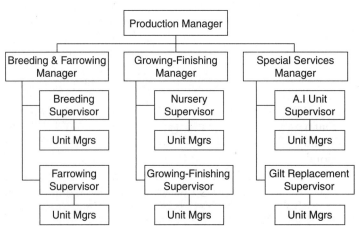

FIGURE A–2
Organizational Chart Level 2a.

FIGURE A–3
Organizational Chart Level 2b.

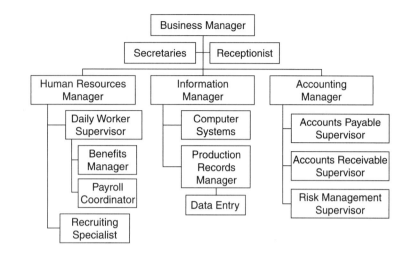

FIGURE A–4
Organizational Chart Level 2c.

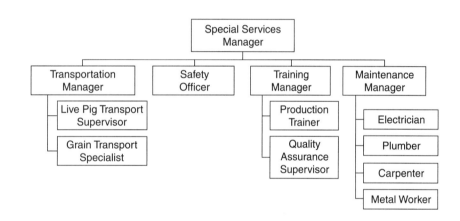

FIGURE A–5
Organizational Chart Level 2d.

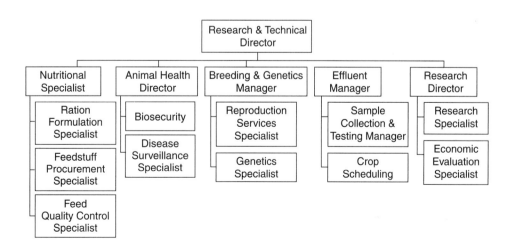

person per job or several people. The days of one person being skilled in all jobs are long gone. If one person runs a smaller farm, he or she will need to hire part-time consultants or specialists to handle some of the specialized tasks such as engineering, buildings and waste handling, veterinary medicine, accounting, and others.

A fully self-contained pig unit is probably impossible. Outside consultation is usually necessary and certainly is desirable for maximum efficiency.

A list of common jobs, by title and responsibility, follows. An organizational chart is included to show how the chain of command might work.

General Manager

The General Manager oversees both the business and biological components of the enterprise and is ultimately responsible for the unit's internal and external operations. The General Manager is usually either the owner or the person that the owner considers ultimately responsible for the production unit.

Communications Director

A pig unit needs to communicate effectively with the local community and other people external to the organization. The company should have a positive image in the community and this can only be established by the purposeful efforts of a skilled communications specialist. If the farm does not have specialists in communication, people should obtain training in effective communication techniques. Good relations with the local community will assure continued success of the pig unit. Alternatively, poor communication with the public will certainly cause the company financial hardships.

Business Manager

The Business Manager is responsible for the accounting, bookkeeping, office administration, human resources, benefits administration, worker's compensation, vacation and sick leave management, and any other nonpig, nonequipment function of the unit. This manager oversees all business and accounting affairs of the company.

Production Manager

The Production Manager is responsible for all of the people and activities related to live-pig production. Animal transport and effluent management may come under the Production Manager's control. All Unit Managers, Farrowing Managers, Breeding Managers, Nursery Managers, and Growing-Finishing Managers report to the Production Manager.

Special Services Manager

The job responsibilities of the Special Services Manager could be split between the Business and Production Managers, but for this illustration they are separate. Examples of departments that might fall under the Special Services Manager include Transport, Waste Management, Safety, Training, Veterinary Care, Feed Milling, Facility

Maintenance, and perhaps others. On mid-sized farms, the jobs of the Special Services Manager are split among workers holding the other job titles.

Research and Technical Director

This job and the workers under this leader might be in-house or they might be part-time consultants. The pig unit must be very large in scale to be able to afford full-time professionals in these jobs. Most of the workers in this division will have professional degrees up to the Ph.D. degree for many slots. With such professionals in this division, the Research and Technical Director often has a veterinary, engineering, or animal science advanced degree.

JOBS UNDER THE PRODUCTION MANAGER

These jobs are the core workers that make the farm run. Workers in this division must have an understanding of pig biology, management skills, and interpersonal and organizational skills. These job descriptions apply to large farms and a three-site production layout (breeding-farrowing, nursery, and growing-finishing). As such, there must be at least three production staffs—one for each site. To have this layout, with clear advantages for biosecurity and labor specialization, requires at least 1,200 sows for efficient operation.

Breeding and Farrowing Supervisor

This individual would oversee the sow herds for the entire production system. If the unit has 10 sow units, each with its own breeding and farrowing unit managers, the Breeding and Farrowing Supervisor would oversee these 10 people.

Breeding Specialist

As an assistant to the person in charge of all sow production, this person specializes in breeding techniques that assure the highest possible farrowing rates. In a large unit, the Breeding Specialist could have advanced training in reproductive biology. This individual is responsible for on-going training and tracking pig performance in the breeding barns across all the units.

Farrowing Specialist

The Farrowing Specialist keeps current on techniques and procedures to maximize the number of pigs weaned/sow. Expertise in causes of pigs born alive, stillbirths, and preweaning mortality is especially critical. This individual is responsible for on-going training and tracking pig performance in the farrowing barns across all the units.

Breeding and Farrowing Farm Managers

The sow unit is the core of the farrow-to-finish production unit and the Breeding and Farrowing Unit Manager is the leader of the core production unit. A typical production unit has five workers/1,200 to 1,400 sows. A 2,400-sow unit may have 10 workers under one Unit Manager. The Unit Managers in these systems oversee the generation of 20,000 to 50,000 pigs/yr. On smaller units, the Unit Manager may have responsibilities for nursery and growing-finishing and, as such, is more like a Production Manager.

Breeding Managers

The Breeding Manager is responsible for heat detection, breeding, and pregnancy checking of the sows in a given unit. Daily care of the sows and boars is also under the direction of the Breeding Manager.

Breeding Worker

This person assists the Breeding Manager in breeding and animal care tasks.

Farrowing Managers

The Farrowing Manager is responsible for sound farrowing barn practices that result in the highest numbers of pigs born alive and weaned and the lowest numbers of stillbirths and preweaning deaths. Daily care of the farrowing sows is under the direction of the person in this position.

Farrowing Worker

This person assists the Farrowing Manager in all farrowing and animal care tasks.

Utility Worker

On some farms, the farrowing and breeding assistants are the same as the Utility Workers. In this model, the Utility Worker takes care of routine cleaning, equipment set-up, and animal movements. The Utility Worker has the responsibility to power-wash the farrowing crates in indoor units. This job title is considered the starting point on many farms and, as such, has a high turnover rate due to people being promoted or terminated, or quitting due to not liking the work.

Nursery/Growing-Finishing Supervisor

This person is responsible for all pigs from weaning until market. Nursery and Growing-Finishing Supervisors report to this supervisor.

Nursery Complex Manager

This person is responsible for all nursery pigs at the various nursery sites.

Nursery Site Manager

A given nursery site may have from 1,000 to 32,000 pigs at one location. This person is the on-site manager with responsibility for the care and feeding of all the pigs at one location. The Nursery Site Manager also has local supervisory responsibilities for workers at this location and is typically a hands-on worker/manager.

Nursery Worker

The Nursery Workers do the manual tasks of the nursery. This includes moving pigs in and out, sorting pigs by size and gender and observing animal behavior for problems with the thermal environment or animal health problems.

Growing-Finishing Complex Manager

This person is responsible for all growing-finishing pigs at the various nursery sites.

Growing-Finishing Site Manager

A given growing-finishing site may have from 1,000 to 50,000 pigs at one location. This person is the on-site manager with responsibility for the care and feeding of all the pigs at one location. This manager also has local supervisory responsibilities for workers at this location. The Growing-Finishing Site Manager is typically a hands-on worker/manager.

Growing-Finishing Worker

The Growing-Finishing Workers do the manual tasks of the nursery. Tasks include moving pigs in and out, sorting pigs by size and gender, and observing animal behavior for problems with the thermal environment or animal health problems.

Special Units Manager

Some of the production units are separate from the core pig production. These are special units, including, but not limited to artificial insemination (AI) center, gilt replacement (and possibly gilt breeding) unit, and pig isolation facility. For biosecurity reasons and to develop specialized staff, these units should have staffers who are physically separate from the core production staff.

AI Manager

On modern farms, AI is used for nearly 100% of the matings. A specialized AI facility typically collects, extends, packages, and delivers semen to the breeding units. The AI manager might oversee from just a few to hundreds of boars.

Semen Collection and Semen Processing Workers

For a boar stud, the core workers feed boars, clean pens, and collect and process semen. If the boar stud has a large number of boars, specialized equipment might be used to process and package semen.

Gilt Replacement Manager

This person oversees the gilt replacement facility and process. In some cases a "gilt breeding program or project" is a part of the gilt replacement facility. A gilt breeding program delivers pregnant gilts to sow units as replacement animals. The gilt breeding program is in contrast to the traditional method of delivering peri-pubertal gilts to the sow units and allowing the sow breeders to inseminate gilts. In the gilt development area, gilts are held from a young age until either mating or until 30 d after mating when they are delivered to the units. The needs of the production unit for gilts determines the size and through-put of the gilt replacement unit.

Gilt Development/Replacement Workers

These workers care for developing and cycling gilts. In units with gilt breeding programs, workers in this area feed and care for the gilts, and possibly inseminate and assist in delivery.

Pig Isolation Manager and Workers

New breeding stock must be brought in to affect genetic improvement. New stock (gilts and boars) should be held in isolation and blood-tested to be sure they do not carry new pathogens to the farm. The manager and workers in this isolation facility must be in strict isolation from the primary pig unit to avoid unwanted contamination of in-house pigs.

JOBS UNDER THE BUSINESS MANAGER

Secretaries and Receptionist

The reception staff influences a visitor's first impression of a company. These individuals should be welcoming, pleasant, and personable to internal staff and visitors. The secretarial staff should support the business needs of the office and farm staff.

Human Resources Director

This job is especially critical because of the high turnover rates on U.S. farms and due to the many laws and regulations that apply to companies. Relevant laws include job discrimination laws and occupational health and safety laws and regulations. The Human Resources (HR) Director has four primary responsibilities that are performed by the Director or, if the unit is large enough, by specialized staff. The next four positions report to the HR Director.

Daily Worker Manager

This person assures each position is filled each day. If people call in sick or need to have time off for a personal reason, the Daily Worker Manager ensures positions are covered or temporarily filled. Daily, weekly, monthly, and annual work schedules and weekend and holiday workers are organized and managed by this person.

Benefits Manager

Full-time workers have several benefits that require management. These benefits include production bonuses, vacation pay, sick leave, disability, retirement, and other business perks.

Payroll Coordinator

This person assures people receive the proper pay for all work. This person combines personnel and accounting services.

Recruiter

This job is critical for success and competition for new employees. The Recruiter travels to local universities, high schools, job fairs, and employment agencies to find prospective employees. The Recruiter must be personable, outgoing, and have good interview skills. Job interviews could be conducted on the farm or in another town during the recruiting process.

Information Services Manager

Internal communication and information flow is critical to success in the modern operation. The following two positions report to this manager.

Computer Systems and Information Specialist. This position is responsible for establishing and maintaining the computer system on the farm.

Production Records Entry. If production records are handled on-site, labor is needed to enter the records and generate reports. Timely reports are needed for proper management of the biological units.

Accounting Manager

This person is in charge of accounting services. This job and related activities could be out-sourced on a smaller farm, but most farms of any size perform the majority of accounting services on-site.

Accounts Payable

This person gathers and pays all bills that arrive on a defined schedule.

Accounts Receivable

This person sends out bills and collects payment. Ordinarily, the income to this farm arrives once/wk (or so), so a full-time employee should not be needed in accounts receivable.

Risk Manager

Many farmers and business people buy, sell, and trade options and futures on both inputs and outputs of the production unit. Grain and pig/pork commodities can be managed by a person on the farm who tracks these markets with an eye to minimizing risk to the farm.

JOBS UNDER THE SPECIAL SERVICES MANAGER

The special services group handles all operations that are not related to pigs, technical matters, or the business office. This department could be spread between the production and business departments. Having a separate special services department allows the production and business staffers to specialize in their work.

Transportation Manager

This department and position could be easily out-sourced. If the farm is large enough to allow efficiency of scale, the transportation of grain and pigs will take place nearly every day.

Live Pig Transport

Pigs are moved from isolation to gilt breeding, gilt breeding to sow units, farrowing to nursery, nursery to growing-finishing, and growing-finishing to slaughter. All but the last transport occur within the farm and should be done with in-house equipment. The shipment of pigs from finishing to slaughter can be out-sourced, with proper biosecurity control measures.

Grain Transport

Even a small-scale production unit needs either complete feed or feed ingredients (grain, etc.) delivered on a regular basis. Grain transport is typically handled by outside vendors.

Safety Officer

Safety is an important part of a quality company and providing a safe workplace is not only good business, but a requirement of law.

Training Manager

The training department could be a part of special services, production, or the business office. Regardless, the Training Manager should work with the staff to train workers either in specialized training farms or on regular production units with close oversight of the training staff. The Training Manager often works with specialized technical specialists (e.g., breeding, farrowing, nursery and grow-finish) to train new staff.

Production Trainer

This person conducts the hands-on and classroom training.

Quality Assurance (QA) Trainer

QA is an important part of every operation on the farm. The QA office is a critical part of new programs using identification of Hazard Analysis Critical Control Points (HACCP) programs.

Maintenance Manager

The physical plant of the farm is in need of repair and upkeep from the moment construction is complete. As the farm ages, the time and money spent in maintaining the farm increase proportionally. The maintenance department will include workers trained as electricians, plumbers, carpenters, and metal workers.

JOBS UNDER THE RESEARCH AND TECHNICAL DIRECTOR

Nutritional Services

Most pig units are not large enough to have their own full-time Pig Nutritionist. Those that are, have the following three main functions, each with distinct job requirements.

Feed formulation. One person is needed to conduct routine ration formulations and reformulations. Ingredients change composition over time and feedstuffs vary in cost. A commercial laboratory is used to assay feedstuffs for major nutrients and this information is used to reformulate diets.

Quality control. A person needs to sample ingredients and complete fees. Changes in feedstuff composition and feed ingredient mixing completeness need to be constantly checked and adjusted.

Feedstuff procurement. Because feed represents from 50% to 80% of the cost of production, obtaining quality feed ingredients at a reasonable cost is a critical job.

Animal Health Director

Pig health is critical for optimum herd performance. A veterinarian usually heads the animal health program. This person could be a part-time consulting veterinarian.

Biosecurity

Keeping disease out of the farm is the major function of the biosecurity officer. Incoming vehicle and people traffic should be closely monitored.

Disease Surveillance

The herd must be monitored for low-level and acute pathogen status. Only with an informed health status can the herd have its health remain at a high level.

Animal Health Product Evaluation and Quality Control

This person monitors the effectiveness of vaccines, stock isolation, and antibiotic use.

Breeding and Genetic Director

Reproductive services is critical for the continued success of the breeding herd. The following three primary workers are in this division.

Reproduction services. This breeding specialist controls boar semen collection and gilt/sow insemination.

Genetics manager. This person oversees the herd genetic improvement and herd replacement programs. The Genetics Manager directs gilt and boar selection, assuming an in-house genetic multiplication unit is in place.

Reproduction and genetics quality control. As with nutrition and animal health, a quality control person is needed to ensure semen is properly processed, matings are made appropriately, and that the general reproduction and genetics program is on line and going according to the master plan of the farm.

Effluent Manager

Pig units produce a large amount of effluent—feces, urine, spilled feed, and dead animals. The Effluent Manager directs disposal of these valuable by-products.

Sampling and Testing Manager

This person samples the effluent and croplands to determine correct application rates.

Crop Scheduling

A full-time crop manager may be needed if the effluent is spread on the farm's land. Crops must be selected, planted, and managed to take up as much of the nutrients from effluent as is possible.

Research Director

Too often, research is half-way done by nonspecialists. Without proper oversight, research studies can be misinterpreted or, worse, studies can even lead to wrong directions for the farm. Large production corporations may spend 1% to 5% of their gross sales on research and product improvement. For a 100,000-sow unit, the research budget should be from $2 to $10 million. Because pig production units do not spend enough ear-marked money on organized research, opportunities are lost to improve productivity. Research of this nature does not cost—it pays!

Research Technicians

Staffers are required to conduct studies so the core production staffers are not interrupted in their daily routines.

Economic Evaluation of Corporate Plans

An economist is needed to evaluate corporate research results. This person can also evaluate the constant flow of new ideas that arise from day-to-day discussions about future directions.

HUMAN RESOURCES LAWS AND REGULATIONS

Laws abound that regulate human resources. Laws are enacted at city, county, state, and federal levels. In the United States, the primary federal laws refer to equal opportunity, lack of discrimination, and minimum wage limits. At the state level, the laws of most concern are worker compensation laws and regulations. Most businesses are required to carry worker compensation insurance in case of on-the-job injury.

Owners should be familiar with federal and state laws to be in compliance with laws that govern retirement and profit-sharing.

CONFLICT RESOLUTION

From time to time, conflicts arise in the workplace. Management's first rule in resolving conflicts is to let the affected people work out the problem on their own. Let's use the example of a worker and unit manager having a conflict and one party consults the Production Manager. The Manager's first option should be to send them home to think about the problem and develop their own solution. Then, failing that solution, the Production Manager must arbitrate the conflict. Getting the facts straight is the next order of business. This is best done with both parties present (to avoid exaggeration). After gathering the facts, the Production Manager should wait a small amount of time to offer a solution. In some cases, the two sides will resolve the conflict during that period. The wait should be less than 24 hr because the problem could be serious enough to warrant conflict resolution.

When the Production Manager makes a decision, he or she should clearly convey the outcome to both parties and give reasons for the decision to both parties. A chance for appeal should usually be granted, but only one chance should be given to avoid a long, protracted problem.

OTHER POTENTIAL CONTENTIOUS AREAS

Drugs and Alcohol

Most firms today perform pre-employment drug screening and random drug testing. Because the work is potentially hazardous, the employer cannot risk having impaired employees on the work site.

Relatives

Because most farms are in small communities, relatives are often employed by the business. Every attempt should be made to have relatives working on different sites and performing different tasks.

Persons Abusing Worker Compensation Insurance

People do get injured from time-to-time on the farm. Some people, however, will fake an injury or fake an extension of the injury to get more time off work with pay. These people should be treated as fairly as possible, but the company must avoid being taken advantage of.

Espionage

Persons from some other companies or activist groups (environmental or animal rights) may get a job at the site to gather information to hurt the company. Checking references and even checking with other industry sources about potential employees is advised. This is becoming increasingly difficult because many companies will not give references on former employees due to past legal action.

FINDING A JOB IN THE PIG INDUSTRY

Students or others interested in a job in the U.S. pig industry or the industry in Asia or Europe will have great opportunities for the foreseeable future. The employer's first criterion is that the potential worker is highly motivated to work in the industry. Once the worker is highly motivated, he or she can be placed in a number of jobs or training positions. In any case, new employees must be trained in the methods and policies of the company—whether family-style, corporate-style, or family corporation-style.

When in search of a job in the pig industry, prospective employees can consult a plethora of sources. Some of the sources are given here for employment in the United States or for companies with U.S. offices and production in other countries.

Personnel Search Firms

Agri-Careers: *http://www.agricareers.com/*
Ag Jobs USA: *http://www.agjobsusa.com/*
PIC USA: *http://www.pic.com*
Iowa Select Farms: *http://www.iowaselect.com/welcomefla.htm*
Newsham Hybrids (USA), Inc.: *http://www.newsham.com/*

Large U.S. Pork Producers

Smithfield Foods, Inc. corporate office: *http://www.smithfield.com/*
Murphy Family Farms (owned by Smithfield): *http://www.u-n-i.net/murphyfamilyfarms/*
Tyson Foods: *http://www.tyson.com/*
Premium Standard Farms: *http://www.psfarms.com/*
Seaboard Farms: *http://www.seaboardpork.com/*

National Organizations

National Pork Producers Council:
P.O. Box 10383
Des Moines, IA 50306
Phone: (515) 223-2600

http://www.nppc.org

QUESTIONS AND ACTIVITIES

1. Prepare an organizational chart and a list of job descriptions for two farms:
   a. a 10,000-sow unit
   b. a 1,200-sow unit
2. What human resources challenges are likely for a farm with 100,000 sows or more?
3. What are the responsibilities that might be in a job description for the owner and sole operator of a 100-sow, farrow-to-finish farm?
4. Discuss routine and more creative solutions to the following problems:
   a. Finding motivated workers
   b. Keeping qualified people on the job
   c. Maintaining productivity on the farm

B
Glossary

Abattoir
See: **Slaughter house.**

Abortion
Termination of pregnancy.

Ad libitum feeding (ad lib feeding)
A system of feeding in which no limit is placed on feed intake.

Age at slaughter
The number of days from a pig's date of birth to that pig's date of slaughter.

Age at weaning
The number of days from a piglet's birth to that pig's removal from the sow.

Arrival date
The date when a breeding company delivers a replacement gilt or boar to a farm or the date when the gilt or boar joins the breeding herd after selection.

Artificial insemination
The deposition of spermatozoa in the female genitalia by artificial rather than natural means.

Assets, current
Items that are used or sold and can be converted into cash within a 12-mo period without disrupting the business. Examples include cash, cash value of bonds, stocks and life insurance, accounts receivable, nondepreciable items such as grain, fresh flowers, forage inventories, feeder and market livestock, and supplies such as vermiculite and fertilizer.

Assets, fixed
Items that have combined, depreciated value; includes the operation's buildings, land, machinery, vehicles, breeding stock, other long-term assets, and any other disposable items of value.

Atrophied piglets
The wasting away or decreasing in size of muscle, fat, or any tissue or organ. May result from weakening disease or from disease.

Availability, of breeding stock
The quantity of breeding stock of any class ready for shipment.

Availability, of nutrients
A feed's nutrient concentration that is absorbed and utilized by the animal.

Average
An arithmetic mean; the sum divided by the sample size.

Average, weighted
A sum divided by the number of contributing units.

Backcross
The breeding of F1 females to a male of the same breed as the F1's parents.

Baconer
A class of finishing pig in the United Kingdom; typically in the weight category of 59 to 70 kg (130 to 155 lb).

Barrow
Male pig that was castrated when young.

Boar
A male pig that has not been castrated (intact).

Boar, breeding
A male, uncastrated intact pig that is used in the breeding herd.
Syn: **Stock boar**

Boar, replacement
A male that will be used in the breeding herd.

Boar, unworked
An intact (uncastrated) male before the date of his first mating.

Boar, working
An intact (uncastrated) male that will be used only for breeding purposes after the first day of mating.

Boar, vasectomised
A male that is made infertile by means of surgery.

Breeding herd
The livestock retained to provide for the perpetuation of the herd or band.

Carcass
The major portion of a meat animal remaining after slaughter. Varies among animals, but usually the head and internal organs have been removed. Skin and shanks are removed from cattle and sheep.

Casualty slaughter
A pig that is euthanized because of injury or disease.

Conception rate
In cattle breeding, the percentage of first services that conceive.

Condemnation
(1) Describing an animal, carcass, or food that has been declared unfit for human consumption. (2) Referring to real estate property acquired for public purposes under the right of eminent domain.

Condition scoring
The assessment of a sow's physical condition; method is based on the sow's shape, level of body fat, and how easily "H" bones can be felt when pressure is applied with the hands.

Congenital
Acquired during prenatal life; certain conditions that exist at birth; often used in the context of birth defects.

Congenital defect
An abnormality of a newborn piglet present at birth; can be fatal.

Contemporary group
A group of cattle of the same breed and sex that are raised in the same management group (same location on the same feed and pasture). Individual animals can then be accurately compared with the others in the group. Contemporary groups should include as many cattle as can be accurately compared.

Controlled feeding
A predetermined amount of feed delivered at each meal.

Correlation
The degree of relation between variables, sets of data, or traits. Positively correlated traits vary in the same direction; negatively correlated traits vary in opposite directions.

Correlated genetic response
A genetic change in one trait caused by selection for another trait.

Creep feed
A system of feeding young domestic animals by placing a special fence around feed for the young. The fence excludes mature animals but permits the young to enter. Or, feed is provided to nursing piglets.

Critical temperature, maximum
Environmental temperature high enough to reduce an animal's performance.

Critical temperature, minimum
Environmental temperature below which an animal must create extra heat to maintain body temperature.

Crossbred
An offspring that results from the breeding of two purebred parents of different breeds.

Criss-cross
A method of breeding crossbred animals. Breed A males are usually bred to females with Breed B and Breed B males are usually bred to Breed A females.

Cull
(1) Anything worthless or nonconforming that is separated from other similar and better items; the act of removing the inferior items; to cull out. (2) The lowest marketing grade of meat carcasses or dressed poultry. (3) Any animal or fowl eliminated from the herd or flock because of unthriftiness, disease, poor conformation, etc.; a reject.

Culling percentage, females
The number of females culled/yr multiplied by 100 and divided by the average female inventory.

Culling percentage, sows
The number of sows culled/yr multiplied by 100 and divided by the average sow inventory.

Cutter
A class of finishing pig in the United Kingdom; typically in the weight category of 50 to 80 kg (110 to 178 lb).

Cycle
The age of a sow calculated by the number of reproductive cycles completed also parity.
   A sow remains in her first cycle from entry date until the first mating after her first farrowing. She then moves into her second cycle.
   This term has advantages over parity in that all periods are of approximately equal length, that is one farrowing interval (though this may vary from herd to herd).
Con: **Parity**

Daily live weight gain
The gross weight of a live animal as compared with the dressed weight after slaughter.

Dam line
A breed selected to contribute the female parents of the slaughter generation in a hybrid breeding program.

Days to market
The number of days from a pig's date of birth to that pig's date of slaughter.

Death percentage, females
The number of females that die/yr multiplied by 100 and divided by the average female inventory.

Death percentage, sows
The number of sows that die/yr multiplied by 100 and divided by the average sow inventory.

Dressing percentage
Carcass weight divided by live weight and multiplied by 100. Usually the cold carcass weight is used. The dressing percentage for cattle averages around 50% to 60%, hogs average around 70%.

Depreciation
The decrease in value of business assets caused by wear and obsolescence.

Emaciated
Backfat less than 0.6". H bone painfully and easily visible. Body shape: backbone, rib cage, hip bones prominent, sharp to touch.

Embryo
Any organism in its earliest stages of development.

Empty days
The number of days between weaning and conception.

Energy—digestible
The proportion of energy in a feed that can be digested and absorbed by an animal.

Energy—gross (of feed)
A measure of the chemical energy of feed by means of a calorimeter.

Energy—metabolizable
The total amount of energy in feed less the losses in feces, combustible gases, and urine. Also called available energy.

Energy—net
The amount of energy that remains after deducting from a feed's total energy value the amount of energy lost in feces, urine, combustible gases, and heat increment. Sometimes called work of digestion.

Enteritis
Any inflammatory condition of the lining of the intestines of animals or people. Characteristics of enteritis are frequent evacuations of a liquid or very thin, foul-smelling stool that may or may not contain blood; and straining, lethargy, and anorexia. In acute cases, there is a rise in body temperature. The condition is seen as a symptom of a number of infectious diseases or it may be caused by specific bacteria or viruses. Other common causes include plant and animal poisons, parasites, overeating, faulty nutrition, and poor environmental factors.

Entry date
The date when a boar is first bred or a gilt is first mated.

Entry date (25-d arrival rule)
In Germany and France, the date when a gilt becomes a sow automatically if she remains in the herd for more than 25 d even if she has been mated.

Entry date (200-d-old rule)
The date when gilts automatically become sows after they reach 200 d of age.

Equity
Net ownership of a business; the difference between the assets and liabilities of an individual or business as shown on the balance sheet or financial statement (assets − liabilities = owner equity).

Estrus
The period of sexual excitement (heat) and a certain hormonal profile, at which time the female will accept coitus with the male.

Estrous
Pertaining to estrus (heat) in animals.

Event date
The date when an event occurs.

Farrowing
Normal-delivery of one or more live, stillborn, or mummified pigs, on or after the 110th day of pregnancy.

Farrowing crate
A crate or cage in which a sow is placed at time of farrowing. The crate is constructed as to prevent the sow from turning around or crushing the newborn pigs as she lies down.

Farrowing date
The date of birth of the first pig of a litter.

Farrowing, failure
Failure to farrow within 120 d after an effective service.

Farrowing index
The number of farrowings per sow per year divided by average sow inventory.

Farrowing interval
The number of days between a sow's two consecutive farrowing dates.

Farrowing rate
The number of farrowings/yr divided by the number of farrowing services.

Fat
Backfat 0.9″. No detection of H bone. Body shape: thickening of trunk behind front legs and in neck region. Rear rounded.

Fat
(1) (a) The tissues of an animal that bear an oily or greasy substance. (b) Any animal or fowl that abounds in fat. (2) The oily substance in milk.

Feed
(1) Harvested forage, such as hay, silage, fodder, grain, or other processed feed for livestock. (2) The quantity of feed in one portion. (3) To furnish with essential nutrients.

Feed consumed/yr—total
The initial feed inventory, plus feed purchases and farm-produced feed; subtract the ending feed inventory, feed sales, and feed eaten by other livestock.

Feed conversion ratio
The rate at which an animal converts feed to meat. If an animal requires 4 lb of feed to gain 1 lb, it is said to have a 4:1 feed conversion ratio.

Feed efficiency
A term for the number of pounds of feed required for an animal to gain 1 lb of weight.

Feeder pig (USA)
A pig 8–12 wk of age destined for finishing.

Feed intake/d
The rate at which animals consume feed.
NOTE: What we really measure is the rate of FEED DISAPPEARANCE, which includes both consumption AND waste. In slatted systems with poorly-designed troughs, waste can easily exceed 5% of allocation.

Female herd
The total number of all maiden gilts on a farm or those selected from the finishing accommodation, plus all mated sows until they are culled and removed.

Female herd turnover rate percentage
A statistic reached by adding the culling percentage and the death percentage for all females in a herd.

Female-to-boar ratio
The average female inventory divided by the average boar (breeding and unworked) inventory.

Fetus (USA)
An unborn animal.

First cross (F1 or F1 cross)
The first-generation progeny that is produced by crossbreeding two different lines.

F2 generation
The second-generation progeny produced by crossing two F1 individuals.

First litter sow
After a successful pregnancy, a sow between the date of the first effective service and the date of the next effective service.

Fixed cross
A crossbreeding or hybrid system that places breeds or lines in fixed positions as sire and dam lines; the final progeny are slaughtered and purebred or crossbred parents are provided as replacements.

Flat deck
Specialized housing with perforated floors and temperature-controlled environment for recently weaned pigs.

Fostering
Transferring piglets from one litter to another.

Full feeding
A feed or ration being fed to the limit of an animal's appetite.

Full siblings
Individuals that have the same parents; can include littermates and full brothers and sisters from repeated services of the same parents.

Gastroenteritis
Inflammation of the stomach and intestines.

Generation interval
The period of time between the birth of one generation and the birth of the next.

Genetic correlation
The extent of similarity between two traits (ranging from −1 to +1) due solely to genetic influences. Two traits can be controlled by the same genes. Positive genetic correlation indicates that selection for one trait will result in a positive, correlated response in another trait.

Genetic lag
The time required for genes to pass from the nucleus to the slaughter generation. Successive generation intervals in the nucleus, daughter nucleus, and multiplier layers in the breeding pyramid comprise this period.

Genetic level
A particular sow or boar's position in a breeding pyramid.

Genetic status
The genetic composition of boars and sows in common terms, such as to purebred or hybrid.
Syn: **Line**

Gestation
Period of pregnancy, or the time between service and the subsequent farrowing or abortion (the day of service is counted as day 0).

Gilt
A female that has arrived in the breeding herd but has not yet been mated. An in-pig gilt is a female that has been served and therefore should properly be described as a sow.

Gilt pool
Replacement gilts used for breeding.

Gilt replacement rate percentage
The number of gilts that enter the herd/yr multiplied by 100 and divided by the average sow inventory in the same period.

Grandparent
The parent stock (boar or sow) in commercial production in a multiplier herd.

Great-grandparent
The grandparent stock (boar or sow) in a nucleus herd.

Grower pig
A pig after transfer to (UK) or from (USA) the nursery to the moment of transfer into the finisher building.
Con: USA and UK terminology
Syn: **Growing pig**

COUNTRY	USA	EUROPE
Start Weight	50 lb (23 kg)	5 kg (11 lb)
End Weight	120 lb (55 kg)	30 kg (66 lb)

Grower pig
Swine from about 40 to 100 lb.

Growth promoters
Non-nutrient feed additives that decrease the feed conversion ratio and increase growth rate.

Growth rate
The per-day rate at which live weight increases between two successive weighings.

Heat period
Estrus; the period during which a female is sexually receptive.

Half-siblings
A half-brother or half-sister.

Half-sibs, maternal
Half-siblings that have the same dam.

Half-sibs, paternal
Half-siblings that have the same sire.

Heavy hogs
A finished pig class in the United Kingdom in the weight category over 82 kg (180 lb) carcass weight.

Hedging gain or loss
A net gain or loss resulting from trading in pig futures.

Herd size, female
The average number of sows and gilts in a herd.
Con: **Herd size, sow**

Herd size, sow
The average number of mated females (but not unserved gilts) in a herd.

Heritability
The proportion of the variation in traits attributable to genetics. Heritability varies from zero to one. The higher the heritability of a trait, the more accurately the individual performance predicts breeding value and the more rapid should be the response due to selection for that trait.

Heterosis
The amount of superiority observed or measured in crossbred animals compared with the average of their purebred parents; hybrid vigor.

Hog
(1) a pig of any size or (2) a British term for a castrated male pig.

Hybrid
An animal produced from the crossing or mating of two animals of different breeds.

Hybrid vigor
The increase of size, speed of growth, and vitality of a crossbred over its parents.

Hysterectomy
(1) Technique used in some specific pathogen-free laboratories to remove unborn pigs from the sow. The entire uterus is removed with the pigs inside. This operation makes the sow useless, and she is slaughtered immediately after the operation.
(2) Surgical removal of all or parts of the uterus.

Ideal
Backfat 0.7″ to 0.8″. H bone can only just be felt. Body shape: tube shaped. "Lean but fit" look.

Indexing
A system for comparing animals with a herd, or area, based on the average of the group; usually the figure 100 is used for an average index; animals receiving an index of 100 or over are the top end while those indexing less than 100 are the bottom end.

Induction of farrowing
A treatment that can cause farrowing in a pregnant sow in late pregnancy.

Insurance
Guarantee against loss for fire, casualty or third-party liability in exchange for premiums paid on an enterprise, buildings, equipment, or livestock.

Intensity of selection
A standardized measure that compares selected individuals and a group.

Interest
Amount of funds paid to a lender for the use of money.

Interference level
A minimum level of performance.

Killing-out percentage
The carcass weight multiplied by 100 and divided by the live weight.

Labor and fringe benefits
Wages and noncash benefits such as staff housing, social security, health insurance, and pension contributions.

Lactation
The process of forming and secreting milk.

Lairage death
In the United Kingdom, a pig that dies between unloading at the slaughterhouse and the point of slaughter.

Late foster sow
A sow that weans her own litter and then weans piglets from other sow(s).
Syn: **Nursemaid sow**

Leak
Leakage of semen from the vulva during or after artificial insemination.

Leaving date
The date at which the sow leaves the herd, i.e. is sold, culled or dies.

Liabilities
Money, goods, and/or services that are owed.

Litter
The multiple offspring born during the same labor; also, a substance used by animals that is appropriate for absorption of waste products.

Litter number
See: **Cycle.**

Litter scatter graph
The percentage of litters with more than and/or less than a specified number of liveborn piglets.

Litter weight
A litter's total weight.

Litters farrowed per crate per yr
The total number of farrowings/yr divided by the average number of farrowing crates.

Litters per female per yr
The total number of farrowings/yr divided by the average total female inventory.

Litters per mated female per yr
The number of litters/sow each year.
Syn: **Farrowing index**

Live births per female per yr
The total number of live births/yr divided by the average total female inventory.

Live births per sow per yr
The total number of live births/yr divided by the average sow herd size.

Live-born piglets
Piglets that are born alive.

Live births/litter
The total number of live births divided by the number of litters farrowed.

Live birthweight, average
The total birthweight of pigs that are born alive divided by the total number of live pigs weighed in the same time period.

Live weight
The gross weight of a live animal as compared with the dressed weight after slaughter.

Losses
Deaths.

Losses, causes preweaning
Common causes for preweaning deaths include congenital defects, low viability, low birthweight, scour, starvation, or death by being laid on or killed by the sow deliberately.

Losses, causes post-weaning
Common causes for post-weaning deaths include scour, dysentery, respiratory and nervous system diseases, stress, and injury.

Maiden gilt
See: **Gilt.**

Maintenance ration
The amount of feed needed to support an animal when it is doing no work, yielding no product, and gaining no weight.

Margin
The sum obtained by the difference between a certain type of revenues and related expenses.

Market hogs
In the United States, finished pigs destined for slaughter.

Market weight, average
The *total* weight of all marketed pigs divided by the number of pigs marketed in the same period.

Marketable pork produced/yr
The weight of ending inventory, plus sales, minus the beginning inventory, minus purchases.

Mating
(1) To pair off two animals of opposite sexes for reproduction. Mating may be for a single season or for life. (2) In plants, to be cross pollinated.

Mated females
The total number of females in a breeding herd that have had at least one mating.

Mated female death loss percentage
The number of sows that die/yr multiplied by 100 and divided by the average sow inventory.

Mated female-to-service boar ratio
See: Sow-to-boar ratio.

Mean
A middle point between two extremes; the average.

Megajoule (MJ)
An SI (Systeme Internationale d'Unities) unit of energy; 4.184 MJ = 1,000 calories = 1 kilocalorie.

Morbidity
The condition of being diseased, or the incidence or prevalence of some particular disease. The morbidity rate is equivalent to the incidence rate.

Mortality rate
The number of overall deaths, or deaths from a specific disease, usually expressed as a rate; i.e., the number of deaths from a disease in a given population during a specified period, divided by the average number of people or animals exposed to the disease and at risk of dying from the disease during that time.

Mortality (preweaning)
Death that occurs between live birth and weaning.

Mortality (post-weaning)
Death that occurs in finisher pigs between weaning and slaughter.

Mummification
(1) In animal reproduction, the drying up and shriveling of the unborn young.

Nonproductive sow days
Total number of days that elapse between the first service date and the date when a sow is culled, dies, aborts, or is repeat served or fails to farrow.
The number of days that a sow is kept in a herd while she is nonproductive.

Not in pig
A sow that fails to become pregnant up to 110 d after a presumed effective service.

Nursery
A place for housing either (a) lactating sows and their piglets (older term) or (b) weaning pigs (newer term).

Nursery, cold
Specialized housing for newly weaned pig that lacks supplementary heating.

Nursery, hot
Specialized housing for newly weaned pigs that has supplementary heating.
Syn: **Flat deck**

Nursery, pre-
Specialized, temperature-controlled housing for early-weaned pigs (3 to 4-wk-old until 6 wk old).

Nursemaid sow
See: **Late foster sow.**

Nursing pig
A pig that has not been weaned.

Oestrus (UK)
See: **Estrus.**

Opportunity cost
The hypothetical cost or benefit of a forgone opportunity.

Overfat
Backfat 1". No detection of H bone. Body shape: excessively thickened trunk behind front legs and in neck region. Bulbous shape.

Parity
The number of times a female has borne offspring.

Parturition
Giving birth; called farrowing or pigging in swine.

Performance testing
The systematic collection of comparative production information for use in decision making to improve efficiency and profitability of beef production.

Phenotypic correlation
The observed relationship between traits, caused by both genetic and environmental effects.

Piglet
A young pig of either sex.

Pigs born live/female
The number of pigs that are born alive/yr divided by the average sow inventory in the same period.

Pigs sold/yr
The total number of pigs of a defined type sold/yr.

Pigs sold per sow per yr
The sales of stated type/yr divided by the average sow inventory.

Pigs weaned per farrowing crate per yr
The number of pigs weaned/yr divided by the average number of farrowing crates during the same time period.

Pigs weaned per female per yr
The number of pigs weaned/yr divided by the average female inventory during the same time period.

Pigs weaned per sow per yr
The number of pigs weaned/yr divided by the average sow inventory during the same time period.

Pigs weaned per sow-place per yr
The number of pigs weaned/yr divided by the average sow capacity during the same time period.

Pork
The meat of pigs.

Porker
Any young hog.

Post-weaning loss
Pigs that die between weaning and slaughter, expressed as a percentage of the number of at-risk pigs.

Post-weaning death loss percentage
The post-weaning death loss multiplied by 100 and divided by the number of pigs weaned in the same group.

Pregnancy
The condition of a female animal having a living fetus in the uterus; occurs after the ovum has been fertilized by the male sperm cell. See: **Gestation period.**

Pregnancy rate, 40-d
The number of sows that become pregnant within 40 d of service divided by the number of sows serviced in the same time period.

Preweaning loss
A pig that is born alive but dies before weaning.

Preweaning death loss percentage
The number of preweaning death losses multiplied by 100 and divided by the number of liveborn pigs in the same group.

Premature farrowing
The birth of fetuses before the 110th day of pregnancy wherein some survive for more than 24 hr.

Premature pigs
Piglets that are born alive before the 110th day of pregnancy and survive for more than 24 hr.

Progeny testing
Determining the breeding value of an animal by studying its progeny.

Purebred
Designating an animal belonging to one of the recognized breeds of livestock. Such animals are registered or eligible for registry in the official herdbook of the breed. Purebred, registered, and pedigree stock are often used interchangeably, and the term "thoroughbred" is often improperly used for "purebred."

Rent and lease
Payments that are made for leased or rented property or land.

Repair and maintenance
The costs of repairing or maintaining buildings, equipment, or machinery.

Repeat breeder
A female that returns to estrus and is served before the anticipated farrowing date.

Replacement gilt
Females that have been selected to be in the breeding herd.

Replacement rate percentage
The number of pigs mated for the first time multiplied by 100 and divided by the average sow inventory during the same time period.

Restricted feeding
A system of feeding pigs (particularly sows) whereby feed is provided only during certain periods of the day and in less than ad libitum quantities.

Rotation cross
Systems of crossing two or more breeds wherein the crossbred females are bred to males of the breed contributing the least genes to that female's genotype. Rotation systems maintain relatively high levels of heterosis and produce replacements from within the system. Opportunity to select replacement heifers is greater for rotation systems than for other crossbreeding systems.

Scours
Loose stools or diarrhea usually caused by an infectious agent.

Selection
Choosing certain individuals for breeding purposes to propagate or improve some desired quality or characteristic in the offspring.

Semen
A fluid substance produced by the male reproductive system containing spermatozoa suspended in secretions of the accessory glands.

Service
For a sow, one or more completed matings within the same estrous period.
For boars, all the matings in a particular service.

Service, date of
The date of the first mating in any one estrous period.

Service, effective
A successful mating until proven otherwise.

Service period
The period of time during when one or more matings can occur.

Service, repeat
An additional service in the same estrous cycle.

Service, return to
Symptoms of heat in a sow more than 5 d after service, measured from the first day of mating.

Service, normal or regular return to
Symptoms of heat in a sow 18 to 24 d after previous service, measured from the first day of mating.

Service, irregular return to
Symptoms of heat in a sow outside the 18 to 24 d interval, measured from the first day of mating.

Service, early return to
Symptoms of heat in a sow prior to 18 d post-service, measured from the first day of mating.

Shoat
A young pig.

Sire line
A breed of males and females in a hybrid breeding program that exclusively contributes to the sires of the slaughter generation.

Slaughter
The killing of pigs for food.

Slaughterhouse
Location where pig slaughter occurs.
Syn: **Abattoir, Packing house**

Slaughter weight
The live weight of a pig at slaughter.

Sow
A female swine, usually one that shows evidence of having produced pigs or one that is obviously pregnant.

Sow death loss percentage
See: **Death percentage, sows.**

Sow inventory
See: **Herd size, sow.**

Sow-to-boar ratio
The average sow inventory divided by the average breeding boar inventory.

Sow-to-gilt ratio
The average sow inventory divided by the average gilt inventory.

Sows weaned/yr
The number of sows that gave birth and whose pigs were removed from them, regardless of how long they lactated.

Stage of production
The phases of the pig production process; includes Breeder, Preweaning, Nursery, Grower, Finisher, and Off-test.

Standing reflex
A sow's response to pressure on her back; useful in AI. This may be elicited by pressing firmly on the sow or actually sitting astride her. A strong positive response is shown by the sow standing still, pushing upwards and giving characteristic grunts. It can be measured on a three-point scale where 3 indicates the strongest positive reaction and 1 means a total lack of interest on the sow's part.

Stig
The identification tag on a pig's ear.

Stillborn
Born lifeless; dead at birth having not taken a breath.

Stillborn percentage
The number of stillborn piglets multiplied by 100 and divided by the total number of pigs born alive in the same group.

Store pig
An animal not yet ready for slaughter.

Substandard pig
A pig at weaning that is low in weight and/or vigor.

Suckling pig
A young pig still nursing its mother. When slaughtered at this stage, it produces a small carcass for roasting whole.

Supplies
Materials used in operation of pig enterprise; includes clothing, cleaning products, small tools, and other materials.

Target
A goal to be achieved; a level of production to aim for.

Taxes
Compulsory charge that is levied by a federal, state, or local unit of government against income or wealth for the common good.

Testing, performance
See: **Performance testing.**

Thin
Backfat 0.6″. H bone easily felt, or visible protrusion. Body shape: hollowness at loin, (flat or "slab" sides). Cavity around tail setting.

Total births/litter
The total number of live births, plus the number of stillborn and mummified pigs/litter, divided by the number of litters farrowed in the same time period.

Total female inventory
The total number of replacement gilts and mated females in a herd.

Total female-to-boar ratio
The average total female inventory divided by the total number of pigs born alive in the same group.

Total pigs born
The total number of pigs born in a litter, including liveborn, stillborn, and mummified pigs.

Transit death
A pig that dies in transport between the farm and the slaughterhouse or final disembarcation point.

Trucking cost (USA)
The total freight charges and associated expenses related transport of pigs.

Transport costs (UK)
See: **Trucking cost (USA).**

Utilities
Water, electricity, gas, telephone, and other fuels used in the pig production enterprise.

Vet and medicine
All veterinary fees and medications directly related to the pig enterprise.

Wasted days
The days between service and an unsuccessful outcome.

Wean
(1) To make a young animal cease to depend on its mother's milk. (2) To accustom partly grown birds to do without artificial heat.

Weaning age, piglet
The number of days from the birth of a litter of piglets to weaning or removal from the sow.

Weaning age, sow
The number of days from a sow's farrowing until she is separated from her piglets.

Weaning, late
The separation and weaning of a late foster sow from her foster piglets.

Weaning, split
The weaning of only part of a litter; allows smaller piglets continued access to the sow.

Weaning percentage
The proportion of pigs of a given population of live pigs born that are weaned.

Weaning to service interval
The time between the weaning date and the date of first service.

Weaning weight
A piglet's weight at weaning.

Yield
Grade in meat animals, referring to the amount of lean meat produced in a carcass.

INDEX

Abortion, 60
Acceptability, of pork, 116
Actinobacillus pleuropneumonia, 329
Additives, 69, 104–07
Addresses, of clubs, 83–85
Adrenal gland, 39
Aflatoxins, 340
African swine fever, (ASF), 321–322
Agalactia. See Hypogalactica
Agriculture 31–32, 258
   community, 61–62
   Engineering publications, 258
   production, 25
Aisles, 236–237
alar scare, 26
Alfalfa, 169–70, 262
alantois, 95
All-in-all-out systems, 232, 314
Alternative finishing systems, 231–235
Amino acids, 125–127
   composition, 114
   synthetic, 170
Amnion, 95
Anatomical drawings, 37–40
Animal welfare, 301
Anatomy and physiology, 34–40, 51–65, 105–107
Anemia, 46

Animal,
   activists, 25
   Handlers, and chutes, 255
   Heat, 197–198
   temperature, 197
   welfare, 219, 301
Antibiotics, 318–319. See Individual diseases
   For Actinobacillus pleuropneumonia, 329
   And Collibacillosis, 331
   And pneumonic pasteurellosis, 330
   For swine dysentery, 331
Antimicrobials, 318–319
Antibiotic-resistant bacteria, 118–119
Antibodies, 48–49
Anti-clotting factors, 42
Anticoagulant, 42–43
Anti-corporate farming laws, 32
Apple scare. See Alar scare
Arteries and veins, 40–45
Arthritis, 322, 329
Ascarids, 335–338
Ascaris suis parasite, 335–336
Aspergillus flavus, 340
Asphyxiation, 317
Atherosclerosis, 45–46
Atrophic rhinitis (AR), 327–328

Artificial insemination building, 222
Artiodetyla, order of, 3
Ash, in feedstuffs, 77
Asian pigs, 4
Atttachment to uterus, 93
Aujeszky's disease, 322, 325
Automated feeding device, 102, 208
Average Daily Lean Gain, (ADLG), 226
Average daily weight gain (ADG), 89–90, 228, 246–248, 252, 300
Average market weights, 226

Babirussa, 2–3
Backfat thickness, 72–76
Back pressure, on sows, 62–63
Bakery products, 162
Bananas, 161
Barrow,
   defined, 51
   and leanness, 79
Basophils, 49
B cells (Blymphocutes), 49
Bearded pig, 2
Bedding, 18, 231–235, 239–240, 309
Behavioral,
   Problems, 342
   Thermoregulation, 4
   Traits, in estrus, 61–62

383

Behaviors. *See also* Individual types
    Feeding, 292–295
    Problems, 301–307
    Of sows, 284–285
    And time cycles, 251–252
    After weaning, 246
Berkshire, 82
    a century ago, 81
    Daily feeding behaviors, 294
    description and address, 84
Best Linear Unbiased Prediction (BLUP), 67
Beta-adrenergic Agonists, 104–105
Biofix, 208
Biomedical research, 34–35
Biosecurity, 306
    And antibiotics, 318–319
    In buildings, 233
    modern systems, 75–76
    Pathogens and herds, 131
    risk, 74
    risk of replacements, 75
    Showering of humans, 313–314
Biotechnology, 128
Biotin, 131
Birth,
    adaptations, 98–99
    canal, 58
Births,
    survivors after, 99
    weight, 97
    weight and survival, 99
Bleeding,
    by syringe, 43
    by vacuum tube, 44
    volumes (safe), 47
Blood, 34
    along dorsal wall, 39
    cells, 164
    clots, 41
    Fetal and maternal, 93–94
    meal, 164
    plasma, 164
    samples, 40–45
    sampling, 41
    sampling from piglets, 44–45
    sampling of older pigs, 45
Bloody scours. *See* Swine dysentery
Blunt trauma, 317

Boars,
    and collecting semen, 62
    Improved performance traits of, 71
    like stimuli, 63
    mating phases, 62
    odor, 117
    in optimal crossbreeding, 75–76
    preference of a sow in estrus, 62
    reproductive tract, 53
    seeking, of a sow, 61
    semen produced by, 61
    sex organs of, 52–56
    sexual functions of, 53–56
    slaughter weight of, 106
    stimulating features, 63
    versus barrows, 117
Body, 35–36
    changes in composition, 105–107
    chemical composition of, 107
    fat content, 125
    growth rates of parts, 91
    temperature at birth, 100–101
    weight gains, 181–182
Breed,
    defined as, 81, 86
    examples of traits of, 86
    Selected for ADG and ADLG, 226
Breeding, 51–65
    To assess proficiency, 273
    areas, 210–224
    behavioral expressions in male, 53
    and boar contact, 63
    early, 4
    and feeding, 144–145
    and the HAL gene, 77
    help from hormones, 64
    Herd goals, 267
    and housing of females, 63
    like-to-like, 67–68
    Management, 275–290
    And pathogens, 314
    Poor versus good rates, 276
    Square, 222
    Stock company website, 274
    terminal systems, 75–76
    traditional scaled back, 86
Breeds,
    to contact rare, 87
    of early America, 6

    for poor meat quality, 76
    major US, 83
    modern, 81
    of pigs, 23
    some pure born in 1990, 82, 86
    rares on the Web, 87
Brewery by-products, 159
British seedstocks, 86
Brucellosis, 313, 324–325
Bulbourethral gland, 54
Buildings, 219. *See also* Facilities; Floors; Individual types
    Cargill-style, 233
    Costs of construction, 233
    Costs of hoop buildings, 235
    designs, 198
    Diagrams, 236
    Economics of, 228–230
    Goals for, 235
    heat, 195–196, 200
    With hoop-style designs, 234
    Indoor-outdoor lots, 233
    Most successful, 232
    And pig handling, 237
    Styles of, 228–231
    Total confinement, 232
    Tunnel-ventilated, 232
Bushpig, 2

Calcium, 132–133, 171–172, 180
Calorie,
    absorption and utilization, 177
    contents, 111–114
    and weight, 191
Cam borough -15 (C15),
    Behaviors of, 293
    Female, 85
    Response to enrichment, 300
Campylobacter, 118
Canadian Swine Breeders Association on the Web, 87
Cannibalism, 301–304
Canola (Rapeseed) meal, 167–168
Carbohydrates, 122, 148–149
    classification of, 124
    in feedstuffs, 177
Carbon,
    Content in manure, 261

Carcasses, 106, 115, 226
Cardiac arrest, 46
Cargill-style buildings, 233–234
Casein, 165
Cassava, 160
Castrated male pig. *See* Barrow
Cause of death, 46
C.outline, 333
CD8 cells, 49
CD4 cells, 49
Cells, of the testes, 53–54
Cellulase, 122
Cellulose, and newborns, 103
Celtic pigs, in the US, 4–5
"Central" in anatomy, 36
Cereal grains, 122, 178
Cervix, 58–59
Changing seasons, 199
Chemical composition of body, 107
Chester, 82
    White breed, 6
    Description and addresses, 84
Chewing/rooting, 296, 299
Chilling, 281–282
Chlorine, 133
Cholesterol contents, 111–114
Choline, 131–132
Classification. *See* Phylogenetic classification
Clipping teeth, of newborns, 103
CL regression, in pregnancy, 60
Clitoris, 58–59
Clostridium, 333
Coat color, 68, 81–82, 85
Coccidiosis, 332
Cold animal house, 198
Coconut meal, 168
Colder climates, 230
Collibacillosis, 331
Colorado Newsham, 86
Colostrum, 101, 227
Commercial,
    farms and pigs suffering, 28–29
    lines, 76, 85
    pig production and anatomy, 34–40
    seedstock, 85
Community benefits, from pig farms, 28
Communities, and pig farms, 28

Computer technology, 178–180
Concentrated Animal Feeding Operations, (CAFO), 259–260
Concrete seats, 232
Condensation, 196, 199
Conduction, 196
Consolidation, 16, 32
Consumer acceptance, 116–118
Construction costs, of buildings, 233
Contracting process, 227
Controlateral, 35–36
Convection, 196
Conventional facilities, 231
Cooking,
    effects on, 111
    effects on palatability, 116–117
    to prevent contamination, 118
Cooling methods, 201, 282
Cooling systems, in buildings, 232
Copper, 135
Corn, 122
    belt and pigs, 6
    gluten feed, 160
    percentage in diets, 179
    and soybean diets, 31
Coronal plane, 35–36
Corporate versus family farms, 31–32
Corpus luteum (CL), 58, 60
Costs,
    Of buildings, 232–236
    formulation for diets, 180
    of production, 10, 13–14, 16–17
Cotswold, 86
Cottonseed meal, 168
Courses on the Internet, 65
Courtship behavior, 62–63
Cowpers gland, 54
Cows' milk, 102
C.porringers Type C, 333
Crates/huts, 209, 271–272, 280–281, 342
Crossbred boars, 71, 74, 76
Crossbreeding,
    effects on performance traits, 71
    purpose of, 83
    systems, 76
    systems and heterosis, 72
    two-way terminal, 75–76

Crushing,
    And different positions, 286–287
    And guardrails, 279–280
    And humans, 288
    And larger pens, 279–280
    And postural adjustments, 285, 288
    By sows, 286–288
Cryptorchidism, 52
C. tetani, 333
Culling, 68
Cured pork, 6
Curtain-sided buildings, 229
    And slatted floors, 239

Daily,
    average weight gain, 105–106
    feeding schedule, 102
    to increase weight gain, 104
    weight gain of growers, 143
Dalland, 86
Danish lines, 86
Danbred, 86
Dark breeds, 82
Dark, firm and dry (DFD), 116–117
Deaths, post-natals, 100
Decreased sperm count, 64
Defecation. *See* Dunging behavior
Deficiencies,
    of minerals, 132–136
    of vitamins, 128–132
Dehydration, 46
DeKalb Swine Breeders, 85–86
Demargination, 47
Demodectic mange mites, 334
Denmark, 9
Dental pattern, 103–104
Depreciation,
    Of buildings, 229
Diarrhea, 78, 102
Diets, 30–31, 175–193
    and amino acids, 126
    for breeders, 144–145
    early, 4–5
    feedstuffs of, 141
    formulation, 176–179
    for growing, 143–144
    New modern, 226
    in the New World, 5
    pork in, 1

by processing, 185–193
starter, 142–143
Supplements to, 180
vitamins and minerals, 128–136
well-balanced, 121–147
Digestables, 122
Digestability, 127, 138–140
Digestion, 261
Digestive,
enzymes, 137–138
tract, 136–140
Diseases of Swine, 312
Diseases, 316–318. *See* Individual diseases
And antibiotics, 318–319
Causes of, 317
Control system, 226
And humans, 330
Lists of, 321–324
organisms and site layouts, 18
And pigs' ages, 241
Surveillance, 316
Threats, 9–10
In the US, 10
Dissection, of fetal pig, 92
Distillery, by-products, 159
Distribution, 7–9
Domestic pigs, 1–2, 34–50
chromosomes of, 4
as descendants of, 3
and the ear carriage trait, 82
of 1500 A.D., 4
and mating, 61
prylogenetic classification of, 3
Of *sus* genus
Domestication, 207
Dominance order, 304–305
Dorsal points, 35–36
Double-curtain buildings, 231. *See also* Curtain-Sided buildings
Drafts, 196
Drinking behaviors, 295
Dry diet. *See* Transition to a dry diet
Dunging behaviors, 295–296
Duroc, 82
boar bred, 85
Daily feeding behaviors, 294
description and address, 83
Dust particles, 326

Ear,
Carriage trait, 82
Chewing, 342
Posture, 85
Early explorers and pigs, 5
Early weaned pigs, and behaviors, 342
E. coli, 78, 118, 331
Economics,
of additives, 105
and birth traits, 99
and boar odor, 117
And buildings, 228–230
of growth hormones, 105
Of hoop buildings, 234
And lean growth, 88
And molds and fungi, 340–341
And productivity measures, 269
of reproduction, 64
And returns for buildings, 228
and the skin color gene, 78
Effects of crossbreeding, on performance traits, 70
Eggs (ovum), 58–60
Electronic sow feeders, 208
Elimination behavior, 296
Embryonic mortality, 97
Embryos, 58
during pregnancy, 60
stages of development, 92–95
Embryology, a reference to, 92
Empty body mass, 116
Energy, 122–125
digestible (DE) versus metabolizeable (ME), 139–140
from fats and oils, 162–163
other sources of, 148–149
utilization, 177
values of feedstuffs, 176–177
Enrichment, effects of, 301
Entertainment, 239, 249–250 *See also* Toys; Enrichment
Enteric diseases, 330–333
Environment,
And collibacillosis, 331
And crushing, 286–287
Exploration of, 296
Physical, 198

And sickness, 296
And stillbirths, 279–280
And swine dysentery, 331
Environmental,
Activism and the North Carolina Hog industry, 266
Standards, 32
Environments, 194–206
Enzootic pneumonia, 328–329
Eosinophils, 49
Epididymis, 54
Ergot, 341
Erysipelas, 338
Estrogen, 58
is increased, 61
during ovulation, 59
during pregnancy, 60
Sulfate Receptor (ESL), gene, 77–78
Estrus,
aid in inducing, 64
cycle, 51–65,59–60
delay in return of, 64
Cycle and economics, 51–65
one sign of, 58
of sows, 58–60
stance, 63
Eurasian wild pig, 2
European,
Union, 9–10
Wild boar, 5
Wild pig, 3–4
Euthanasia, 317. *See* Individual diseases or conditions
EXP-94,
Behaviors of, 293
Response to enrichment, 300
Experimental lines, 85
Exsanguination, 317
Extensively-kept pigs, 194–206
Evaporation, 196

Facilities,
Costs of 231–232
To determine the cost of , 229
Effects on performance, 231
For growing, 228
Low-investment, 238
Fallopian tubes, 58
Family-style farms, 31–32

Fans, in housing, 232
Farming laws. *See* Anticorporate farming laws
Farms,
   features of, 18
   layouts of, 19–20
   pig flow of, 22–23
   production phases, 18–19, 22
   productivity, 22–23
   site selection, 18–19
Farrowing,
   Behaviors of sows, 284
   And cold air temperature, 282
   Environments, 280–281
   Good versus bad rates, 277
   Management, 275–290
   rates of, 211, 270–274
   And sow behavior, 284
   Successful units, 288
   synchronizing of, 60
Fats,
   contents, 111–114
   in feedstuffs, 177
   and oils, 123–125, 162–163
   as percent of carcass, 106
Fattening, with corn, 5
Fatty acids, 122–12
Fear of humans, 298–300
Feed,
   Budgets, 183–184
   Efficiency, 226, 235
Feeders, 247
   Design of, 244
   Materials for, 243
   Methods of production, 225, 227
   Moved to dry, 246
   And pig performance, 248
   Purposes of, 244–245
   Space required for, 244–247
   Wet versus dry, 247–248
Feeding, 140–145
   Behaviors, 282–285, 294–295
   Competition, at, 246
   devices, 102
   Of early weaned pigs, 243
   On the ground, 243
   intake lowered in estrus, 61
   Patterns, 294
   Of sows during lactation website, 290
   space, 204–205
   Space required for, 243
   And watering stations, 254
Feed,
   Intake, 61, 295
   Preservation, 181
Feedstuffs, 148–174
   composition of, 171–172
   And contaminants, 340–341
   effects on processing, 181–182
   major classes of, 176–177
   processing methods, 180–181
Female pig, 61–62. *See also* Gilt; Sow
Fences,
   In buildings, 237
   Heights of, 241–243
   Materials, 241–243
Feral population, 5
Fertility, 4, 272–273
Fertilization, defined as, 93
Fetal,
   growth and death, 97–98
   membranes, 95–96
   pig dissection, 92
Fetus, supplies to, 95,
Fibrin, 1
Fighting, 305–306
Fights, by housing, 310
Finishing,
   Alternative systems, for, 231–232
   Buildings for, 233
   Floors for, 242
   Most popular buildings, 232
   Out-of-favor systems, 232
First domesticated. *See* Domesticated
Fish,
   meal, 165
   solubles, 165–166
5–alpha-and rostenone, 117
Flavor, in pork products, 71, 79
Fleas, 335
Flies, 335
Flooring, 239
Floors,
   Characteristics of, 241
   Concrete slatted, 240–242
   In indoor-outdoor lots, 233
   Materials for, 240–241
   Materials for slotted, 242
   And pigs' ages, 241, 243
   Slatted, 239–242
   Space needs for, 253
   And tent designs, 235
   For transporting pigs, 307
Folacin, 131
Follicles, in females, 54, 57–58
Food,
   additives, 29
   after birth, 101
Foot-and-mouth disease, (FMD), 9–10, 321–322, 324
Foot support, 242
Formula,
   for absolute growth rate, 89
   for genetic improvement, 72
Fortifying, the diet, 180
Free space, 202–203
Frontal plane, 35–36
FSH,
   in follicular phase, 59
   in gilt, 58
   is increased, 61
Full-fat soybeans, 174

Genes, 91
   effects, on average daily gain (ADG), 69
   estrogen sulfate receptor (ESR), 77–78
   for ovulation, 77 *See also* Ovulation
   for skin color, 78
Genera, of pigs, 3
General pig information, 11
Generation interval (GI), 72
Genes, 91
Genetic improvement,
   of commercial companies, 85
   to effect the rate of, 72
   factors and formulas, 72
   Improvement in meat quality, 71
   objective of improvement, 68, 78
   Progress, to effect the rate of, 72
   in traits of, 78
Genetics,
   In disease resistance, 78
   New superior system, 226
   And productivity goals, 272
   And purebreds, 82
Genotype,
   and average daily gain (ADG), 69

distribution formulas for, 68
improvement, 67
Gestation,
   diets, 189
   and farrowing systems, 211–217
   length, 97
   period, 93
   systems, 208–224
Giant forest pig, 2
Gilts, 56
   Breeding project, 269
   commercials in the 1970s, 85–86
   and fat, 79
   follicles in, 58
   Herds and mating, 269
   inducing estrus in, 64
   odor, 117
   and presence of a boar, 63
   replacements, 75
Glans penis, 54
Glucose, 100, 122–124
Glycolysis, 42
GnRH hormone, 64
Grains, 17, 30–31, 149–158
   by-products of, 158–160
   as feedstuffs, 173
Groundnut meal, 168
Group,
   Pens, 209
   Sizes, 254
"Growers," as pig producers, 225, 226, 228
Growing and handling pigs, 311
Growth,
   And body composition, 89, 104–107
   Curve, 89
   defined, 88–89
   equations, 90
   And hormones, of fetus, 97
   lean, 88
   and nutrient deficiencies, 91
   postnatal, 99
   rate, 89–91
   regions, in the US, 14–15, 17
   and steroids, 125
Gunshot, 317
"Gut closure," 48, 101
"Gut edema," 331

HAL-1843 gene, 46, 77
Halothane, ("Hal") gene, 46, 117
Hampshire, 82
   description and address, 84
   effect, 76
   and a gene, 117
Handling facilities, 258, 342
HCG, 64
Head sizes, 245
Health,
   And antibiotics, 318–319
   Of herds, 312–345
   Of pork, 118–119
   In pork production, 342–343
   Websites on, 345
Heart,
   failure, 46
   valve, 45
Heat,
   and humidity, 200
   increment, 177
   Index, 206
   tress, 309
   stress, and sperm, 56
Heaters, in housing, 232
Heating goals, 195–196
Heat stress, 56, 309
Heavier carcasses, 226
Hematocrit, 46
Hematoma, 41
Hemophilus pleuropneumonia, 329
Heparin, 42
Herds,
   And antibiotics, 318–319
   avoiding contamination of, 75
   Best versus worst, 269
   Diseases, and humans, 312–314
   An Example of well-managed, 268
   goals of, 75–76
   and HAL gene, 77
   And health programs, 312–345. See also
   Individual diseases; Diseases; Swine diseases
   Health, and PRRS, 326–327
   to make more lean, 79
Hereford, 85
Heritabilities, and genetics, 69–72
Hernias, 341–342

Heterosis, 75–76
   Amount of, 74
   definition of, 70
   and heritability, 69–70
   types of, 71
High-investment facility, 207–208
High-moisture contents, 340
History of pigs, 12
   in North America, 5–7
   traits of interest, 81
   Hog cholera, 321–322
H1N1. See Swine influenza
Hog and Pig Report, 23
Hogs,
   cholera in, 10
   defined, 51
   of North America, 82
   purebreds, 83
Hominy feed, 160
Hoop-style facility, 230–232, 234–236, 238
Hormones, 125,
   during courtship, 62
   during the estrus cycle, 60
   for fetal growth, 97
   for increased farrowing rates, 62
   to induce estrus, 64
   to be injected, 104–107
   in parturition, 61
Housing,
   of females, 63
   Systems as influences on production, 230
   websites of environments, 206
H3N2. See Swine influenza
Human,
   Handling, 300
   pig interactions, 301
   Use of pig organs, 35
Hunting, of pigs, 4–5
Hybrid pig, 4
Hybrid vigor. See Heterosis
Hydrolyzed feather meal, 165
Hyperplasia, 91
Hypertrophy, 91
Hyogalactica, 286

Iberian pig, 4–5, 79
Illeitis, 332

Illinois requirements, 18
Immune system, 48–49
Immunity,
　After birth, 101
　feed utilization, 104
　reproductive performance, 62
Inbred genetic line, 68
Indoor-Outdoor lots, 233–234
Indoor systems, 229–239–240, 253
Inducing estrus, 64
Infections, 47, 49
Inhibin, 54
Initiation of parturition, 60
Injections, 318, 320
Iodine, 135
Insemination, 51, 56
　described, 93
　and reproductive rates, 62
　and sow's stance, 63
Intensively-kept pigs, 194–206
Internal organs, 37
Internet resources, 33. *See also*
　Websites;
　Individual topics
Intestines, 37–38
Intramuscular fat percentages, 71
In vitro fertilization, 51–52
In vivo reproduction, 52
Ipsilateral, 35–36
Iron, 134
Ivermectin, 334–335

Javanese warty pig, 2
Joint problems, 338
Jugular veins, 40–41, 45

Kansas farming laws, 32
K88 antigen/protein, 78
Kjeldahl procedure, 176–177
Kidneys, 39

Lactation, 61, 190
Lameness, 24
Landrace, 82
　bred to get a Camborough, 85
　description and address of, 83
Lateral regions, 35–36
Lean,
　capacity for growth, 127
　content increased by, 104–105
　growth, 181–182
Leanness, 114–116
Legal requirements, of farms, 17–18
Legislation. *See* Restrictive legislation
Leg problems, 232
Legumes, 262
Leptospirosis, 333–334
Lesions, 316, 324
　Of bones, 328
　On foot pads, 232
　In ileitis, 332
　Of lungs, 328
　Of mange mites, 334
　on toe pads, 240
Less water-holding capacity, 117
Leutinizing Hormones (LH), 53
LEVIS System, 220–221
Leydig cells, 53
LH,
　in follicular phase, 59
　in gilt, 58
　is increased, 61
Lice, 335
Life-cycle feeding, 121–147
Liner, T. Eriel, 85
Linoleic acid, 123, 125
Linseed meal, 168
Lipids, 123, 125
Litters,
　adding piglets to 77–78
　first and last, 99
　To increase, 273–274
　numbers increased by hormones, 62
　Per sow per year, 268–269
　performance of, 211–217
　Processing of, 288
　sizes of and survivals, 60, 99
　sizes of in early gestation, 95
　in 1990, 82
　space during farrowing, 281
　Stillbirths, 99
Loading chutes, 254–255
Lordosis stance, 63
Lowering costs, 16–17
Low-investment facility, 207–208
Lubbock Swine Breeders, 85–86, 210–224
Lungs, 37–38

Lymphocytes, 49
Lysine,
　content in diet, 187
　levels of, 190
　percentage in corn, 179

Macrominerals, 132–134
Magnesium, 133
Maintenance behaviors, 292
Major exporting countries, 13–14
Male pig. *See* Boar
Malignant
　Hypertension, 46
　Hypothermia, 339
Manganese, 135
Manure, 27–28, 204
　Buildup on floors, 239
　Categories of nutrients in, 261
　Defined, 260
　And dry bedding, 234
　Dry versus wet, 263–265
　To encourage elimination, 241
　And flooring, 239–241
　Handling systems, 264–265
　Management of, 259–265
　Nutrient collection, 264
　And odor of rotten eggs, 262
　Production per animal unit, 261
　Removal of, 235
　Storage and fuel, 265
　Technologies, 261
　Volume produced, 260
　Wallowing in, 296
　And water, 249
　And young pigs, 243
Marbling, 111
Marker-assisted selection, 77–79
Market weights. *See also* Average market weights
　body/carcass weight of hogs, 114–116
　to-breeders ratio, 15
　for pork, 13
　Weight, 22, 105, 225–236, 230
Mastitis-metritis galactic (MMA), 339
Mating,
　Behavior of a sow, 62
　and fertilization, 93

Meat,
  and bone meal, 165
  consumption, 31
  meal, 164–165
  quality and breeds, 71
  quality for poor, 76–77
  quality traits, 71
Mechanically-operated buildings, 229
Mechanically-ventilated buildings, 231
medical,
  plans, 35–36
  points, 35–36
Megafarms, 16–17, 32
Meishan, breed, 85–86
  Activity of, 292
  Behavior, 293
  As a purebred line,
  And teat fidelity of, 246
  Traits of, 85
Mendelian genetics, 67–69
Metabolism,
  of proteins and amino acids, 130
  and vitamins and minerals, 128–136
Microorganisms, 313–314, 317–318
Middsaggital section, 35–36
Midventral laparotomy, 35–36
Midwest agriculture, 6
Milk, 48
  dried, 166
  Production, and diseases, 339
  Provided to piglets, 61
  Replacer diet, 102
Milling by-products, 158–159
Minerals, 132–133, 171–172
  Absorption and excretion, 140
  Concentrations of, 125
  Elements of, 112–114
  Sources of vitamins, 171
  Traces of, 134–136
Modern,
  Farms, 20–23
  Legal requirements for pork production, 17
  Schedules of pork production, 23–30
Modified open-front building, 230
Molasses, cane, 161–162
Molds, in feedstuffs, 340–341
Molecular genetics, 76–78

Monsanto, 86
Mortality, in pre-weaning, 100, 211, 215. See Individual diseases
Mummified fetuses, 98
Mycoplasmal pneumonia, 328–329
Mycotoxins, in feed, 340
Myths,
  About pigs, 26
  About pork, 28–32

Native to the Americas, 2
Necropsies, 34, 37–40
Needle, in blood sampling, 44–45
Nest-building behaviors, 342
Neurotic behaviors, 28
Neutrophils, 49
New antibiotics, 118–119
Newborns, 99
  Digestive system of, 102–103
  Without a sow, 102
  Teeth of, 103
Newsham, 86
Niacin, 130
Nitrogen,
  Balance and digestibility, 127
  Content in manure, 261
  And pollution, 128
Nitrogenous compounds, 176
NK cells, 49
Nonadditive gene action, 70–71
Nonproductive days, 269
Normal,
  Blood supply, 46–47
  Distribution, of individual traits, 68
Noses ringed, 5
NPD seedstock, 86
Number of pigs,
  Per pen, 204
  By farm sizes, 16, 23
  And regulations, 18
  By state in the US, 15
Nursing, 102, 108, 285–286
  Patterns of, 292–293
  Problems with, 286
Nutrients,
  Absorption versus digestability, 138
  Ability, 138
  Carbohydrates, 122–124

Classes of, 121
Composition of, 110
Composition of sources, 171–172
Fats and oils, 162–163
Lipids, 125
Minerals and vitamins, 171–172
Proteins, 163–170
Required by pigs, 121, 124–125
Requirements of, 147, 176
transport of, 138–139
Use, by crop, 263
Websites, 192–193
Nutrition, 136–147
  During lactation, 145
  Websites, 147, 174
Nutritional Secondary Hyperparathyroidism, (NSHP), 327–328

Ochratoxicosis, 340
Odors,
  Of buildings, 229–241
  Of manure, 263
Offspring trait, 69
Oilseed neals, 167
One-size production, 18
Open-air bedded building, 232
Open-front buildings, 233–234, 239
Organs,
  Of the abdomen, 38
  Of fetuses, 93
Orient, The, 31
Origins, of pigs, 1–2, 4
Outdoor production, 208–209
Ova, 93
Ovaries, 56–61
Overeating, in newborns, 102
Oviducts, 58–59
Ovulation, 58–59
  Gene for, 77
  And hormones for, 60
Oxygen, supply at birth, 98
Oxytocin, 62–63
  To stimulate milk and let-down, 286
  And uterine contractions, 284

Packed cell volume (PCV), 46
Packing plants, 21, 116
Palatability factors, 116

Pale, Soft, and Exudative (PSE) pork, 77, 116–117
Pantothenic acid, 130
Parasites, 316, 323, 334–338
Parasitic infection, 49
Parturition, 61
Passive immunity,
    Of Piglets, 101
    Of SEW pigs, 226
Parvo virus, 322, 334
Pathogens, 48, 313–345
Peanut meal, 168
Peccary, 2
Penetrating captive bolt, 317
Penis, 54, 58
Penna's pig production, 15
Pens, 90, 208
    Within buildings, 237
    Economics of, 229
    Layouts of, 239–254
    And mortality, 280
    For sick pigs, 237
    Space in, 281
    Well-designed, 295
Per capita consumption, 110
Performance,
    Effects of diets, 180
    Influenced by nutrition, 175–193
    Maximizing, 188
    Of sows and litters, 218
    Traits and crossbreeding, 70
    Traits and progeny tests, 71
    Up to potential, 127
Peritoneal cavity, 39
PGFa, 64
PG600, 64
Phosphorus, 133
    And content of manure, 261
    Requirement, 179
    In supplements, 171–172
Phylogenetic classification, 3
Pickled pork, 6
PIClit gene, 78
PICment, and skin color gene, 78
PICUSA,
    Database, 212–214
    Description of, 85
    And gene for ovulation, 77–78
    Position in the 1990s, 85–86

Pietrain, 85–86
    As a purebred line, 85
Pig,
    Anatomy of, 36–40
    Biology and growth of, 88–109
    Breeds on the web, 87
    Density, 8–9
    Desired locations of production, 7
    Distribution in the US, 14–15
    Early pens, 14
    Early production, 1–4
    Expansion of production, 6–7
    Facilities, and disease, 328
    Farms, 15–23, 26–28, 30–33
    Flow, 22–23
    Genome Project, 67. See also US Gene Mapping Program
    Genomes on the web, 80
    And growth of, 88–109
    Health, 34. See also Health, of herds
    Heart, 45
    And hog inventories in the USA on a Website, 274
    Injuries and diseases, 37–40
    Inventory by US state, 15
    Nature, 207
    Operations, 16–24
    Output, 16, 22–23
    Production 4, 8–9, 33, 98–100, 140–141
    Production units. See also Societal Concerns
    Reproduction of, 51–65
PigChamp, 212–214
Piglets,
    And antibodies, 317–318
    Behaviors, 286–287
    At birth, 58
    And blood sampling, 44–45
    Crushing of, 279–280
    Deaths, 277
    And a delivered/sow, 277
    And diarrhea, 333
    And heat, 283
    How to keep them warm, 283
    Immune systems, 48, 281
    Management website, 290
    And mortality, 268, 288–289

    Numbers of, 14–16
    Numbers in litters, 60
    Nursing of, 282, 331
    On PPDS sows, 339
    And preferred temperature, 280
    Processing, 6
    "Producer" defined, 15
    Remained from the sow, 227
    SEWs, 227
    Stillborns and parvo, 334
    And TGE, 330
    Thermal needs of, 281
    And urine-drinking, 342
    Weaned in the wild, 226
    Weaning weights of, 277
Pigs,
    Anatomy of the heart of, websites, 50
    Biology of, websites, 109
    Business of, website, 258
    Canadian performances, 230
    Carcasses appraisals website, 120
    Concerns of citizens, 26–33
    Costs of production, 267
    Courses on websites, 65
    Critical measures of, 280
    As curious animals, 243
    Diet components websites, 174, 193
    As donors, 35
    In early America, 5–7
    On early voyages, 5
    Effects of mixing and sorting, 305–306
    Evolution of, 292
    Entertainment for, 239
    Examples of traits, 79
    Farms and odors of, 263
    Farms and PPSYs, 267–268
    Feet and toes, 240
    Fear of humans, 287–288
    Feeder spaces for, 247
    Fetal dissection, of websites, 50
    Fighting in hoop buildings, 235
    Gangs of, 254
    Genetics on the web, 80, 120
    genomes websites, 80
    Growing patterns for, 251, 109

Growing in medieval times, 229
Growth effected by social states, 305
Handling websites, 311
Health websites, 345
History of, 4
history of housing of, 228–230
Hoop structures website, 258
and housing and waste management
Human factors in production, 287–288
humans' organs sizes, 34
Industry of, websites, 24
Information websites, 11
Inventories of, website, 274
Marketed per year/per week, 21–23
Measures of, 276
Moderns' weights, 107
And mortality in hoop buildings, 235
Mortality and pigs per pen, 253
North Carolina industry websites, 266
(Numbers of) Produced Per Sow Per Year (PPSY), 267, 276–290
Nutrition websites,109, 174, 192
Pharmaceutical websites, 345
Production of, websites, 33
Rare breeds information, 87
Reproduction fact sheets websites, 65
Pigs' Peformance
Agricultural engineering websites, 258
In breeding and farrowing, 276–290
And building costs, 232–235
And buildings, 229
Compared by facilities, 230–231
On early voyages, 5
And economics, 228, 231
Education websites, 258
And feeder space, 244–248
And free spaces, 252
Handling pigs website, 224, 258
Heat Index Chart website, 206
In hoop-style designs, 234–235
Housing and environment websites, 206
Modern practices, 226–258
Parasites, 334–338. *See also*

Antibiotics; Diseases; Swine diseases
Per sow, 16, 21–23
Pharmaceutical websites, 345
Pig chutes website, 224
And pollution, 30
Pork Industry website, 223
Produced per week, 21, 23
recognition, 254
socialization, 28–29
Sows and piglets websites, 290
space per pig, 238
Standards, 256
"stink," 26–27
Suffering on farms, 28
Systems, 225–258
total numbers born, 268
Traditional stages, 225
Trivia website, 12
Wastes management website, 266
Wastes historically
And water pollution, 30
Weaning ages, 269
Weaning rates, 272
Placenta, 93
Growth of, 95
Shown, 96
Weight of, 97
Placentation,
On the Internet, 65
Plantains, 161
Plant sources of energy, 161–162
Plasma, 42–43
Fluid, 41
Glucose concentrations, 46
Paoge, Roy, 85
Playing, 296
Pneumonic pasteurillosis, 330
Poland China breed, 6, 82–84
Pole shed, 233
Pollution. *See* Pigs; Pollution; Manure-related, 259, 263–264
Polysacceharides, 122–124
Ponds, and manure, 264
Poor countries, 31
Porcine respiratory corona virus (PRCV),326, 339–340
Porcine Stress Syndrome (PSS), 77, 339

Pork,
Over beef, 8
"causes cancer," 29
Consumed locally, 8
And the consumer, 110
And consumer acceptance, 116
Consumption, 8, 31
Contents of, 111–114
Costs of production, 10
Early center of, 6
In early diets, 1, 4
Exporters, 8–10
Fat and calorie composition, 111–114
For food, 110–119
Industry Handbook, 20
Industry Institute website, 223
Numbers of processing plants, 23
Per capita, 7–9
Processing in Kansas, 32
Production,, and health, 312–345
Prohibited by religions, 1, 4
Pork quality,
Effect of the production system, 257
And genes, 71
Of products, 71
Postpartum dysgalacttica syndrome (PPDS), 339
Post-weaning, and feeding behaviors, 293
Potassium, 133, 261
Potatoes, 160–161
Poultry by-product meal, 166
PPSYs,
Achieveable rates of, 277–278
Contributing factors to, 271
And farrowing rate, 270
Goals for, 271
Herds with highs, 270
Increased by, 268
Influential factors, 273
And litters sizes, 270
Rates of, 275–290
To reach top levels, 268
Pregnancy, 60
Pregnant mare serum, 64
Prenatal development, 92–98
Preservatives, 29
Prenatal development, 92–98
Prenatal mortality, 97

Pressure sprayers, 314
Preweaning,
  And crushing, 286–287
  Mortality, 270–271, 277
  Rates of, 278–280
Pricing, 106
Principles of nutrition, 121–147
Problems of the industry, 11
Processing,
  Plants, 23
  And disease surveillance, 316–319
Production,
  Methods, 13, 17, 22–23
  Phases on farms, 18–19, 22
  Role of environments, 194–206
  Schedules, 20–22
  Size, 20–23
  Systems, 204–258
  Systems of indoor versus outdoor, 18
  Three critical measures of, 276
Productivity,
  Was improved, 228
  And measures of herds, 269
  And pigs per pen, 254
  Problems and areas to improve, 270–273
Progesterone, 58–61
Prolapse, 341
Prolificacy, 5
Proliferative enteropathies, 332
Prostaglandins, 60
Prostrate gland, 54
Proteins, 125–126, 163–170
  Contents, 114
  In feedstuffs, 172–173
  Or lysine (amino acid), 178, 180
  The proximate analysis formula, 176
  Requirements of, 141–144
Proximal regions, 35–36
Pseudorabies, 313, 322, 325
PST injections, 105
Purebreds, 82

Quantitative Trait Loci (QTL), 77

Radiation, 196
Rare breeds contact information, 87

Record-keeping program about pigs: a website, 274
Red blood cells, 41
Redistribution of population, 7
Regulations, examples of, 24
Relationship of foot dimensions to body rate, 90
Religions, prohibiting pork, 4, 8
Reproduction, 51–65, 211–224
  Courses on the Internet, 65
  Efficiency (PPSY), 267
  Environments for, 207–224
  Stages of, 56
  Steroids, 125
  Vitamins, 128
Reproductive,
  Diseases, 333–334. *See also* Individual diseases
  Management, 60
  Performance, 63, 211
  Performance and heterosis, 69–70
  Problems, 287–288
Reproductive,
  Rates, 207–224
  Rates and courtship behavior, 62–63
  Tract sizes, 64
Respiratory,
  Diseases, 321–330
  Symptoms, 329–337
Restrictive legislation, 32
Retail cuts, 112–113
Rolling in mud, 4
Rotational systems, 74
Roots and tubers, 160–161
Rotaviral diarrhea, 333
Rotavirus, 322

Safety of pork, 118–119
Safflower meal, 168
Saliva, and boar pheromones, 63
Salmonella, 118, 332
Salt pork, 5–6
Sampling blood, 44–45
Sanitation, 205, 312–315, 331
Sanitizing, 314–315
Sarcoptic mange mite, 334
Seaboard farms, 213
Seasonal infertility, 273
Seaweed, 170

Seghers, 86
Segregated early weaning (SEW), 226
  Defined, 227
  To produce healthy piglets, 227
  Thermal needs of, 256
Selecting,
  For ADG and ADLG, 226
Selection,
  And genetic progress, 71
  For growth and leanness, 116
  For leanness, 114–116
  Marker-assisted, 77
  Methods, for pig traits, 66–80
  Natural, 81
  And polygenic traits, 68
  Technologies for, 77
Selective breeding, 67–69
Selenium, 135
Semen, 56
  Collection, 62
  Produced, during ejaculation, 61
Serosa, 95
Semi-trucks, 130
Sending pigs to market, 306. *See also* Semi-trucks
Semi-trucks, 230
Sending pigs to market, 306. *See also* Semi-trucks
Sertoli cells, 53–54
Service. *See* Breeding
Sesame meal, 169
Set-back requirements, 17–18
Serum, 41–43
Sexual behavior, of sows, 61, 210–224, 272
Sexual development of the male, 53–54
Sheltered lots on dirt, 231
Shipping pigs. *See* Transporting pigs
Showering-in practices, 313–314
Sickness behaviors, 296
Signoret, Jean-Pierre, 62–63
Similarities, of humans and pigs, 34–50
Single cell protein, 170
Size of trucks, 21
Skeleton, 37
Skin,
  Color gene, 78
  Tumors, 49

Slats,
  And tail-biting, 303
Slats or slots, 240
Slats, used in buildings, 232
Slaughter,
  And antibiotics, 318–319
  Checks, 316, 319
  Weight, 228
Small versus large farms, 15–17, 23, 30–32
Snares, 45
Snouts, twisted and diseased, 327–328
Social,
  Behavior, 302
  Facilitation, 294
  And early pigs, 292
  Order, 243
Sodium fluoride, 42
Software for livestock websites, 274
Somatotropin, 105
Sorting boards, 308
Sows,
  Behavior of, 284–286
  Bred for market, 85
  Cooling of, 282
  And crushing, 286–287
  Dynamic space for, 279–280
  Effects of warm weather on, 282
  In estrus, 58–60
  In estrus, near a boar, 62
  Feeding the outdoor, 243
  Giving birth, 284–285
  Herd productivity, 16, 22–23, 268–271
  And hormones, 62
  Housing systems, 217
  Improved performance, 71
  Inventory of, 208
  Land requirements for, 261
  Location of herds, 19
  Lying/standing, 285
  Management websites, 290
  Milk replacer, 100
  Mounting other sows, 61
  Numbers to breed, 271–273
  Numbers needed, 23
  Numbers in a litter, 60
  In optimal crossbreeding, 75–76
  Preferred temperatures, 280
  Problems in nursing, 286
  And standing reflex, 62–63
  And suffering, 28–29
  Units needed, 21–23
Soybeans. See Full-fat soybeans
Space,
  Allowances, 202
  Dynamics for, 279
  Effects of free, 252
  Needs, 201–205, 251, 253
  Occupied formula, 251
  Requirements, 238, 253
  Utilization and building designs, 229–241
  In wean-finish concept, 228
Species, 3
Spermatids, 53
Spermatocytes, 53
Spermatogenesis, 54–55
Sperm count, 64
Spot, 6, 82, description and address for, 84
  Spotted Poland China, 6
Staff. See also Workers; Stockpersons; Stockmanship; Farrowing, 276–277, 285–286
Stalls, 209–210
Starches, 122, 124
Statistics, on the industry, 24
Stillborn piglets, 98
  And behaviors, 284–285
  And birthing hormone, 284
  "normal" number of, 276
  And sows kept outdoors, 279–280
Stockmanship, 297
Stockpersons, 296, 302
  Behaviors of, 287–288
  Duties of, 283–287
  Features of, 288
  In moving pigs, 306–309
  And pig sickness, 296
  And tail-biting, 302
Stomach lesions, 181
Stress, 46, 77
  And birthing hormone, 284
  And netrophils, 47
  And reproduction, 211
  Signs of, 342
  Of sows, 219
Stunting effect, 91–92, 317
Suffering, By pigs on farms, 28

Sugar, cane and beet, 161
Suidae family, 3–4
Suina, suborder of, 3
Sulawesi warty pig, 2
Sulfur, 134, 261
Sunflower meal, 169
Survival, after birth, 100
Sus genus, 4
Sweat, 122
"Sweating like a pig," 26
Sweet potatoes, 161
Swine,
  Diseases, 312–334. See also Injections
  Dysentery, 331
  Farms, 6, 274. See also Pigs
  Fever, 321–322
  Influenza, 322
  Pox, 322, 339
Symptoms, of swine diseases, 321–334
Syringe use, 41–43

Tail-biting,
  And bedding, 232
  And ear chewing, 301–304, 306
  Resolution of, 342
Taiwan, 8–10
Tamworth,
  A century ago, 81
  Traits of, 85
T-catatonic suppressor cells, 49
Teat fidelity, 246
Technology,
  Efforts by Monsanto, 86
  To improve feeding, 104
  For selection, 77
Temperature, of environment, 197
Tenderness, in pork products, 71
Tent designs, 235
Terminal,
  Crossbred boar, 85
  Crossing systems, 74
Testicles, 32, 54
Testosterone, 53–54, 117
Tethers, 209
TGE virus, exposure to, 281
T-helper cells, 49
Thermal,
  Excess, 309
  Needs, 280–283

Three-breed terminal cross, 75–76
Ticks, 335
Tools, for moving pigs, 308
Top 10 US farms, 17
Total confinement buildings, 232–233
Total head in US, 15
Total meat consumption, 110
Toxicity,
    Of minerals, 128–136
    Of vitamins, 128–136
Toxo. *See* Toxoplasmosis
Toxoplasmosis, 332
Toys, 293, 299–301
Traits,
    Of economic importance, 68
    For the future, 83
    And individual genes, 70–71
    Of major breeds, 83–85
    Sought-after, 83
Transition to a dry diet, 102
Transmissible Gastro Enteritis (TGE) Virus, 281
Transplants, 33, 45
Transporting, of pigs, 307–309
Transverse planes, 35–36
Trichinal parasite, 29
Trichinella spiralis, 335, 337–338
Trichinosis, 29, 118
Trickle feeding, 208
Triglycerides, 123–124
Trough, for piglets, 44–45
Truckloads, (semi-trucks), 20–21
Trucks, and supplies, 314
Tryptophan, 130
Tunnel-ventilated buildings, 232
Turkey buildings, 232
Tusks, 103
Two-bred rotational cross, 75–76
Two-site or three-site facilities, 19
Two-way terminal cross, 75–76

Ulcers, 181
Umbilical cord, 99
United Kingdom, 86
Urea, 127, 261
Urethra, 54
Urination. *See* Dunging behavior
US,
    expansion sites, 15
    Gene Mapping Program, 67

Pig industry, 14–17
Pork production competition, 13–14
Sow herd, 16, 22–23
Swine industry, changes in, 230
Uterine horns, 58–60
Uterus, 58–59, 96
    In estrus cycle, 60
    In lactation, 61
    During pregnancy, 60
    Responds to weaning, 61

Vaccinations, 313, 317, 328
Vaccinia virus, 339
Vaccines, 329–331. *See* Individual diseases
Vacutainer, 43–44
Vagina, 58–59
Vas deferens, 54
Vena cava, 45
Venipuncture, 41
Ventilation,
    Of buildings, 229
    Goals of, 200–201
    Natural and beddings, 232
Ventral sites, 35–36
Vesicular
    Glands, 54
    Stomatitis (VS), 324
Veterinarians, 316
Vibrionic dysentery. *See* Swine dysentery
Vitamins A through K, 113–114, 128–131
    Absorption and excretion of, 140
    Deficiencies, 128–132
    Fat-soluble, 125, 128–130
    In feedstuffs, 177
    Water-soluble, 130–132
Vitamin premixes, 174
Vocalization of sows, 61
Vulva, 58–59, 342

Wallowing, 4
Warm animal house, 199
Warthog, 2–3
Waste management, 132
Water,
    Alternatives for, 250
    Contaminants, 250

Content in manure, 261
Deprivation, 248
Devices, 249–250
Drip, 282
Intake per pig size, 249
As a nutrient, 122
Weaning,
    Ages, 226–227, 269–270
    In diet, 184–185
    And TGE virus, 281
    Versus slaughter weight, 105–107
    Weight at weaning, 22
    In the wild, 226
Wean-to-finish buildings, 19, 227–228, 232
Weather Safety Index, 309
Weekly numbers, of pigs, 22–23
Weights, 80, 235, 245–248, 306
    Gain, 30, 205, 318
    Weaning to market, 104–107
Whey, 166
White,
    Blood cells, 41
    Breeds, 82
Wild pigs, 2–3
Windchills, 198–206
Workers,
    Job duties of, 291–292
    Job satisfaction, 234–235
    And Zoonoses, 321
    Website for, 290
Worms, as internal parasites, 335–336
Woven-wire floors, 241

Xenotransplantation, 35

Yolk sac, 95
Yorkshire, 82–85
"Your room is like a pig sty," 26

Z-axis plane, 35–36
Zinc, 135–136
Zone-cool/heat, 280–281
Zoonoses,
    To combat, 321
    Lepto, 334
    Swine fever, 329
Zoonotic diseases, 312, 338

DATE DUE

7-4-19

DEMCO, INC. 38-2931

Hartness Library
Vermont Technical College
One Main St.
Randolph Center, VT 05061

DISCARD